REMOTE SENSING
PRINCIPLES AND APPLICATIONS

REMOTE SENSING
PRINCIPLES AND APPLICATIONS

B. C. Panda
Former Head & Professor
Division of Agricultural Physics
Indian Agricultural Research Institute
New Delhi

Viva Books

New Delhi | Mumbai | Chennai | Kolkata | Bangalore | Hyderabad | Kochi | Guwahati

Publisher's note

Every possible effort has been made to ensure that the information contained in this book is accurate at the time of going to press, and the publisher and author cannot accept responsibility for any errors or omissions, however caused. No responsibility for loss or damage occasioned to any person acting, or refraining from action, as a result of the material in this publication can be accepted by the editor, the publisher or the author.

Every effort has been made to trace the owners of copyright material used in this book. The author and the publisher will be grateful for any omission brought to their notice for acknowledgement in the future editions of the book.

Copyright © Viva Books Private Limited

All rights reserved. No part of this book may be reproduced, stored in a retrieval system, or transmitted in any form or by any means, electronic, mechanical, photocopying, recorded or otherwise, without the written permission of the publisher.

First Published 2005
Reprinted 2006, 2008, 2009, 2011, 2018, 2019

Viva Books Private Limited

- 4737/23, Ansari Road, Daryaganj, New Delhi 110 002
  Tel. 011-42242200, 23258325, 23283121, Email: vivadelhi@vivagroupindia.net
- 76, Service Industries, Shirvane, Sector 1, Nerul, Navi Mumbai 400 706
  Tel. 022-27721273, 27721274, Email: vivamumbai@vivagroupindia.net
- Megh Tower, Old No. 307, New No. 165, Poonamallee High Road, Maduravoyal, Chennai 600 095
  Tel. 044-23780991, 23780992, 23780994, Email: vivachennai@vivagroupindia.net
- B-103, Jindal Towers, 21/1A/3 Darga Road, Kolkata 700 017
  Tel. 033-22816713, Email: vivakolkata@vivagroupindia.net
- 194, First Floor, Subbarama Chetty Road Near Nettkallappa Circle, Basavanagudi, Bengaluru 560 004
  Tel. 080-26607409, Email: vivabangalore@vivagroupindia.net
- 101-102, Moghal Marc Apartments, 3-4-637 to 641, Narayanguda, Hyderabad 500 029
  Tel. 040-27564481, Email: vivahyderabad@vivagroupindia.net
- First Floor, Beevi Towers, SRM Road, Kaloor, Kochi 682 018
  Tel. 0484-2403055, 2403056, Email: vivakochi@vivagroupindia.net
- 232, GNB Road, Beside UCO Bank, Silpukhuri, Guwahati 781 003
  Tel. 0361-2666386, Email: vivaguwahati@vivagroupindia.net

www.vivagroupindia.com

ISBN: 978-81-7649-630-8

Published by Vinod Vasishtha for Viva Books Private Limited, 4737/23, Ansari Road, Daryaganj, New Delhi 110 002.
Printed and bound in India.

Dedicated to my

Revered Parents

SUKRITA AND KANHU CHARAN

Contents

Acknowledgements	*xv*
Preface	*xvii*
Foreword	*xix*
Picture/Image Gallery	*xxi-xxx*

1. Introduction   **1–3**

    1.1 Concept of Remote Sensing   1
    1.2 Essential Components of Remote Sensing   1
    1.3 Natural Remote Sensing   2
    1.4 Artificial Remote Sensing   2
    1.5 Passive and Active Remote   2
    1.6 Applications of Remote Sensing   2
    Suggestions for Supplementary Reading   3

2. The Signals   **4–16**

    2.1 Disturbance in a Force Field   4
       2.1.1 Gravitational Field   4
       2.1.2 Electric Field   5
       2.1.3 Magnetic Field   5
    2.2 Acoustic Signals   5
    2.3 Particulate Signals   7
    2.4 Electromagnetic Signals   7
    2.5 Generation of Electromagnetic Signals   8
    2.6 Emission of Electromagnetic Signals from a Material Body (Thermal/Black-body Radiation)   10
    2.7 Some Characteristics of Black-body Radiation   12
       2.7.1 Kirchhoff's Law   12
       2.7.2 Planck's Radiation Law   13
       2.7.3 Stefan – Boltzmann Law   13
       2.7.4 Wien's Displacement Law   13
       2.7.5 Radiant Photon Emittance   14
       2.7.6 Radiation Terminology   15
    Suggestions for Supplementary Reading   16

3. Target–Signal Interaction in Optical Region   17–36

    3.1 An Overall View   17
    3.2 Solar Irradiation above the Atmosphere and at the Earth's Surface   25
    3.3 Interaction of Electromagnetic Radiation with Atmospheric Constituents   26
        3.3.1 Rayleigh Scattering   26
        3.3.2 Mie Scattering   27
        3.3.3 Non-Selective Scattering   27
    3.4 Target-Signal Interaction Mechanisms   27
    3.5 Electromagnetic Signals Useful for Satellite Remote Sensing (the Atmospheric Windows)   28
        Suggestions for Supplementary Reading   34

4. Target–Signal Interaction in Microwave Region   37–55

    4.1 Radar Return from Characteristic Surfaces   39
        4.1.1 Facets   39
        4.1.2 Spheres   41
        4.1.3 Cylinders   42
        4.1.4 Corner Reflectors   42
    4.2 Effect of Dielectric Properties of Targets on Radar Return   43
        4.2.1 Surface and Volume Scattering   45
        4.2.2 Influence of System and Target Parameters on Imagery Gray Tones   47
    4.3 Microwave Radiometry   49
        4.3.1 Passive Microwave Emission from Target Surfaces   50
        Suggestions for Supplementary Reading   55

5. The Sensor and Sensor Platforms   56–89

    5.1 Sensor Materials   56
    5.2 Sensor Systems   57
        5.2.1 Framing Systems   59
        5.2.2 Scanning Systems   62
    5.3. Scale, Resolution and Mapping Units   66
        5.3.1 Scale   66
        5.3.2 Resolution   67
        5.3.3 Mapping Units   72
    5.4 Remote Sensor Platforms   73
        5.4.1 Ground Observation Platform   73
        5.4.2 Airborne Observation Platform   73
        5.4.3 Space-borne Observation Platforms   76
    5.5 Ground Systems   78
    5.6 Satellite Launch Vehicle   78

5.7	Remote Sensing Satellite Orbits	78
	5.7.1 Geosynchronous Orbit	78
	5.7.2 Sunsynchronous Orbit	86
	5.7.3 Shuttle Orbit	86
	5.7.4 Coverage	86
	5.7.5 Passes	86
	5.7.6 Pointing Accuracy	87
	Suggestions for Supplementary Reading	88

6. Remote Sensing Data Acquisition and Dissemination — 90–98

6.1	Radio Communication and Data Transmission	90
6.2	Preprocessing of Satellite Digital Data	92
	6.2.1 Radiometric Correction of Remote Sensor Data	93
	6.2.2 Geometric Correction of Remote Sensor Data	94
6.3	Data Formatting and Archival	95
6.4	Remote Sensing Data Dissemination	95
	Suggestions for Supplementary Reading	98

7. The Sensing — 99–158

7.1	Requirements of Digital Image Processing (Hardware and Software)	99
	7.1.1 Characteristics of Digital Image Processing System	99
7.2	Data Loading and Image Restoration	100
	7.2.1 Data Loading	100
	7.2.2 Image Restoration/Display	101
	7.2.3 Image Reduction	101
	7.2.4 Roam Mechanism	102
	7.2.5 Image Magnification	103
	7.2.6 Transects	103
7.3	Image Rectification and Registration	104
	7.3.1 Image Rectification	104
	7.3.2 Image Registration	106
7.4	Image Statistics Extraction Using Radiometric Data	106
	7.4.1 Univariate Image Statistics	106
	7.4.2 Information obtained from Univariate Image Statistics	108
	7.4.3 Multivariate Image Statistics	109
	7.4.4 Correlation Between Pixel Values in Two Bands k and l	110
	7.4.5 Statistical Evaluation of Image Quality Parameters	112
7.5	Image Enhancement using Spectral Transforms	113
	7.5.1 Linear Contrast Stretch	114
	7.5.2 Saturation Stretch	115

	7.5.3 Non-linear Contrast Stretch	115
	7.5.4 Histogram equalization	115
	7.5.5 Normalization Stretch	116
	7.5.6 Reference Stretch	116
	7.5.7 Thresholding	116
	7.5.8 Color Images	117
7.6	Image Enhancement Using Spatial Transforms	125
	7.6.1 Convolution Filters	126
	7.6.2 Linear Filter	126
	7.6.3 High-Boost Filters (HBF)	127
	7.6.4 Directional Filters	127
	7.6.5 Cascaded Linear Filters	127
	7.6.6 Statistical Filters	129
	7.6.7 Gradient Filters	129
	7.6.8 Fourier Transforms	130
	7.6.9 Power Spectrum	131
7.7	Ground Truth Collection to Support Image Classification	131
	7.7.1 Data Calibration/Correction	131
	7.7.2 Interpretation of Target Properties	131
	7.7.3 Training	131
	7.7.4 Verification	131
7.8	Thematic Image Classification and Information Extraction	132
	7.8.1 Spectral Pattern Recognition	132
	7.8.2 Spatial Pattern Recognition	132
	7.8.3 Temporal Pattern Recognition	132
	7.8.4 Hard and Soft Classification	133
	7.8.5 Supervised Classification	134
	7.8.6 Unsupervised Classification	135
	7.8.7 Minimum-distance-to-means Classifier	136
	7.8.8 Level-slice Classifier	136
	7.8.9 Parallelepiped Classifier	138
	7.8.10 Maximum Likelihood Classifier	138
	7.8.11 K-means Clustering Algorithm	142
	7.8.12 Hybrid Supervised-Unsupervised Classification	143
	7.8.13 Artificial Neural Network (ANN) Classification	143
7.9	Sub-pixel Classification	144
	7.9.1 Fuzzy Set Classification	145
7.10	Hyper-spectral Image Analysis	145
	7.10.1 Visualization of Hyperspectral Image	145
	7.10.2 Training for Classification	146

		7.10.3 Feature Extraction from Hyperspectral Data	146

       7.10.3 Feature Extraction from Hyperspectral Data   146
       7.10.4 Spectral Fingerprints   146
       7.10.5 Absorption-band Parameters   146
       7.10.6 Spectral Derivative Ratio   146
       7.10.7 Classification Algorithms for Hyperspectral Data   147
       7.10.8 Binary Encoding   147
       7.10.9 Spectral-angle Mapping   147
       Suggestions for Supplementary Reading   148

8. Radar Remote Sensing   159–173

   8.1 Introduction   159
   8.2 Parameters Affecting the Radar Return Signals   160
       8.2.1 Radar System Parameters   160
       8.2.2 Terrain/Target Parameters   160
   8.3 Radar Wavelength   161
   8.4 Smooth Surface Criteria   162
   8.5 Terrain Moisture Content and Depth of Penetration   162
   8.6 Depression Angle   162
   8.7 Spatial Resolution   164
       8.7.1 Range Resolution (R_r)   164
       8.7.2 Azimuth Resolution (R_a)   165
       8.7.3 Synthetic Aperture Radar (SAR)   166
   8.8 Satellite Radar Systems   167
       8.8.1 Seasat   167
       8.8.2 Shuttle Imaging Radar (SIR-A)   169
       8.8.3 Shuttle Imaging Radar (SIR-B)   170
       Suggestions for Supplementary Reading   171

9. Global Positioning System (GPS)   174–184

   9.1 What is GPS?   174
   9.2 Functional Segments of the Global Positioning System   175
       9.2.1 The Space Segment   175
       9.2.2 The Ground Control Segment   176
       9.2.3 The User Segment   176
   9.3 How does GPS work?: The Basics of GPS Functioning   176
       9.3.1 Satellite Ranging   176
       9.3.2 Synchronization of Codes   176
       9.3.3 Computation of the 3-D Location of the User GPS Receiver:
             The Method of Triangulation (Geometrical Method)   177
       9.3.4 Algebraic Computation of the 3-D Location of the GPS Receiver   178

		9.3.5 Accuracy of Positional Measurement by GPS Receiver	179

      9.3.5 Accuracy of Positional Measurement by GPS Receiver   179
      9.3.6 Errors in GPS   180
      9.3.7 Improving Efficiency in Location Finding   181
  9.4 Applications of GPS   181
      Suggestions for Supplementary Reading   184

10. Principles of Geographic Information System (GIS)   185–211

  10.1 Introduction   185
  10.2 Spatial Data Representation   187
      10.2.1 Raster   187
      10.2.2 Vector   190
  10.3 Space and Spatial Objects   190
  10.4 Representing Reality   192
  10.5 Spatial Data As Coverage   193
  10.6 Spatial Objects   194
      10.6.1 Points   194
      10.6.2 Lines   194
      10.6.3 Polygons   195
  10.7 Relationships Between Spatial Objects   196
      10.7.1 Adjacency   197
      10.7.2 Connectivity   200
      10.7.3 Containment   202
  10.8 GIS Functions   203
      10.8.1 Data Input Functions   203
      10.8.2 Data Management Functions   204
      10.8.3 Data Manipulation and Analysis Functions   204
      10.8.4 Data Output Functions   207
  10.9 Linkage between Remote Sensing and GIS   208
      10.9.1 Remote Sensing Inputs to Geographic Information System   209
      10.9.2 GIS Input to Remote Sensing System   209
  10.10 Image Based Information System   209
      Suggestions for Supplementary Reading   210

11. Application Areas of Remote Sensing and Geographic Information System   212–246

  11.1 Agricultural Applications   212
      11.1.1 Crop Area Estimation   212
      11.1.2 Crop Stress Monitoring   212
      11.1.3 Crop Production Forecasting   212
  11.2 Water Resources Related Applications   213

	11.2.1 Assessment of Land-surface Water	213
	11.2.2 Assessment of Subsurface Water	213
11.3	Weather and Climate Related Applications	214
	11.3.1 Desertification	214
	11.3.2 Deforestation	215
	11.3.3 Urbanization	216
	11.3.4 Land Use/Land Cover Changes	216
11.4	Forest and Rangeland Related Applications	217
	11.4.1 Forest Fire Hazard Zone Mapping	218
	11.4.2 Rangeland Resource Monitoring	219
11.5	Engineering Applications	220
	11.5.1 Dam Site Selection	220
	11.5.2 River Crossing Site Selection for Bridges	221
11.6	Geologic Hazard Zone Monitoring/Mapping	221
	11.6.1 Seismic Hazard	223
	11.6.2 Geomorphic Hazard	224
	11.6.3 Volcanic Hazard	224
	11.6.4 Glaciologic Hazard	226
11.7	Human-induced Geologic Hazard	226
	11.7.1 Coal Mine Fires	227
	11.7.2 Dam Failure	227
	Suggestions for Supplementary Reading	228

Appendix I : Indian Space Odyssey : The Milestones	247–248
Appendix II : Acronyms	249–255
Appendix III : Glossary	256–278
Appendix IV : Index	279–288

Plate 1

Plate 2

Plate 3

Plate 4

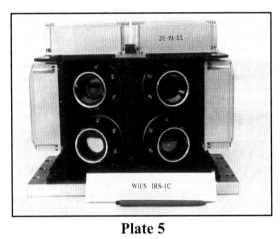

Plate 5

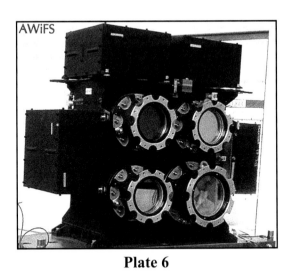

Plate 6

Plate 7

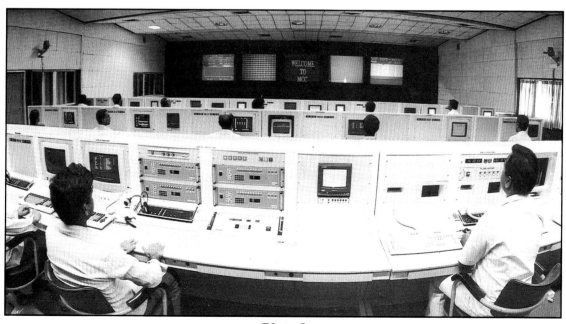

Plate 8

Plate 9

Plate 10

Plate 11

Plate 12

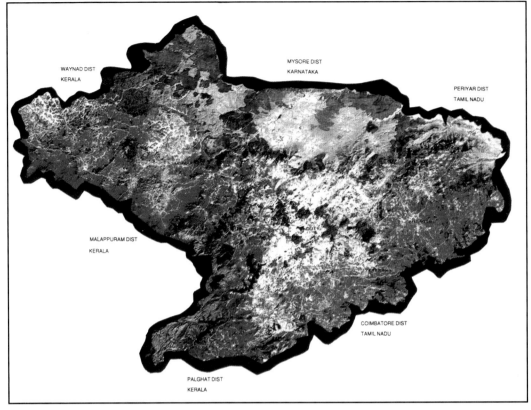

Plate 13

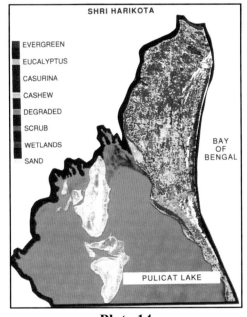

Plate 14

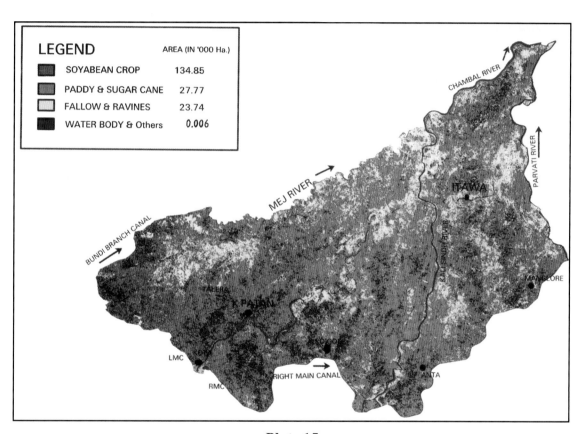

Plate 15

Plate 16

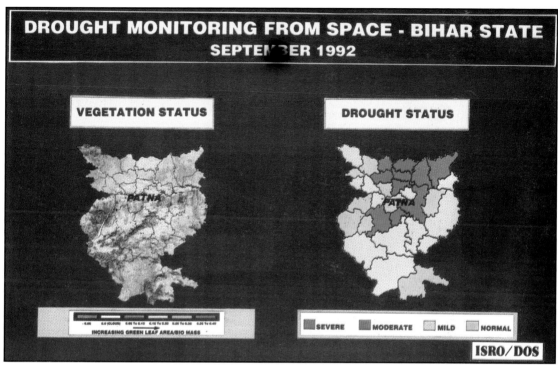

Plate 17

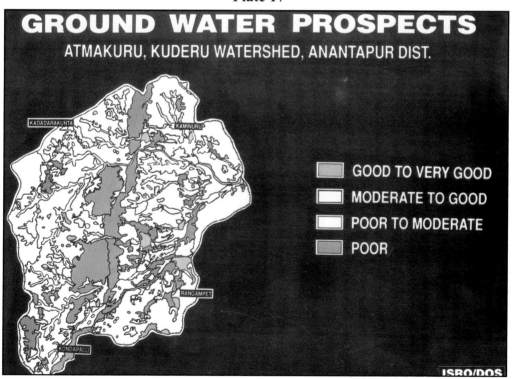

Plate 18

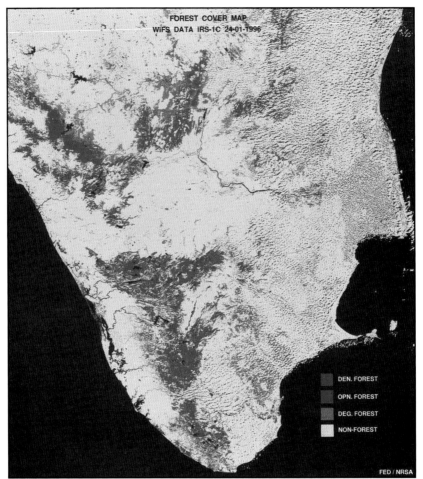

Plate 19

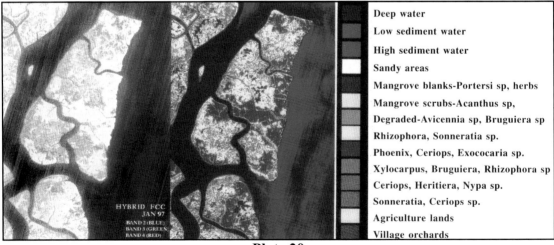

Plate 20

Plate 21

Plate 22

(A)

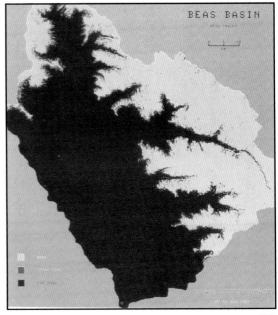

(B)

Plate 23

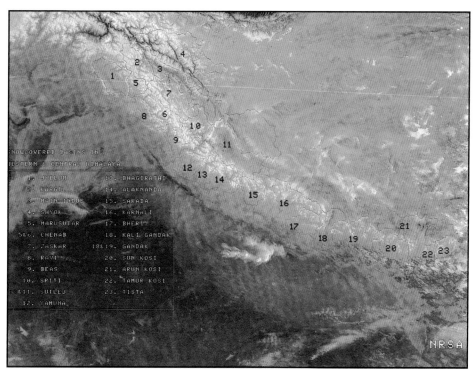

Plate 24

Plate 25

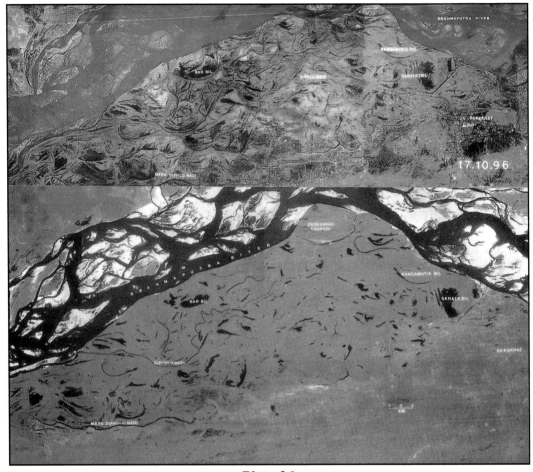

Plate 26

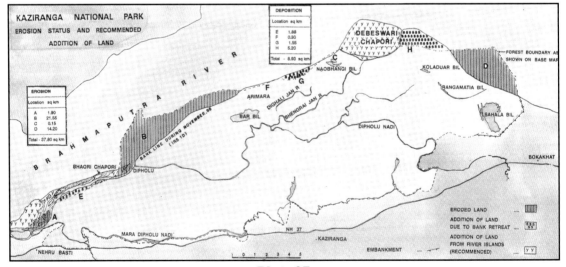

Plate 27

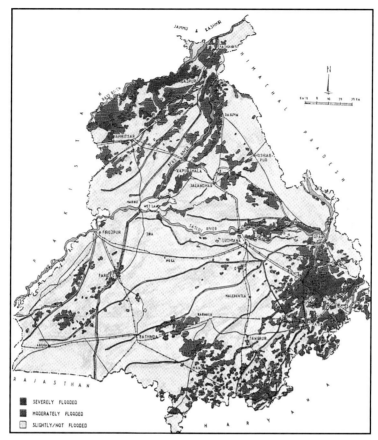

Plate 28

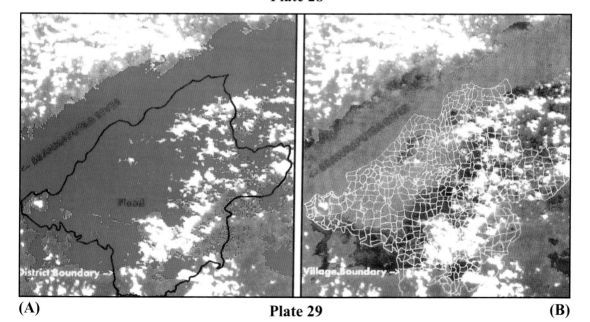

(A)   **Plate 29**   (B)

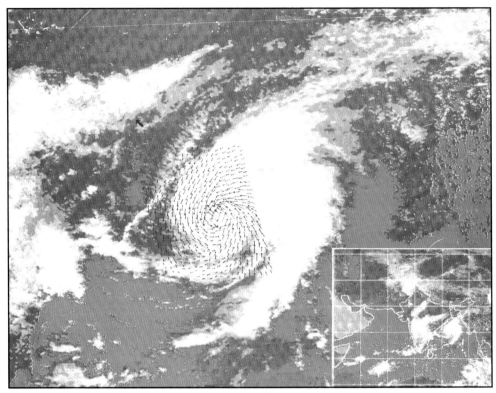

Plate 30

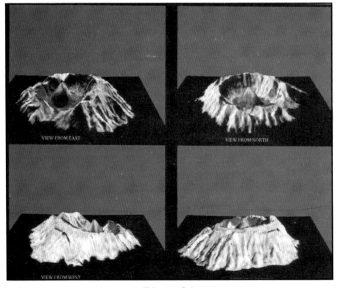

Plate 31

Plate 32

Acknowledgements

At the outset let me say how indebted I am to all the authors of the numerous books and research publications, cited under *Suggestions for Supplementary Reading*, for their classic and innovative contributions to the field of Remote Sensing.

I have consulted these literature, either fully or partly, appreciated their work, picked up the relevant materials, synthesized them and put them in an organized manner in my present book mainly for the benefit of the student community. I wholeheartedly express my gratitude to these great teachers and researchers who have contributed enormously to the quality and information content of this book.

I am deeply grateful to the Indian Space Research Organization, Department of Space, Government of India for adding enough colours to this book as can be found in an organized "Picture/Image Gallery". I am particularly thankful to the Publications & Public Relations Unit of ISRO Headquarters, Bangalore for permitting me to use these colour photographs. I deeply appreciate the kind gesture of Dr. V. Jayaraman, Director, Earth Observation System, ISRO/DOS who not only agreed to go through the manuscript but also to write a Foreword to this book. I thank my affectionate student Dr. S. K. Srivastava, Deputy Director, Disaster Management System, ISRO/DOS for his technical help at the time of my need. My thanks are also due to Mr. Jim Anderson who presented to me a lot of useful illustrative material on geographic information system to be used freely for my students. The constant reminder from a large number of my students to write a text book on Remote Sensing remained with me as a sustained inspiration during the entire period of preparation of the manuscript of this book. I thank them one-and-all for their affection.

I am extremely grateful to Mr. Nishikant Choudhary of Viva Books Private Limited, Delhi for offering me a proposal to publish this book which encouraged me to complete this task in time.

Lastly I would like to appreciate the endurance, love and cooperation of my dear family members – Chhabi, Parthasarathi, Samhita, Bindu and Siddhanta, for whom I could not spare enough time, during the preparation of the manuscript of this book, to be with them and share their moods.

B. C. Panda

Preface

Seeing is believing – it is said. But seeing by an artificial eye through the electromagnetic waveband to which our eyes are not at all sensitive is all the more exciting, as this allows us to explore an entirely new set of information on the earth's surface features using remote sensors from suitable platforms.

I had the opportunity to formulate four courses on remote sensing which were introduced to the Post-Graduate School Syllabus of Indian Agricultural Research Institute, New Delhi. We started offering these courses, in Trimester Mode, from 1987 for the benefit of our Post-Graduate students. Morning shows the day. The number of students from different agricultural and related disciplines registered for these courses on a sustainable basis showed that our remote sensing courses became extremely popular. Exposure to this new subject also opened up job opportunities to our students in Indian Space Research Organization and related governmental and non-governmental organizations including in oversea countries.

Although our students love to possess some text books and reference books on remote sensing, the high cost of these books which mostly come from western countries debarred a majority of these students to have them. Probably it is for this reason that my students used to request me to write a text book on this subject which they can all afford to keep with them. But being occupied in varied activities like teaching, research, extension, administration and external funded project management, I could hardly get enough time to pay attention to this job.

However, after taking superannuation from my service, I thought it proper to concentrate on writing a text book on remote sensing for the benefit of the students in general, and for the Indian students in particular. The generous offer of Viva Books Private Limited, Delhi to publish such a book added enormous encouragement to me to undertake this work sincerely. As a result *Remote Sensing : Principles and Applications,* a text book for graduate and post-graduate students, took its present shape.

Honestly speaking, I do not consider a text book to be a piece of original work. According to me, original work is confined to the field of research – in which the researcher generates his own concept/idea/theory, designs his own experiments for taking observations for verification, writes the necessary program for computation/analysis of the data and also for information extraction, interpretation and applications. But for a text book, the author has to conceptualize the design of the book and accordingly go through the existing literature, understand them, collect the relevant materials, organize them as per the design of the book, put them in a communicable language illustrated with necessary figures and plates, and impart it an appealing get-up. We must also recognize that a text book acquires special flavour and taste if the author of the book is also a teacher who loves teaching and consciously tries to improve his teaching performance to the best of the satisfaction of his students. Moreover, a good text book must possess the quality to inspire the imagination of the students to be enquiring, innovative and aspiring to work on the frontiers of the field of study.

In a modest way, I have tried to design the present book in the pattern of *Physics by the Physical Science Study Committee* – a text book for school students in USA, and the Berkeley Course of Physics. This forced me not to place the references in the body of the text which is allowed to flow naturally as I used to teach in the class. However, the books and research references are given under *Suggestions for Supplementary Reading* after each chapter, from which the materials for the chapter were meticulously chosen.

It will give me great pleasure if this book becomes useful to the student community. To make the best use of this book, I would only advise the stuents and other users to undertake as much of hands-on-training as possible to gain confidence in the technique and speed to further embark upon innovative applications. I shall appreciate constructive suggestions from the readers which will improve the quality of the book in the subsequent editions.

I have put the manuscript directly to my PC. Hence any omission and commission errors that have still escaped our notice will be shared by me and partly by the editorial board of the publisher. Any intimation regarding this will be appreciated which will help us to add a column of Errata to this book.

B. C. Panda

Foreword

Remote Sensing science has always been a fascinating topic over the years, and with the advent of the Earth Observations satellites and a host of advanced instruments with the capability to monitor closely the land-air-ocean interactions, the field has expanded dramatically covering almost all the areas, say, from cartography to climate. The advances in the imaging optics, devices, signal processing and materials, not to speak of the developments in modeling and algorithms, remote sensing, as a science, has seen a quantum jump in the recent times. Further, the developments in the enabling tools such as the image processing, photogrammetry, Geographical Information System, Global Positioning System as well as the emergence of powerful computing systems have further pushed the science, technology and applications of remote sensing to hitherto unexplored areas. The explosive developments calling for 'everything digital' has enabled the integration of the multi-dimensional requirements into a single workstation. There have been newer applications coming up everyday with such enhanced possibilities. Obviously, keeping track of such newer developments has become imperative for the academic community at large.

It is no wonder that with so much happening in remote sensing field, Dr. B.C. Panda, the doyen of agricultural physics, with his passion for the topic, has taken up the onerous task of writing a textbook on *Remote Sensing: Principles and Applications*, bringing all the essence into a single place. The lucid manner in which Dr. Panda has explained the nuances of the advances in the field is to be greatly appreciated and I am sure, this reader-friendly book will not only directly serve the needs of the students as a textbook, but also serve as a reference manual for the professionals working in the field.

(V. Jayaraman)
Director, Earth Observations System
Indian Space Research Organisation
Bangalore

Picture/Image Gallery

This section is intended to take the students and the interested readers round the exciting and illustrated Picture-cum-Image Gallery, and explain to them, at a cultural level, the essential aspects of Remote Sensing and its capability to handle the complex and daunting human problems at local, national and international levels.

The Gallery is not exhaustive, but contains typical pictures and satellite images related to the various components of Remote Sensing covering sensors, sensor platforms, satellite launch vehicle, launch control system, spacecraft control system, telemetry tracking and command network, and a number of application areas covering agriculture, forestry, fishery, soil degradation, land-use/land-cover classification, water resources and snow-cover monitoring, and natural disaster like drought, flood, cyclone, earthquake and volcanic events monitoring and management.

It is expected that by the time the visitors come out of the Gallery, they would have gathered enough of confidence and liking for the subject to enter into serious business with the rest of the book, targetting to acquire a sound knowledge and a satisfying career option in this high-tech field for themselves.

Plate 1. Indian Remote Sensing Satellite (IRS)

Presently India has the largest constellation of operational remote sensing satellites providing continuity of services both at the national and global levels from 1988 onwards. Data from IRS series of satellites are available in a variety of spatial resolutions. The sensors of IRS spacecraft can also collect data of the earth's surface features in several spectral bands. The state-of-the-art Panchromatic Camera of IRS-1C launched in 1995 achieved a ground resolution of 5.8m. IRS-P6 (Resourcesat-1) launched in October 2003 included LISS-IV Camera extending the ground resolution to 5.8m in multi-spectral mode. In addition, it has Advanced Wide-Field Sensor (AWiFS) operating in multispectral mode in visible - near infrared (VNIR) and short wave infrared (SWIR) regions. The forthcoming Remote Sensing Satellites : Cartosat-1, Cartosat-2 and RISAT (Radar Imaging Satellite) will provide imageries of much better spatial resolution with more spectral bands. Plate-1 presents an artistic view of IRS-1C in space.

Plate 2. Charge-Coupled Detector (CCD)

Indian Remote Sensing Satellites collecting earth surface spectral signatures in the visible and near infrared bands of the electromagnetic spectrum employ Charge-

Coupled Devices (Detectors) as sensors. Silicon photo-diodes used as the sensor elements impart stability and long life to the Charge-Coupled Detectors. In Charge-Coupled Detectors, the sensor elements are arranged in linear arrays and hence the Along-Track (Push-Broom) Scanners used in Indian Remote sensing Satellites are called the Linear Imaging Self Scanners (LISS). Plate-1 shows one Charge-Coupled Detector Camera used in IRS-1C.

Plate 3. Linear Imaging Self Scanner (LISS)

IRS-1C Spacecraft carries LISS-III Camera (Plate-3) which is a multi-spectral camera operating in four spectral bands, three in the visible – near infrared (VNIR) range and one in short wave infrared (SWIR) band. LISS-III provides a ground resolution of 23.5m in VNIR bands and 70.5m in SWIR band. This camera operates in the push-broom scanning mode as it employs linear array charge-coupled devices. Construction of LISS-I/II Cameras launched on IRS-1A/1B, and LISS-IV Camera on board the IRS-P6 (Resourcesat-1) is similar to that of LISS-III Camera.

Plate 4. Panchromatic (PAN) Camera

This Plate shows a Panchromatic Camera launched on IRS-1C in 1995. This camera operates in a single panchromatic spectral band (0.50 – 0.75 mm) and provides a spatial resolution of 5.8m at nadir. The Camera works in the push-broom scanning mode as it employs linear array charge-coupled devices and covers a ground swath of 70 Km which is steerable upto ± 26° from nadir in the across-track direction. This off-nadir viewing provides the capability to acquire stereo-pairs from two different orbits and an ability to revisit any given site with a maximum delay of 5 days.

Plate 5. Wide Field Sensor (WiFS) Camera

This WiFS Camera on board the IRS-1C spacecraft has a spatial resolution of 188m and covers a swath of 804 Km. The wide swath helps in achieving a temporal resolution (repetitivity) of 5 days. WiFS Camera operates in two spectral bands of LISS-III, namely 0.62 – 0.68 µm and 0.77 – 0.86 µm. It also operates in the push-broom scanning mode as it employs linear array charge-coupled devices.

Plate 6. Advanced Wide Field Sensor (AWiFS) Camera

IRS-P6 (Resourcesat-1) launched in October, 2003 has an AWiFS Camera on board, as shown in this plate, along with two other sensors : LISS-III and LISS-IV. The Advanced Wide Field Sensor Camera was realized in two electro-optic modules, namely AWiFS-A and AWiFS-B, and provided a combined swath of 740 Km. It provides enhanced capabilities compared to the WiFS Cameras on board the IRS-1C/1D spacecraft in terms of spatial resolution (56m vrs. 188m), radiometric resolution (10 bits vrs. 7 bits) and spectral bands (4 bands vrs. 2 bands), with the additional feature of on-board detector calibration using LEDs.

Picture/Image Gallery

Plate 7. Polar Satellite Launch Vehicle (PSLV)

Indian Space Research Organization started its Satellite Launch Vehicle (SLV) programme with SLV-3 in 1979, followed by the Augmented Satellite Launch Vehicle (ASLV). Now ISRO has commissioned its Polar Satellite Launch Vehicle (PSLV) and Geosynchronous Satellite Launch Vehicle (GSLV). The four stage PSLV is a cost effective vehicle for launching upto 1300 Kg Class satellites into an 800 Km polar orbit. Alternately, it can take two piggyback satellites each weighing upto 100 Kg, along with the main payload of 1000 Kg; or four satellites weighing 300 Kg each. It can also launch 1050 Kg Class satellites into geostationary transfer orbit (GTO). ISRO's PSLV has successfully placed IRS-P2 (1994), IRS-P3 (1996), IRS-1D (1997), IRS-P4 (Oceansat) (1999), India's TES along with Belgian PROBA and German BIRD (2001) and IRS-P6 (Resourcesat-1) (2003) in orbit. Plate-7 shows a close-up view of the PSLV and its lift-off.

Plate 8. PSLV Launch Control Centre (LCC)

PSLV Launch Control Centre is situated in Sriharikota Satellite Launch Complex about six kilometers away from the Launch Tower. The Integrated Launch Vehicle is checked out using an automatic check out system employing four mini-computers and forty micro-computers located in the Launch Control Centre.

Plate 9. Telemetry Tracking and Command (TTC) Antenna, Bangalore

Telemetry Tracking and Command Network for the Indian Remote Sensing Satellites comprise of ISRO Telemetry Tracking and Command (ISTRAC) stations and the External Agency Ground Stations, providing the complete ground segment support of the spacecraft operations, tracking and control activities. Plate-9 shows the TTC Antenna which is an ISTRAC station located at Bangalore.

Plate 10. Main Spacecraft Control Centre (SCC), ISTRAC, Bangalore

The main Spacecraft Control Centre, as shown in this Plate, is situated at Bangalore and has all the mission control facilities. The Centre uses mission computers, the mission-specific software and displays to monitor the spacecraft health and control operations.

Plate 11. Ahmedabad Earth Station

ISRO Telemetry Tracking and Command (ISTRAC) Network performs the spacecraft control operations and TTC functions for the IRS Mission. Ahmedabad Earth Station, shown in this Plate, is one of the earliest established ISTRAC stations.

Plate 12. S-Band Antenna, Sriharikota

For TTC functions, the Spacecraft Control Centre (SCC) is supported by a network of ground stations in S-band. The SCC computer system with its attendant software and displays provides the environment for conducting flight operations and mission analysis. It also supports flight dynamics software for orbit determination and attitude computation. Network communication software in conjunction with dedicated communication links connects all the TTC stations and work centres. Plate-12 shows an S-band Antenna at Sriharikota (SHAR), where India's Satellite Launching Station is situated.

Plate 13. Nilgiri District as viewed by IRS-1B

NRSA Data Centre (NDC) supplies remote sensing data free from geometric and radiometric errors. Using a Digital Image Processing System, the user restores the image and registers it with respect to a map or an existing imagery, and thus produces what is called an FCC. Plate-13 shows a standard FCC of Nilgiri District, Tamil Nadu as obtained by IRS-1B data.

Plate 14. Land Cover Map of Shriharikota

Shriharikota, an island in Pulicat Lake where India's Satellite Launch Complex is situated, is about 80 Km north of Chennai. It is protected from the direct influence of the Bay of Bengal by a strip of main land of Andhra Pradesh. A classified land cover map of Shriharikota and its environs is showed in this Plate.

Plate 15. Crop Area Estimation

At national level, crop production forecasting assumes great importance for policy planners. Crop production being the product of crop yield and crop acreage, it is difficult to run yield models to achieve the state or even the district level yields satisfactorily. While in dry land region, both crop yield and crop acreage vary dramatically with weather and input (specially water) availability, the command areas produce stable and assured yields. Thus as a thumb rule it is possible to assess crop production from command areas with a satisfactory crop acreage estimation. Plate-15 demonstrates the ability of remote sensing technique for soybean crop area estimation in Chambal Command Area.

Plate 16. Land Degradation Monitoring

Development of soil salinity causes land degradation and reduced soil productivity. Left unchecked, this salinization process transforms our arable land into wasteland. Similarly ravinous land, resulting from soil erosion by running water, belongs to the wasteland category. Thus monitoring and reclamation/management of ravines and salinity affected land becomes important for sustainable agricultural production. In Plate-16, the satellite imagery of a part of Uttar Pradesh shows vast stretches of salinity affected land on the sides of Yamuna river and the ravines in the vicinity of the river banks. Also one can find the famous Taj Mahal which stands on the bank of Yamuna in Agra as a white spot in the imagery.

Plate 17. Drought Monitoring from Space

Drought monitoring is an important activity to take corrective measures right at the stage of onset. NOAA/AVHRR data in conjunction with Landsat/IRS data are used to monitor vegetation dynamics in different parts of India during cropping season, and drought bulletins are regularly issued for various districts/regions. Plate-17 shows the drought status of Bihar State during September, 1992.

Plate 18. Ground Water Prospects by Remote Sensing

During 7th Five Year Plan, Government of India launched the National Drinking Water Mission to meet the basic needs of rural population for potable water. Under this Mission, ISRO/DOS prepared hydrogeomorphological maps of the whole country using satellite remote sensing data to prioritize areas potential for ground water prospecting. The success rate of ground water targetting was improved by hydrogeological and geophysical investigations. District-wise maps were made on 1: 250,000 scale. Plate-18 shows ground water prospect map of Atmakuru, Kuderu Watershed, Anantapur District, Andhra Pradesh. The success rate of striking ground water using this approach now stands at 90% compared to less than 50% using the conventional technique.

Plate 19. Forest Cover Map generated from WiFS data

Forest Survey of India (FSI) monitors the forest cover of India biennially using IRS-1B LISS-I data and generates maps on 1 : 250,000 scale, providing information with respect to dense, open and mangrove forests. With 22 days revisit cycle of IRS-1B, this work was found to be difficult to complete in time. However, WiFS on board the IRS-1C spacecraft, with its high temporal resolution (5 day revisit time) wide swath (804 Km) data amply demonstrated its capability to provide rapid information on forest resources for their surveillance, monitoring and mapping. Plate-19 shows a classified output of forest cover of the entire South India developed from WiFS coverage of January 24, 1996.

Plate 20. Classified Output of Sunderbans Mangrove Forest by Remote Sensing

Sunderbans is the largest single block of mangrove in the world consisting of about 64 species of which 20 are important. However, only two species, namely *Heritiera fomes* (Sundri) and *Exococaria agallocha* (Genwa) constitute the major part of the Sunderbans mangrove forest. Plate-20 shows a classified output image of Sunderbans Mangrove Forest obtained from IRS-1D LISS-III data of January 1997.

Plate 21. Man-made Forest Fire for Shifting Cultivation (Jhuming) in Nagaland

Shifting cultivation, alternately known as jhuming is practiced in the North-eastern Hill Regions of India by the local inhabitants. This is clearly demonstrated in the satellite imagery of a part of the Himalayan State of Nagaland (Plate-21) developed from IRS-1A data. While jhuming sites are vividly manifest by greyish and light brownish patches, the red backdrop implies the presence of dense forest in the imagery. At least three category of jhums can be identified in the imagery, namely freshly burnt areas (dark gray tone), recent cropped (pinkish tone) and recently abandoned (gray tone). The ongoing slashing and burning operation (preparatory to cultivation) is evidenced in the Plate by the smoke spreading with the prevailing wind. This practice causes accelerated soil erosion, loss of precious forest wealth and environmental degradation.

Plate 22. Satellite Imagery of Snow-Covered Himalayas

A large quantity of water is locked up in the form of ice and snow in the high Himalayas ranging across 2500 Km from west to east. The permanent snow fields at higher altitudes form glaciers. Some of the mightiest rivers of Asia, like the Ganga, Brahmaputra and Indus are glacier-fed. Plate-22 shows the satellite imagery of a part of the snow covered Himalayas developed from IRS-1B data. The southern part of the imagery shows good vegetation, while the northern Tibetan plateau, devoid of any vegetation, stands out prominently with its glaciated valleys and lakes.

Plate 23. Classified Satellite Image of Snow-Capped River Basins

During summer months when the demand for water is very high, the accumulated ice and snow on different Himalayan basins melt feeding the snow-melt runoff water to the rivers originating from them. Plate-23A shows the Landsat MSS image of the snow-covered Beas and Parbati basins. In Plate-23B, this satellite image of the snow-covered Beas and Parbati basins has been classified into snow, transition (melting snow) and non-snow areas.

Plate 24. Himalayan Basins where Snow-Cover Monitoring is being done regularly

A large number of rivers flowing in the Indo-Gangetic plain originate from their snow-covered basins. Therefore, snow-cover monitoring and mapping are undertaken by NRSA/DOS, Government of India for planning and management of snow-melt runoff water, particularly in multi-purpose projects (like Bhakra Nangal Project) for hydroelectric power generation, irrigation and drinking. Plate-24 shows NOAA-AVHRR imagery of the western and central Himalayas where 23 large basins are demarcated for snow-cover monitoring on a regular basis.

Plate 25. Chlorophyll Image of Parts of the Bay of Bengal and the Arabian Sea

Ocean color remote sensing is a well recognized tool for monitoring optically active bio-geo-chemical parameters like phytoplankton pigments (e.g. chlorophyll), suspended particulate matter and the yellow substance. Routine monitoring of the regional and temporal variability of ocean chlorophyll is essential for the subsequent assessment of the secondary production processes such as marine fisheries. Ocean color information on a global scale is also of importance in studying the bio-geo-chemical cycles of carbon, nitrogen and sulphur. Plate-25 shows the chlorophyll image processed from

IRS-P4 OCM (Ocean Color Monitor) data of June 3,1999 covering parts of the Bay of Bengal and the Arabian Sea. Chlorophyll-a concentration is expressed in mg/m^3. Blue lines represent the rivers.

Plate 26. Kaziranga National Park, Assam as seen from IRS-1C/1D

The oval shaped Kaziranga National Park, Assam – the home of the Indian one-horned Rhinoceros is about 45 Km long east-west and about 15 Km wide at its broadest point. It is bounded on north by the mighty river Brahmaputra and on south by the Karbi Anglng Hills. The National Park suffers from floods and erosion almost every year. Assessment of erosion occurring in Kaziranga National Park being difficult following ground based survey methods, attempts were made by ISRO/DOS to extract this information using multi-date satellite data during 1967 to 1998. Kaziranga National Park as viewed by IRS-1C on October 17, 1996 and by IRS-1D on November 11, 1998 is presented in Plate-26.

Plate 27. Erosion Status of Kaziranga National Park

The analysis of multi-date data of American and Indian Remote Sensing Satellites during 1967 to 1998 has revealed the progressive encroachment of the Brahmaputra river over the Kaziranga National Park, putting pressure on the wildlife of the park. During this study period, the total loss of area was found to be 37.80 square kilometers. It was also found that during this period, at places of the river bank, deposition has taken place on a total area covering 8.93 square kilometers. Moreover, during this period, the Brahmaputra indicated significant changes in its configuration. The river developed braids of different dimensions and orientations in its stream course. Some of these river islands became stable as could be seen from their existence during almost all over the period of study. One such river island called Debeswari Chapori having an extent of about 14.55 square kilometers is identified along the northern boundary and another of about 9.00 square kilometers at the western boundary downstream of Dipholu. Thus in addition to the engineering measures of bank protection, it is necessary to reclaim these two river islands and thereby adding another 23.55 square kilometer area to the Kaziranga National Park, providing a great relief to the wildlife of the park in meeting their primary needs for survival. Plate-27 presents the results of the final analysis on the erosion status and areas recommended for extension of the Kaziranga National Park.

Plate 28. Extent of Flooding in Punjab during 1993

The state of Punjab covering an area of 50,360 square kilometers is a part of the Indo-Gangetic alluvial plains. It has three major rivers – Ravi, Beas and Satluj. These rivers feed a vast network of canals. The unique feature of the canal system is the interlinking achieved for transfer of surplus supplies from one river basin for the use in the other. The flash floods in July 1993 inundated 9,59,940 ha of land, i.e. 19.06 % of the total geographical area of Punjab as compared to 9,22,122 ha

(18.31 %) in September 1988 floods. The flood situation was monitored almost in real time using IRS-1A (LISS-I) and ERS (SAR) data. The classified image map presented in this Plate shows severely flooded, moderately flooded and slightly flooded areas. It is found that Fatehgarh Sahib, Patiala, Rup Nagar, Mansa and Gurdaspur districts are the worst affected ones.

Plate 29. Flood Inundation Study using Remote Sensing and Geo-Database

During south-west monsoon period large parts of eastern and north-eastern India experience severe flooding, causing miseries to people and damage to agricultural crops and properties. On behalf of the Department of Agriculture and Cooperation, Government of India, every year since 1989, NRSA is carrying out flood incidence monitoring, inundation mapping and affected area assessment in the Brahmaputra river valley and the Ganga river basin. They use multi-date WiFS data (804 Km swath in two to three days interval) of IRS-1C/1D/P3 along with Radarsat Scan SAR Wide data (500 Km swath and 5 day revisit period) whenever necessary for this purpose. Plate-29A shows flood inundation area in Marigaon District of Assam based on the analysis of IRS-1D WiFS data of September 10, 1998. In Plate-29B, the geo-database of village boundaries of Marigaon District is superimposed onto the IRS-1D WiFS image of September 13, 1998 using GIS software, clearly indicating a large number of villages which are severely affected by floods. This technique helps in efficient flood management through the supply of relief to the needy, the rescue and rehabilitation of the marooned, damage assessment and provision of security measures for the future. Using multi-date historical remote sensing data, this technique can characterize the terrain in terms of inherent flood vulnerability, and target those who live with different degrees of risk. Thus using Remote Sensing and GIS techniques, it is possible to map the population living with different risk/vulnerability zones in a district, namely in zones flooded every year, twice in 3 years, once in 2 years, once in 3 years and so on. Such studies form an Early Warning System for the flood disaster management.

Plate 30. Cyclone Tracking by INSAT

Tropical cyclone is one of the most destructive natural disasters that affect many countries around the globe and inflicts colossal annual loss in life and property. Its impact is the greatest over the coastal areas, which bear the brunt of the strong surface winds, squalls and flooding from storm rainfall and storm surge as well as ocean wave action. Most cyclonic storms are formed over the ocean where surface meteorological observations are few. Under the circumstance, Geostationary Satellites and NOAA series of meteorological satellites have greatly helped in monitoring and tracking these cyclones over such data-sparse regions. The INSAT imagery of November 19, 1992 (0600 GMT) presented in this Plate shows Bangladesh Cyclone in the Bay of Bengal. ERS-1 Scatterometer derived surface winds beneath the cyclone are superimposed on it.

Picture/Image Gallery

Plate 31. 3-D Perspective Views of Barren Island Volcano

Barren Island Volcano situated in the Andaman Sea of the Bay of Bengal is the only known active volcano in India. After nearly 200 years of dormancy, the volcano erupted in March 1991. Four different 3-D perspective views of this volcanic event, as shown in the imagery, were processed using multi-date Landsat TM (Bands 4, 5, 7) and DEM data. This work clearly demonstrated the utility of remotely sensed data for volcanological studies.

Plate 32. SAR Interferogram for Identifying Coseismic Slip during Earthquakes

Some very creative applications of Synthetic Aperture Radar (SAR) data using interference of images from different epochs have revolutionized our ability to monitor ground deformation at the level of radar fringe widths of a couple of centimeters. Possibility of visualizing such fine structure progressions of the earth's dynamic systems create deep insights into the yet inscrutable processes leading to catastrophes, far in advance of their actual happening. Plate-32 exhibits an interferogram of two SAR imageries obtained by ERS-1 in April and August, 1992, showing the coseismic slip that occurred during the Landers Earthquake of June 1992. Here red lines indicate the trace of the field mapped earthquake rupture, and the white lines show the other known faults in the area, some of which experienced triggered slip during the rupture sequence. Such information is crucial for constraining the strain regimes along faults and estimating their strain reservoir capacities for prognosticating their future behavior. Extensive, routine use of SAR Interferometry, can thus revolutionize the way we study ground deformation and greatly advance our understanding of the mechanics of active fault systems in the country, which would, in turn, help continually refine hazard quantification and risk assessment in the respective regions, as new additional data accrue.

References

Plate 1. Current Science **61** (1991) : 150.
Plate 2. Annual Report, DOS, Govt. of India (1993-94) : 45.
Plate 3. Current Science **70** (1996) : 513.
Plate 4. Current Science **70** (1996) : 513.
Plate 5. Current Science **70** (1996) : 514.
Plate 6. Remote Sensing with Resourcesat-1, NRSA (2004) : 2.
Plate 7. SPACE india (July – September,1993) : 3.
Plate 8. SPACE india (July – September, 1993) : 4.
Plate 9. ISTRAC (Pamphlet, undated), ISRO H.Q., Bangalore.
Plate 10. ISTRAC (Pamphlet, undated), ISRO H.Q., Bangalore.
Plate 11. Annual Report, DOS, Govt. of India (1993-94) : 77.

Plate 12. ISTRAC (Pamphlet, undated), ISRO H.Q., Bangalore.
Plate 13. Annual Report, DOS, Govt. of India (1993-94) : 29.
Plate 14. Annual Report, DOS, Govt. of India (1993-94) : 32.
Plate 15. NNRMS(B)-**24** (January, 2000) : 21.
Plate 16. Current Science **61** (1991) : 218.
Plate 17. ISRO-NNRMS-**TR-98-93** (1993) : Plate-III.
Plate 18. ISRO- NNRMS-**TR-98-93** (1993) : Plate-II.
Plate 19. Current Science **70** (1996) : 608.
Plate 20. NNRMS(B)-**24** (January, 2000) : 35.
Plate 21. Current Science **61** (1991) : 237.
Plate 22. SPACE india (July – September, 1993) : 14.
Plate 23. SPACE india (July – September, 1993) : 15.
Plate 24. SPACE india (July – September, 1993) : 16.
Plate 25. NNRMS(B)-**24** (January, 2000) : 9.
Plate 26. NNRMS(B)-**24** (January, 2000) : 15.
Plate 27. NNRMS(B)-**24** (January, 2000) : 18.
Plate 28. NNRMS(B)-**18** (March, 1994) : 40.
Plate 29. NNRMS(B)-**24** (January, 2000) : 49.
Plate 30. NNRMS(B)-**18** (March, 1994) : 49.
Plate 31. NNRMS(B)-**18** (March, 1994) : 12.
Plate 32. NNRMS(B)-**28** (February, 2003) : 17.

*(**Courtesy:** ISRO, DOS, Govt. of India)*

Introduction

$\boxed{1}$

1.1 CONCEPT OF REMOTE SENSING

Remote sensing as it has been accepted today in the scientific literature is comparatively a young branch of science. But the act of remote sensing is perhaps as old as the origin of life on our planet.

Remote sensing is the sensing of an object or a phenomenon from a remote distance. But then, how remote is remote for remote sensing? Consideration of this type is highly relative and depends on the characters of the signals and the sensors, and also on the attenuation properties of the signal transmission channel. One may also like to know whether there can be sensing without physical contact between the sensor and the object. You may say yes, but the real answer to this question is a big NO. This is because of the fact that under all circumstances the objects to be sensed and the sensor are always intimately bathed in an interacting field, namely the gravitational field, the electromagnetic field and/or the pressure field. These fields are not fictitious but are as real as a lump of matter is. Thus whatever may be the distance between the sensor and the sensed, they are always in contact with each other through the field. What then is the special significance of the contact sensing? In fact, during the so-called contact sensing no true material contact is established since the planetary/surface electrons of the two bodies can not touch each other, being both negatively charged they repel one another and keep themselves at a distance. What happens in reality is that during the so-called contact sensing the fields of both the bodies influence each other so markedly that it results in an appreciable amount of interaction force/pressure, which is sensed or measured. Now, without going any further with the passionate philosophical discussion on what is remote sensing, we may content ourselves saying that practically remote sensing is the science and technology for acquiring information about an object or a phenomenon kept at a distance. Basically it is a nondestructive physical technique for the identification and characterization of material objects or phenomena at a distance.

1.2 ESSENTIAL COMPONENTS OF REMOTE SENSING

Essentially remote sensing has three components :

  The Signal (from an object or phenomenon)
  The Sensor (from a platform), and
  The Sensing (acquiring knowledge about the object or the phenomenon after analysis of the signals, received by the sensor at the user's laboratory)

However, the interaction of the signal with the object by which we obtain information about it, and the interaction of the signal with the transmission channel which reduces the signal strength are given due considerations for detail information extraction. Remote sensing is a branch of Physics, namely Reflectance Spectroscopy which has now found extensive applications in almost every field of human activity.

1.3 NATURAL REMOTE SENSING

Sensing in general and remote sensing in particular can be taken as a measure of life and activity of all living organisms, from microbes to man. Any living organism whose sense organs are well developed can interact with its environment better, live a better life and can protect itself better from its enemies and hostile environments. Taking the example of man, we have five well developed sense organs (natural sensors) : eye, ear, nose, skin and tongue along with highly developed sensing systems – the brain and the nervous system.

1.4 ARTIFICIAL REMOTE SENSING

Seeing is believing. But we human beings can see only through the visible light which forms only a very narrow band out of the extremely broad electromagnetic spectrum. This is because our eye is not sensitive to the wavelengths below the violet and above the red region of the electromagnetic spectrum. Similarly our ear is insensitive to the infrasonic and the ultrasonic frequencies – it can sense only in the audible frequency range. Thus the knowledge about objects obtained through our eyes and ears is but partial. In the field of artificial remote sensing, man merely tries to imitate the already existing remote sensors in us and improve upon them to include wider information channels so that they can be used efficiently to collect detailed information about objects or phenomena.

1.5 PASSIVE AND ACTIVE REMOTE SENSING

A remote sensing system that possesses only a sensor and depends on an external (natural) source to irradiate the target to be sensed is called a passive remote sensing system. As for example, in visible light remote sensing system, the sun, an external natural source, irradiates the target and the reflected light from the target is detected by the sensor. An active remote sensing system, on the other hand, possesses both the sensor and the source to irradiate the target. As for example, the radar, which is an active remote sensing system, transmits microwave pulses from its transmitter antenna to irradiate the target and receives the radar returns by its receiver antenna.

1.6 APPLICATIONS OF REMOTE SENSING

In the sections that will follow, the principles of remote sensing will be developed at a cultural level with the objective to elucidate how the technique of remote sensing can help in different facets of human activities. Any technique of science, the remote sensing technique is no exception, generates new knowledge and it is this knowledge that gives us power. As far as the applications of the scientific technique is concerned, this power may be exercised either for the good or the bad of the humanity. Satellite remote sensing techniques which can collect and deliver information on all parts of the globe may thus be used for constructive purposes in natural resources exploration and management, or for destructive purposes in military applications. Of course, the final choice of its applications should be dictated by the human need, the user's conscience and the policy of a nation.

Introduction

Suggestions for Supplementary Reading

These suggestions are not exhaustive but are limited to the works that will be found especially useful for developing interest and creative imagination in the field of Remote Sensing.

Books

American Society of Photogrammetry, 1983. Manual of Remote Sensing, Vol.I, R. N. Colwell (Ed.), Falls Church, Va.

American Society of Photogrammetry, 1984. Multilingual Dictionary of Remote Sensing and Photogrammetry, Falls Church, Va.

Avery, T. E. and G. L. Berlin, 1992. Fundamentals of Remote Sensing and Airphoto Interpretation, 5th ed., Macmillan Publishing Co., New York.

Barrett, C. E. and Curtis, L. F. 1976. Introduction to Environmental Remote Sensing, John Wiley & Sons, New York.

Campbell, J.B. 1996. Introduction to Remote Sensing, 2nd ed., The Guilford Press, New York.

Curran, P. J. 1985. Principles of Remote Sensing, Longman, London.

Feynman, Richard P., Robert B. Leighton and Matthew Sands, 1963. The Feynman Lectures on Physics, Vol.I, Addition-Wesley Publishing Company, Reading, Mass., USA.

Harper, D. 1983. Eye in the Sky: Introduction to Remote Sensing, 2nd ed., Multiscience, Montreal.

Lintz, Joseph, Jr. and D. S. Simonett, 1976. Remote Sensing of Environment, Addition-Wesley, Reading, Mass.

Schanda, E. 1986. Physical Fundamentals of Remote Sensing, Springer-Verlag, New York.

Toselli, F. (Ed.), 1989. Applications of Remote Sensing to Agrometeorology, Kluwer Academic Publishers, Dordrecht, The Netherlands.

Research Papers

Chandrasekhar, M. G., V. Jayaraman and Mukund Rao, 1996. Future Prospectives of Remote Sensing, Current Sci. (Special Issue), 70 (No.7): 648-653.

Deekshatulu, B. L. and George Joseph, 1991. Science of Remote Sensing, Current Sci. (Special Issue), 61 (Nos. 3 & 4): 129-135.

Hartl, Ph., 1989. Fundamentals of Remote Sensing, In: Applications of Remote Sensing to Agrometeorology, F. Toselli (Ed.), Kluwer Academic Publishers, Dordrecht, The Netherlands. p. 1-18.

Kasturirangan, K., George Joseph, S. Kalyanraman, K. Thyagarajan, M. G. Chandrasekhar, D.V.Raju, S. Raghunathan, A. K. S. Gopalan, K. V. Venkatachari and S. K. Shivakumar, 1991. IRS Mission, Current Sci. (Special Issue), 61(Nos. 3 & 4): 136-152.

Kasturirangan, K., R. Aravamudan, B. L. Deekshatulu, George Joseph and M. G. Chandrasekhar, 1996. Indian Remote sensing Satellite IRS-1C : The Beginning of a New Era, Current Sci. (Special Issue), 70 (No.7): 495-500.

Merideth, R. W., Jr. and A. B. Sacks, 1986. Education in Environmental Remote Sensing: A Bibliography and Characterization of Doctoral Dissertations, Photogramm. Eng. Remote Sensing, 52 : 349-363.

Rao, U. R. 1991. Remote Sensing for National Development, Current Sci. (Special Issue), 61 (Nos. 3 & 4): 121-128.

The Signals 2

Signals are carriers of information. So before we try to analyze the signals to get the information on the object from which they have come, we should like to spend some time to study the characters of the signals at some depth. For meeting the requirements of remote sensing, in general, we can recognize the following types of signals :

Disturbance in a force field

Acoustic signal

Particulate signal, and

Electromagnetic signal

2.1 DISTURBANCE IN A FORCE FIELD

Disturbances in the long range force fields such as the gravitational field, the electric field and the magnetic field are helpful for remote sensing purposes.

2.1.1 Gravitational Field

The gravitational field originates due to the presence of mass. When a test body (of mass m) enters the gravitational field of another body (of mass M) it creates a disturbance in the force field of the latter and experiences a force of attraction whose magnitude is given by

$$F = G \frac{Mm}{r^2} \qquad (2.1)$$

where r is the distance between the centers of the two masses and G is the universal constant of gravitation. In SI units M and m are measured in Kg, r in meter and F in Newton. Therefore, G is given in units of $(N\ m^2 / Kg^2)$. The numerical value of the universal gravitational constant is

$$G = 6.670 \times 10^{-11}\ N\ m^2 / Kg^2 \qquad (2.2)$$

The gravitational force field extends up to infinite distance whose strength falls off according to the inverse square law mentioned above (Eqn. 2.1). The possibility of sensing (measuring) the earth's gravitational field leads to gravity survey in the field of geological prospecting. The gravity survey helps in conceptualizing the geophysical theories of the solid earth and explains the facts concerning the nature and deposition of its crust.

2.1.2 Electric Field

A stationary electric charge establishes a static electric field in space surrounding the charge, so that when a test charge enters the field of this charged body it disturbs the force field of the latter and experiences a force which is either attractive or repulsive depending on whether both the charges are of opposite sign or of the same sign. Quantitatively it is given in the vector form as:

$$\mathbf{F}_2 = (k\ q_1 q_2\ \hat{\mathbf{r}}_{21})/r_{21}^2 \tag{2.3}$$

where q_1 and q_2 are scalars giving the magnitude and sign of the respective charges, $\hat{r}_{21}$ is the unit vector in the direction from charge 1 to charge 2, r_{21} is the distance of separation of the two charges and $\mathbf{F}_2$ is the force acting on charge 2 due to charge 1. This equation expresses the fact that like charges repel and unlike charges attract, and that the force is Newtonian, i.e. $\mathbf{F}_2 = -\mathbf{F}_1$. The constant of proportionality k takes care of the units. If one measures r_{21} in cm, F in dynes and the charges in CGS electrostatic units, then k becomes exactly equal to 1. But in SI units, charge is expressed in coulombs, distance in meters and force in newtons, imparting k a value of 8.9875×10^9. A charge of 1 coulomb = 2.9982×10^9 esu. The electric field extends up to infinite distance, but the field strength falls off according to the inverse square law. Sensing of electric field of a charged body has potential applications to map the electric field of a living organism to elucidate its state of health and developmental functions.

2.1.3 Magnetic Field

A magnet has two poles, conventionally called the north and south poles. A stationary magnet establishes a static magnetic field in the space surrounding it so that when a magnetic test charge enters this field it creates a disturbance in the field and experiences a force of attraction or repulsion depending on whether the interactions are between the unlike or like magnetic charges (poles). Although magnetic poles interact just like the electric charges, since the isolated magnetic charges are not found in nature, the magnetic field strength does not strictly follow the inverse square law that we have found for electric fields. Not only a magnetic pole attracts the opposite pole of another magnet, it can also induce magnetism in some materials present in the earth's crust such as iron, cobalt and nickel, and hence attracts them towards it. This property is utilized in sensing through the magnetic force field. It is made use of in the prospecting of commercial deposits of magnetic materials in geomagnetic surveys. Slow changes in the magnetic field of the earth accompany the wanderings of the magnetic poles, while wilder variations are caused by the fluctuations in the solar wind of charged particles arriving from the sun. The earth's magnetic field extending to a great height above its surface interacts with the incoming charged cosmic ray particles and channelizes them towards the polar regions, thereby screening the terrestrial living organisms from damage by these high energy cosmic ray particles.

2.2 ACOUSTIC SIGNALS

Acoustic signals are generated by a time varying pressure field in a material medium and this is brought about by a vibrating body. The vibration of the body produces compression and rarefaction in a fluid medium, i.e. produces temporal variations in the pressure field, and thus a longitudinal compression wave starts propagating from the vibrating source outwards in the fluid medium. The frequency of the vibrating body determines the frequency of the compression wave in the medium. If the frequencies are less than 20 Hz, the waves are called infrasonic which are not audible. If the frequencies are between 20 and 20,000 Hz then they are audible as sound, and are called sonic waves. The frequencies above 20,000 Hz become inaudible again and are called ultrasonic waves. At normal temperature and pressure, the compression wave travels in air with a speed of 332 m/s. The speed of compression wave is higher in a liquid medium and higher still in a solid medium. However, in a solid medium a sudden bang not only generates a longitudinal compression wave but also a sheer wave which is transverse in character and moves with a speed lower than that of the longitudinal wave in the solid. Due to the transverse character of the sheer wave it can be polarized. During an earth quake or detonation, both the compression and the sheer waves are generated which travel at two different speeds from the epicenter to the tracking stations on the earth's surface. Sensing of acoustic signals helps us in sound ranging for locating the source of the sound., in ultrasonic range for directional signaling and depth sounding for ocean floor mapping, in sonar (sound navigation and ranging) for detection of aircraft and submerged submarines, and for flaw detection in metals. The strength of interaction of acoustic waves and the atmosphere is far greater than that for the electromagnetic waves, while the operational range and the speed of propagation of acoustic waves are far less than those for the electromagnetic waves. Sounding rockets (acoustic echo sounders) are used for upper atmospheric studies in operational meteorology.

2.3 PARTICULATE SIGNALS

So far as the remote signals are concerned , we really get a whole class of energetic particles coming from the distant stars and galaxies, from the deeper regions of outer space. These particles and their antiparticles can be classified into four groups :

Photons : *quanta of electromagnetic energy, rest mass = 0*

Leptons : *light weight particles such as neutrinos, electrons and muons*

Mesons : *medium weight particles such as π mesons, K mesons, ρ mesons and ω mesons, and*

Baryons : *heavy particles such as protons, neutrons, and a whole range of strange particles heavier than the proton*

Out of these particles some are the elementary constituents of ordinary (terrestrial) matter, and yet some others are very very strange to us. From the information these so-called cosmic ray particles bring to us from the deeper layers of the outer space, physicists are in a position to deduce : out of what the matter is made and how it is made. The cosmic ray particles also tell us about the structure and activity of the distant stars and galaxies. The massless and chargeless particle neutrino which can penetrate through large chunks of matter like a planet or a star without being lost, holds great promises in the study of neutrino astronomy to probe the activities in the very core of the distant stars including our sun. Intense solar activity during which bursts of high energy particles come down towards the earth produces magnetic storms, ionospheric disturbances and change in auroral activity.

Apart from the cosmic ray signals, we have another form of particulate signal : the molecular signal, which helps the living organisms to get information about a distant object and its activity. When molecules of volatile essential oil from a flower is received by the sensing organ (sensor) of a honey bee or a butterfly, or by the membrane of our nose, information about the flower is obtained through the process of molecular recognition. Following the molecular signals ants move on a trodden path. Molecular signals : feromones, bring together the opposite sex in the insect and animal kingdom for mating, leading to procreation. Even we can find in the plant kingdom the phenomena of mass flowering, fruiting and senescence in action, may be, as a result of a possible molecular signal recognition.

2.4 ELECTROMAGNETIC SIGNALS

In the foregoing sections we have seen that a stationary electric charge produces a static electric field and a stationary magnet produces a static magnetic field in the surrounding space. Now if we oscillate the charge, it produces oscillations in the electric field which propagates as a transverse electric wave from the source outwards. It is interesting to see that the oscillating electric charge not only produces a transverse electric wave but also simultaneously establishes an oscillating magnetic field which propagates a transverse magnetic wave of the same frequency. Both the electric and magnetic waves propagate with the same velocity, equal to the velocity of light, and have the same frequency and wavelength. Thus they can not be separated from each other, and so the resulting wave is called the electromagnetic wave. The electromagnetic wave is a transverse wave in which the electric and magnetic field vectors vibrate perpendicular to the direction of motion, being always inclined at right angles to each other. We can also generate electromagnetic waves by oscillating a magnet which not only produces oscillations in the magnetic field but also induces an oscillating electric field at right angles to the magnetic field. It is worthwhile recollecting at this point, the historic experiments of Oersted (1820) and Faraday (1791-1831) in which they showed that a current carrying coil induces magnetism on a magnetic material , and that an oscillating magnet induces a fluctuating current on a closed electric circuit respectively, thereby establishing the phenomenon of electromagnetic induction (Fig.2.1 and 2.2).

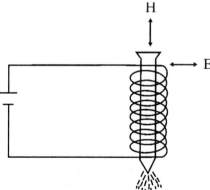

Fig. 2.1 : Hans Christian Oersted (1820), a Danish School teacher demonstrated that a current carrying coil induces magnetism in an iron nail. Oersted's discovery revealed the relationship that the origin of megnetic fields is tied to the motion of electric charges.

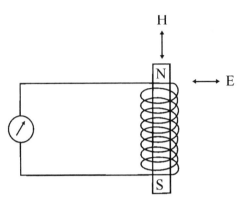

Fig. 2.2 : Michael Faraday (1791-1831) discovered that a magnet when jiggled inside the coil of a conducting wire in a closed circuit, it induces an electric current that flows in the circuit. Oersted and Faraday's observations clearly demonstrated that electricity and magnetism are not only closely related but that they are two different aspects of the same phenomenon. Thus the phenomenon of Electromagnetic induction was firmly established.

James Clerk Maxwell (1831-1879), on the basis of the great discoveries of Oersted and Faraday, succeeded in formulating the laws of electrodynamics through a set of four partial differential equations called the electromagnetic field equations. These equations revealed that the dynamic electric and magnetic field vectors exist together directed perpendicularly to each other being, in turn, perpendicular to the direction of propagation of the field, giving rise to the concept of polarized waves (Fig.2.3). It was also shown that these polarized waves travel with a velocity of 186,000 miles per second – the same velocity at which light travels. Heinrich Hertz (1857-1894) discovered the electromagnetic waves, predicted by Maxwell, which are today called the radio waves and are now so extensively used in wireless telegraphy. Hertz generated the waves with electric sparks and demonstrated their wave nature by forming standing wave patterns. In fact, Hertz showed that Maxwell's electromagnetic waves have all the properties of light and that light itself forms but a very small part of the vast electromagnetic spectrum.

The electromagnetic waves generated by oscillating electric charges all travel with a velocity which is the highest signal velocity that can be attained. This high signal velocity of electromagnetic radiation coupled with its low atmospheric attenuation confers a unique positive advantage to the electromagnetic waves to be used as signals in remote sensing in general and in satellite remote sensing in particular.

2.5 GENERATION OF ELECTROMAGNETIC SIGNALS

The electromagnetic waves generated by the oscillation of electric charges or fields can have various frequencies. Thus the only way these waves distinguish themselves from one another is through their frequencies of oscillation. If we jiggle the charges or fields with higher and higher frequencies, we get a whole spectrum of electromagnetic waves generated. For example, when the frequency of oscillation is of the order of 10^2/s as is the case in AC generators, electromagnetic waves are generated of the frequency 10^2 Hz which are recognized as electrical disturbances. Increasing the frequency of oscillation to 10^5- 10^6 Hz, as is achieved in the radio

The Signals

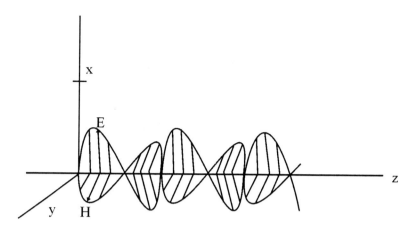

Fig. 2.3 : James Clerk Maxwell (1831-1879), on the basis of the discoveries of Oevsted and Faraday, succeeded not only in formulating the laws of electrodynamics but also in putting forward the concept of polarized electromagnetic waves.

transmitter antenna, we get the electromagnetic waves known as the radio broadcast band. Similarly by a TV transmitter antenna we get FM-TV waves when the frequency of oscillation reaches 10^8 Hz. In electronic tubes such as klystrons and magnetrons, charges can be oscillated with a frequency of the order of 10^{10} Hz, and we get microwaves, radar. For still higher frequencies we go over to the jiggling of atomic and molecular electrons. In the range of frequency from 10^{14} to 10^{15} Hz we sense the electromagnetic waves as visible light. The frequencies below this range are called infrared and above it, the ultraviolet. Shaking the electrons in the innermost shells of atoms produces electromagnetic waves of frequency of the order of 10^{18} Hz, which are known as X-rays. Increasing the frequency of oscillation to 10^{21} Hz, the nuclei of atoms emit electromagnetic waves in the form of γ-rays. From high energy particle accelerators we can get artificial γ-rays of frequency 10^{24} Hz. At times we also find electromagnetic waves of extremely high frequency, as high as 10^{27} Hz, in cosmic rays. Thus we find that the electromagnetic radiation forms a very broad spectrum varying from very very low frequency to very very high frequency, or from very very long wavelength to very very short wavelength. Table-2.1 shows the electromagnetic spectrum with its broad spectral regions and their rough behavior.

Table 2.1 : Electromagnetic spectrum with its broad spectral regions and their rough behavior

Frequency Hz	Wavelength	Name of spectral region	Rough behavior
10^0	300,000 Km	Electrical disturbance/AC	Fields
10^1	30,000 Km	- do -	- do -
10^2	3,000 Km	- do -	- do -
10^3	300 Km	Audio	Waves
10^4	30 Km	- do -	- do -
10^5	3 Km	Radio	- do -

10^6	0.3 Km	- do -	- do -
10^7	30 m	FM – TV	- do -
10^8	3 m	- do -	- do -
10^9	0.3 m	Microwave	- do -
10^{10}	3 cm	- do -	- do -
10^{11}	3 mm	- do -	- do -
10^{12}	0.3 mm	Sub-millimeter / IR	- do -
10^{13}	30 μm	- do -	- do -
10^{14}	3 μm	- do -	- do -
	0.7 μm	Visible	- do -
	0.4 μm	- do -	- do -
10^{15}	0.3 μm	UV	- do -
10^{16}	30 nm	X-rays	- do -
10^{17}	3 nm	- do -	- do -
10^{18}	0.3 nm	- do -	- do -
10^{19}	30 pm	γ- rays (natural)	Particles
10^{20}	3 pm	- do -	- do -
10^{21}	0.3 pm	- do -	- do -
10^{22}	3×10^{-2} pm	γ - rays (artificial)	- do -
10^{23}	3×10^{-3} pm	- do -	- do -
10^{24}	3×10^{-4} pm	- do -	- do -
10^{25}	3×10^{-5} pm	γ -rays (cosmic)	- do -
10^{26}	3×10^{-6} pm	- do -	- do -
10^{27}	3×10^{-7} pm	- do -	- do -

2.6 EMISSION OF ELECTROMAGNETIC SIGNALS FROM A MATERIAL BODY (THERMAL/BLACK-BODY RADIATION)

We have already discussed about the fact that a charge oscillating with a certain frequency (i.e., an accelerated charge) radiates electromagnetic waves of the same frequency. Now let us take an example of an electron oscillating in an atom. This is our charged quantum oscillator which is capable of radiating electromagnetic radiation of frequencies characteristic of the oscillator. In an open system, this electromagnetic radiation never comes to thermal equilibrium with the oscillators since as the oscillators gradually lose energy in the form of radiation, their kinetic energy diminishes and hence the jiggling motion slows down. But when these charged quantum oscillators are imprisoned in an isolated cavity having perfectly reflecting walls, then although initially the oscillators lose energy in the way of radiation and thereby decrease their temperature, in course of time the quantum oscillators maintain an equilibrium temperature. This is because the radiation which does not escape from the cavity gets reflected back and forth and fills the cavity. Under this condition the energy lost by the oscillators in radiation is gained by absorbing energy from the radiation field since they are illuminated by it. We may say that in the cavity electromagnetic radiation has attained thermal equilibrium with matter (oscillators). Now if we make a hole in the cavity and look at it when theoretically its temperature is 0° K, it looks perfectly black. It is for this reason that the cavity is called a black body, and the cavity radiation the black body radiation (Fig.2.4). We have seen that characteristically the black body is a perfect absorber of electromagnetic radiation and also a perfect emitter. On the contrary, a white body is a non-absorber and non-emitter, it is a perfect reflector. As a matter of fact, in the natural world we do not come across a perfect black body or a white body. Natural bodies show a behavior intermediate between a perfect black body and a perfect white body – the gray body.

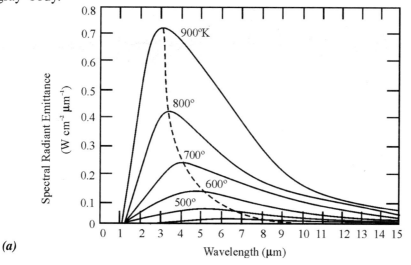

(a)

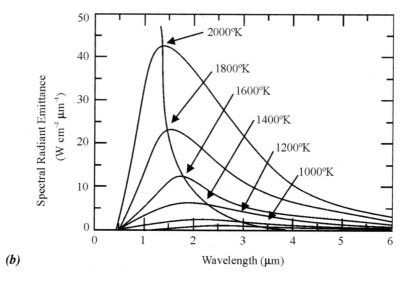

(b)

Fig. 2.4 : Black-body radiation spectra at different temperatures. With the rise in temperature of the black-body, the peak of the black-body spectrum shifts towards smaller wavelengths. This is clearly demonstrated in the two sets of graphs here.

2.7 SOME CHARACTERISTICS OF BLACK-BODY RADIATION

2.7.1 Kirchhoff's Law

Since no real body is a perfect emitter, its emittance is less than that of a black body. Thus the emissivity of a real (gray) body is defined by

$$\varepsilon_g = M_g / M_b \qquad (2.4)$$

where M_g is the emittance of a real (gray) body, and M_b is the emittance of a black body (perfect radiator).

The emissivity of a black body is $\varepsilon_b = 1$, while the emissivity of a white body $\varepsilon_w = 0$. Between these two limiting values, the grayness of the real radiators can be assessed, usually to two decimal places. For a selective radiator ε varies with wavelength. Fig. 2.5 shows the spectral emissivity and spectral radiant emittance as a function of wavelength.

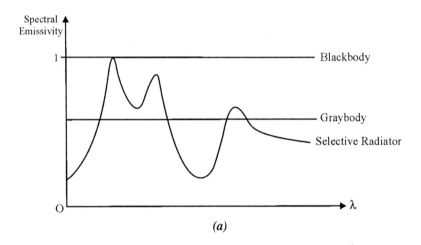

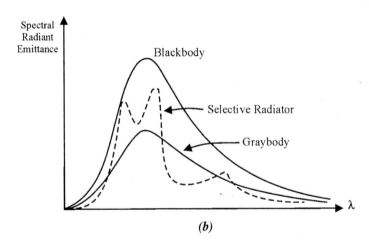

Fig. 2.5 : Characteristic curves of spectral emissivity (ε_λ) and spectral radiant emittance (M_λ) as a function of wavelength (λ) for black-body, gray body and selective radiator.

2.7.2 Planck's Radiation Law

For a radiating body with emissivity ε, the spectral radiant exitance (emittance) is given by Planck's radiation Law as

$$M_\lambda = \frac{\varepsilon\, 8\pi hc}{\lambda^5} \frac{1}{(e^{hc/\lambda kT} - 1)}, \qquad W/(m^2 - \mu m) \qquad (2.5)$$

where ε = emissivity, dimensionless
h = Planck's constant = 6.2559×10^{-34} Js (joule. second)
c = velocity of light = 2.9979×10^8 m/s = 2.9979×10^{14} μm/s
λ = wavelength of radiation
κ = Boltzmann's constant = 1.38×10^{-23} J/K
T = absolute temperature of the radiating body in degree Kelvin (K)

Using the relationship between frequency and wavelength of the electromagnetic radiation

$$\nu = c/\lambda \qquad (2.6)$$

Planck's radiation law (eqn. 2.5) can also be written in terms of frequency as

$$M_\nu = (\varepsilon\, 8\pi h/c^3)\, \nu^3/(e^{h\nu/\kappa T} - 1), \qquad W/(m^2 - Hz) \qquad (2.7)$$

where ν = Radiation frequency, Hz

A useful feature of Planck's law is that it enables us to compute the amount of total spectral radiant emittance which falls between two selected wavelengths or frequencies. This can be useful in remote sensor design, and also in the interpretation of remote sensing observations.

2.7.3 Stefan – Boltzmann Law

If we integrate M_λ or M_ν over all wavelengths or all frequencies, the total radiant emittance M will be obtained for a radiating body of unit area, i.e.

$$M = \int_0^\infty M_\lambda\, d\lambda = \int_0^\infty M_\nu\, d\nu = (\varepsilon\, 8\pi^5 \kappa^4 / h^3 c^3)T^4$$
$$= \varepsilon\, \sigma\, T^4, \qquad W/m^2 \qquad (2.8)$$

where σ is the Stefan–Boltzmann radiation constant, and has the numerical value

$$= 5.669 \times 10^{-8} \ W/(m^2 - K^4)$$

Equation 2.8 is known as the Stefan–Boltzmann radiation law.

2.7.4 Wien's Displacement Law

To obtain the peak spectral radiant emittance, we differentiate M_λ with respect to λ and set it equal to 0, and the resulting equation is solved for λ_{max}. Thus we get

$$\lambda_{max} = \frac{a}{T}, \qquad \mu m \qquad (2.9)$$

where a ≅ 3000 μmK, and T is the absolute temperature of the radiating body.

This shows that as the temperature increases, the peak of M_λ gets displaced towards the shorter wavelengths and the area under the curve increases. Take the examples of the sun and the earth as radiating bodies. For the sun, with a mean surface temperature of 6000 K, $\lambda_{max} \cong 0.5$ µm, and for the earth, with a surface temperature of 300 K, $\lambda_{max} \cong 10$ µm. Thus the wavelength of the maximum radiant emittance from the sun falls within the visible band of the electromagnetic spectrum, whilst that from the earth falls well within the thermal infrared region. We experience the former dominantly as light and the latter as heat.

Another useful parameter, namely the value of emittance at $\lambda = \lambda_{max}$ can be obtained from Planck's radiation law (Eqn. 2.5) by substituting λ by λ_{max} (= a/T as in Eqn. 2.9 above) which yields

$$M_{max} = b\, T^5 \qquad (2.10)$$

where $b = 1.29 \times 10^{-5}$ W/(m^3-K^5). For example, for T = 300 K and a narrow spectral band of 0.1 µm around the peak, the emitted power is about 3.3 W/m^2.

2.7.5 Radiant Photon Emittance

Dividing M_λ or M_v by the associated photon energy $E = hc/\lambda = h\nu$, Planck's radiation law (Eqn.2.5) can be written in terms of the radiant photon flux density. Thus the spectral radiant photon emittance is given by

$$Q_\lambda = (2\pi/\lambda^4)\,[1/(e^{hc/\lambda \kappa T} - 1)] \quad,\quad \text{photons}/(m^3 - s) \qquad (2.11)$$

Integrating over the whole spectrum thus provides the total photon flux emitted by the radiating body of unit area. The Stefan–Boltzmann law for photon flux thus becomes

$$Q = \int_0^\infty Q_\lambda\, d\lambda = \sigma'\, T^3 \qquad (2.12)$$

where $\sigma' = 1.52 \times 10^{15}$ m^2 s^{-1}T^{-3}. This shows that the rate at which the photons are emitted by a radiating body varies as the third power of its absolute temperature. For example, for T = 300 K, the total number of emitted photons becomes equal to 4×10^{22} photons/(m^2– s).

Thus far we have described the generation of electromagnetic radiation. Among its essential properties, it does not require a material medium to propagate. In empty space it travels with the highest signal velocity

$$c = 2.9979 \times 10^8 \text{ m/s}$$

In material medium (non-absorbing or partially absorbing, i.e. transparent or translucent) the velocity of electromagnetic wave (v) is less, depending on the refractive index (µ) or the dielectric constant (k) of the medium :

$$v = \frac{c}{\mu} \qquad (2.13)$$

The dielectric constant of the medium is given by

$$k = \mu^2 \qquad (2.14)$$

The frequency and wavelength of electromagnetic radiation are related by the formula

$$\nu\lambda = c \qquad (2.15)$$

The electromagnetic radiation being transverse in character, it can be polarized and its state of polarization can be changed by reflection, refraction, scattering, absorption and emission.

2.7.6 Radiation Terminology

For the measurement of light we commonly use photometric unit which is based on the assumption that human eye is the ultimate sensor of radiation, and thus the sensitivity of human eye is the basis for the formulation of these units. In remote sensing, however, sensors other than the human eye are used to detect radiation in many other parts of the electromagnetic spectrum to which the human eye is not sensitive. Thus the useful units for measuring radiation in remote sensing are the radiometric units. Table- 2.2 lists the most important of these radiometric quantities along with their symbols, defining expressions and commonly used units.

Table 2.2 : Radiometric and Photometric Quantities

Quantities	Symbols	Defining equations	Commonly used units	
Radiant energy (Spectral radiant energy)	Q	–	joule	(J)
Radiant energy density (Spectral radiant energy density)	W	$W = dQ/dV$	joule/m^3 erg/cm^3	(Jm^{-3})
Radiant flux / Radiant power (Spectral radiant flux)	Φ	$\Phi = dQ/dt$	watt (joule/sec) erg/sec	(W)
Incident flux	Φ_i	$\Phi_i = dQ_i/dt$	watt	(W)
Reflected flux	Φ_r	$\Phi_r = dQ_r/dt$	watt	(W)
Absorbed flux	Φ_a	$\Phi_a = dQ_a/dt$	watt	(W)
Transmitted flux	Φ_t	$\Phi_t = dQ_t/dt$	watt	(W)
Radiant flux density at surface :				
Irradiance (Spectral irradiance)	E	$E = d\Phi/dA$	watt/m^2 watt/cm^2	(W m^{-2}) (W cm^{-2})
Radiant exitance / Radiant emittance (Spectral radiant exitance)	M	$M = d\Phi/dA$	watt/m^2 watt/cm^2	(W m^{-2}) (W cm^{-2})
Radiant intensity (Spectral radiant intensity)	I	$I = d\Phi/d\omega$	watt/steradian	(W sr^{-1})
Radiance (Spectral radiance)	L	$L = dI/(dA\cos\theta)$	watt / (sr. m^2)	(W sr^{-1}m^{-2})
Reflectance	ρ	$\rho = \Phi_r/\Phi_i$	–	–
Transmittance	τ	$\tau = \Phi_t/\Phi_i$	–	–
Absorptance	α	$\alpha = \Phi_a/\Phi_i$	–	–
Hemispherical reflectance (Spectral hemispherical reflectance)	ρ	$\rho = M_r/E$	–	–
Hemispherical transmittance (Spectral hemispherical transmittance)	τ	$\tau = M_t/E$	–	–
Hemispherical absorptance (Spectral hemispherical absorptance)	α	$\alpha = M_a/E$	–	–
Emissivity (Spectral emissivity)	ε	$\varepsilon = M/M_{black\ body}$	–	–

Suggestions for Supplementary Reading

These suggestions are not exhaustive but are limited to the works that will be found especially useful for developing interest and creative imagination in the field of Remote Sensing.

Books

Colwell R. N. (Ed.), *1983. Manual of Remote Sensing,* Vol. I, American Society of Photogrammetry, Falls Church, Va.

Born, M. and E. Wolf, 1964. *Principles of Optics,* Pergamon Press, Oxford.

Campbell, J.B. 1996. *Introduction to Remote Sensing, 2nd ed.,* The Guilford Press, New York.

Cook, C. and M. Bernfeld, 1967. *Radar Signals: An Introduction to Theory and Application,* Academic Press, New York.

Elachi, Charles, 1987. *Introduction to the Physics and Techniques of Remote Sensing,* John Wiley & Sons, New York.

Feynman, Richard P., Robert B. Leighton and Matthew Sands, 1963. The Feynman Lectures on Physics, Vol. I, Addition-Wesley Publishing Company, Reading, Mass., USA.

Harper, D. 1983. Eye in the Sky: Introduction to Remote Sensing, 2nd ed., Multiscience, Montreal.

Papas, C. H., 1965. Theory of Electromagnetic Wave Propagation, McGraw-Hill, New York.

Physical Science Study Committee, 1966. Physics, 2nd ed., D. C. Heath and Company, Boston.

Sabins, Floyd, F., Jr., 1987. Remote Sensing–Principles and Interpretations, 2nd ed., W. H. Freeman and Company, New York.

Toselli, F. (Ed.), 1989. Applications of Remote Sensing to Agrometeorology, Kluwer Academic Publishers, Dordrecht, The Netherlands.

Valasek Joseph, 1956. Introduction to Theoretical and Experimental Optics, 2nd ed., John Wiley & Sons, New York.

Research Papers

Duggin, M. J. and T. Cunia, 1983. Ground Reflectance Measurement Techniques: A Comparison, Appl. Optics, 22: 3771-3777.

Oppenheim, A. V. and J. S. Lim, 1981. The Importance of Phase in Signals, Pro. IEEE. 69: 529-541.

Suits, Gwynn H. 1983. The Nature of Electromagnetic Radiation, In : Manual of Remote Sensing, Vol. I, 2nd ed., R. N. Colwell (Ed.), American Society of Photogrammetry, Falls Church, Va., p. 37-60.

Target–Signal Interaction in Optical Region

3.1 AN OVERALL VIEW

The very fact that electromagnetic radiation incident on a target (object) is partly reflected, partly absorbed and partly transmitted, clearly demonstrates the interaction of electromagnetic radiation with matter. This fact is summarized as

$$\rho_\lambda + \alpha_\lambda + \tau_\lambda = 1 \tag{3.1}$$

where ρ_λ is the reflectance or reflectivity, α_λ is the absorptance or absorptivity, and τ_λ is the transmittance or transmissivity of the object with respect to the electromagnetic radiation (λ) under consideration. The defining equations of the radiometric quantities are given in Table 2.2. The spectral signatures of various earth surface features presented in Figures 3.1 to 3.13 give credence to the above interactions.

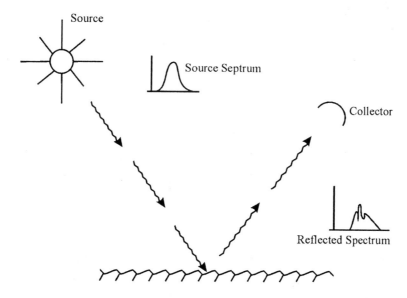

Fig. 3.1 : Basics of Remote Sensing: Radiation modification by target-incoming radiation interaction gives rise to spectral signatures, characteristic of the target, which are essential for remote sensing.

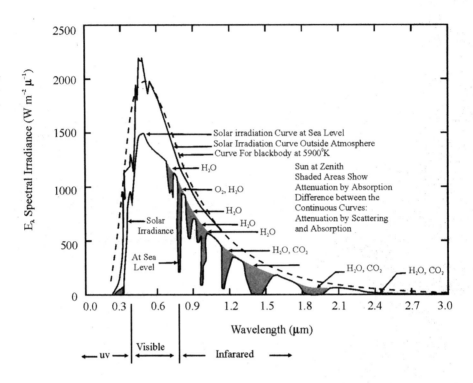

Fig. 3.2 : Insolation spectral irradiance above the atmosphere and on the surface of earth (Adapted from Chahine et al. 1983). This helps in choosing suitable spectral bands not only for monitoring the earth's surface features but also the atmospheric constitutents.

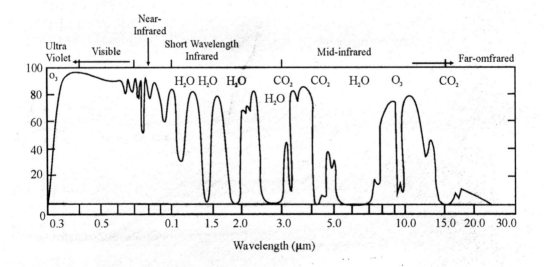

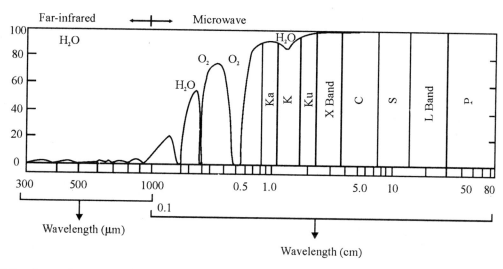

Fig. 3.3 : Atmosphere transmission/absorption spectra of electromagnetic radiation. Atmospheric windows for remote sensing of earth's surface features and atmospheric gases responsible for characteristic absorption bands are indicated. (Adapted from Elachi, 1987)

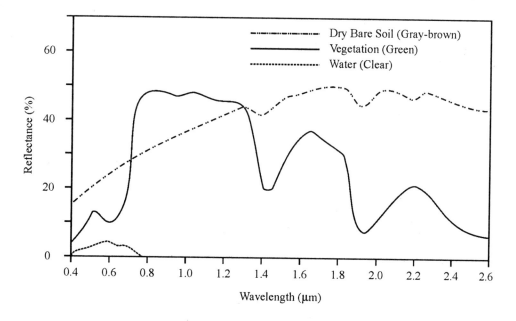

Fig. 3.4 : Typical spectral reflectance signatures of green vegetation, dry soil and clear water (Adapted from Swain and Davis, 1978)

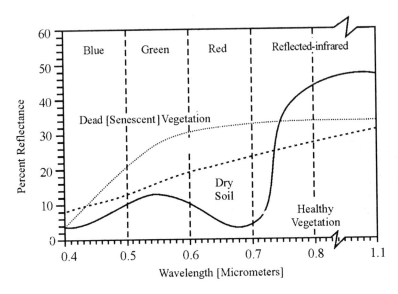

Fig. 3.5 : Typical spectral reflectance signatures of healthy vegetation, dead vegetation and dry soil (from Radhakrishnan et al. 1992)

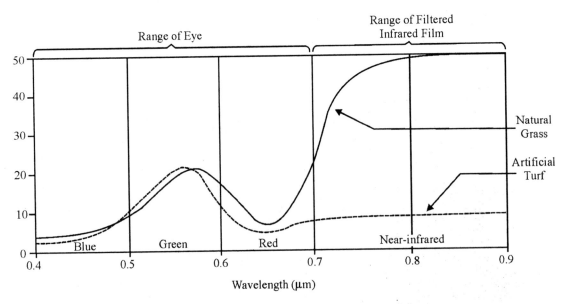

Fig. 3.6 : Generalized spectral reflectance signatures of natural grass and artificial turf (from Lillesand and Kiefer, 1987)

Target — Signal Interaction in Optical Region

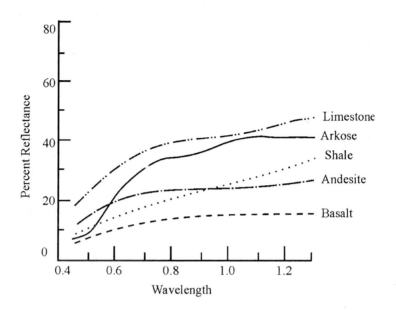

Fig. 3.7 : Typical spectral reflectance signatures of volcanic and sedimentary rocks (from Sabins, Jr. 1987)

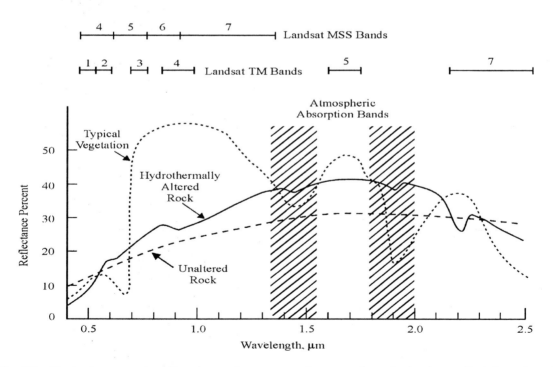

Fig. 3.8 : Typical spectral reflectance signatures of vegetation, hydrothermally altered and unaltered rocks. Also shown the landsat MSS and TM spectral band positions (from Sabins, Jr. 1987)

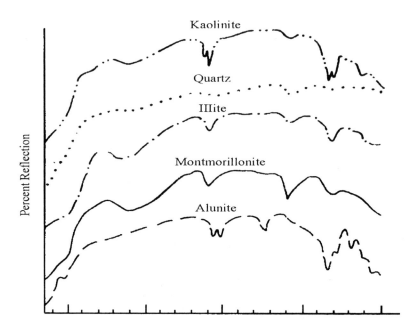

Fig. 3.9 : Spectral reflectance signatures of alteration minerals (Laboratory study). Spectra are offset vertically for clarity. (from Sabins, Jr. 1987)

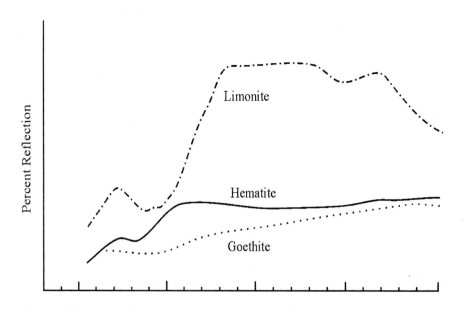

Fig. 3.10 : Spectral reflectance signatures of iron oxide minerals (Laboratory study). Spectra are offset vertically for clarity. (from Sabins, Jr. 1987)

Target — Signal Interaction in Optical Region

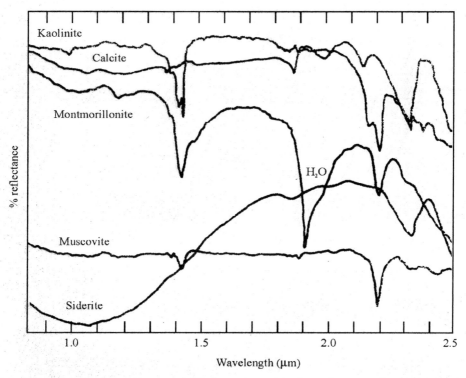

Fig. 3.11 : Spectral reflectance signatures of three dry minerals and two carbonate minerals (Laboratory study) (from Elachi, 1987)

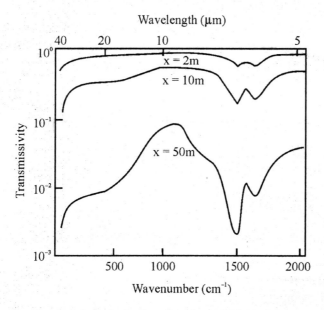

Fig. 3.12 : Spectral transmissivity of clouds of various thicknesses versus wavenumber or wavelength (Yamamoto et al. 1970)

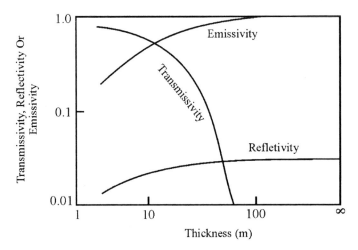

Fig. 3.13 : Emissivity, reflectivity and transmissivity of clouds for the infrared region from 5-50 μ versus cloud thickness in meters (Yamamoto et al. 1970)

If the object (target) is opaque to the radiation, as for example, the earth's crust, then $\tau_\lambda = 0$, and under this condition equation (3.1) reduces to

$$\rho_\lambda + \alpha_\lambda = 1, \quad \text{or} \quad \alpha_\lambda = 1 - \rho_\lambda \tag{3.2}$$

Now, applying the concept that an ideal absorber is an ideal radiator (emitter), we identify absorptance with emittance, and write

$$\rho_\lambda + \varepsilon_\lambda = 1, \quad \text{or} \quad \varepsilon_\lambda = 1 - \rho_\lambda \tag{3.3}$$

This is Kirchhoff's law in a different version which shows that for opaque terrestrial material, the emissivity (absorptivity) and reflectivity (reflectance) are complementary to each other.

Many common terrestrial materials are found to be similar in their thermal emission spectrum in the 8 – 14 μm infrared range with a spectral emissivity between 0.85 – 0.95. Thus for the measurement of temperature of terrestrial materials by remote sensing technique (with the help of an IR thermometer), this 8 – 14μm band is invariably used where it is assumed that the emissivity of all these materials are similar with an estimated emissivity of about 0.9. Table 3.1 summarizes the measured emissivity values of a large number of terrestrial objects.

Table 3.1 : Emissivities of some terrestrial materials in 8-14μm band

Material	Emissivity
Sea water	0.98
Sweet water	0.97
Distilled water	0.96
Snow	0.85
Ice	0.96
Corn canopy	0.94
Cotton canopy	0.96
Tobacco canopy	0.97
Sugarcane canopy	0.99
Meadow grass (dry)	0.88

Meadow grass (green)	0.95
Cactus	0.96
Green tree canopy	0.96
Green conifers (pines)	0.97
Wood	0.90
Granite rock (rough)	0.89
Basalt rock (rough)	0.94
Sand (large grains)	0.91
Sand (large grains, wet)	0.93
Sand (small grains)	0.92
Soil (dry)	0.92
Soil (wet)	0.95
Soil (loam)	0.97
Clay	0.98
Brick	0.93
Concrete	0.92

3.2 SOLAR IRRADIATION ABOVE THE ATMOSPHERE AND AT THE EARTH'S SURFACE

The incoming solar radiation (insolation) that reaches the earth's surface is less than 4μm in wavelength. Fig. 3.2 represents the solar spectral irradiance curves as a function of wavelength at the top of the atmosphere and on the earth's surface at the sea level. It illustrates the atmospheric effects on the solar radiation during its passage to the earth's surface. The extra-terrestrial solar spectrum is comparatively very smooth, strongly resembling the black body spectrum at a temperature of approximately 6000 K. In Fig. 2.4 the black body radiation curves for some terrestrial temperatures are presented.

The total solar irradiance at the mean earth-sun distance was found out to be approximately 1370 W/m^2 above the earth's atmosphere. This irradiance is inversely proportional to the square of the distance from the sun. For the sake of information, Table 3.2 presents the solar irradiance at the surfaces of different planets of our solar system.

Table 3.2 : Solar irradiance at the surface of its different planets

Planets	Mean distance from the Sun (AU)*	Solar irradiance (W/m^2)
Mercury	0.39	9000
Venus	0.72	2640
Earth	1.00	1370
Mars	1.52	590
Jupiter	5.19	50
Saturn	9.51	15
Uranus	19.13	3.7
Neptune	30.00	1.5
Pluto	39.30	0.9

*1 AU (Astronomical Unit) = 150 Million Km

3.3 INTERACTION OF ELECTROMAGNETIC RADIATION WITH ATMOSPHERIC CONSTITUENTS

Electromagnetic radiation from the sun interacts with the atmospheric constituents and gets absorbed or scattered. Essentially two types of scattering takes place : elastic scattering in which the energy of radiation (υ or λ) is not changed due to the scattering, and inelastic scattering in which the energy of the scattered radiation (υ or λ) is changed (Compton scattering and Raman scattering). Inelastic scattering usually takes place when high energy photons are scattered by free electrons in the ionosphere or by the loosely bound electrons of atoms and molecules in the atmosphere.

As regards the elastic scattering, we recognize three types of such scattering phenomena in the atmosphere : Rayleigh scattering, Mie scattering and non-selective scattering depending on the size of the scatterers in relation to the wavelength of radiation (λ) being scattered. The atmospheric scattering processes are summarized in Table 3.3.

Table 3.3 : Atmospheric scattering processes

Scattering process	Wavelength dependence	Approximate particle size (μm)	Kind of particles
Rayleigh	λ^{-4}	$<< 0.1$	Air molecules
Mie	λ^0 to λ^{-4}	0.1 to 10	Smoke, Fumes, Haze
Nonselective	λ^0	> 10	Dust, Fog, Cloud

3.3.1 Rayleigh Scattering

Rayleigh scatterers have size $<< \lambda$. Mostly molecules of the atmosphere satisfy this condition. In Rayleigh scattering the volume scattering coefficient σ_λ is given by

$$\sigma_\lambda = [(4\pi^2 N V^2)/\lambda^4] \, [(\mu^2 - \mu_o^2)^2/(\mu^2 + \mu_o^2)^2] \cong \text{const.}/\lambda^4 \qquad (3.4)$$

where N = number of particles per cm^3
  V = volume of scattering particles
  λ = wavelength of radiation
  μ = refractive index of the particles, and
  μ_o = refractive index of the medium

Rayleigh scattering causes the sky to appear blue. Since the scattering coefficient is inversely proportional to the fourth power of the wavelength, radiation in the shorter blue wavelengths is scattered much more strongly than radiation in the red wavelengths. The red of the sunset is also caused by Rayleigh scattering. As the sun approaches the horizon and its rays follow a longer path through the denser atmosphere, the shorter wavelength radiation is comparatively strongly scattered out of the line of sight, leaving only the radiation in longer wavelengths, red and orange, to reach our eyes. Because of Rayleigh scattering, multispectral remote sensing data from the blue portion of the spectrum is of relatively limited usefulness. In case of aerial photography, special filters are used to filter out the scattered blue radiation due to haze present in the atmosphere.

3.3.2 Mie Scattering

Mie scatterers have size $\approx \lambda$. Water vapor and fine dust particles in the atmosphere satisfy this condition. As a general case, in which there is a continuous particle size distribution, the Mie scattering coefficient is given by

$$\sigma_\lambda = 10^5 \pi \int_{a_1}^{a_2} N(a) \, K(a,\mu) \, a^2 \, da \qquad (3.5)$$

where σ_λ = Mie scattering coefficient at wavelength λ
$N(a)$ = number of particles in interval of radius a and a+da
$K(a,\mu)$ = scattering coefficient (cross section) as a function of spherical particles of radius a and the refractive index of the particles μ

Mie scattering may or may not be strongly wavelength dependent, depending upon the wavelength characteristics of the scattering cross sections. In remote sensing, Mie scattering usually manifests itself as a general deterioration of multispectral images across the optical spectrum under conditions of heavy atmospheric haze.

3.3.3 Non-selective Scattering

Non-selective scatterers have particle size $\gg \lambda$. Water droplets with diameters ranging from 5 – 100 μm scatter all wavelengths of visible light with equal efficiency. As a consequence, clouds and fog appear whitish because a mixture of all colors in approximately equal quantities produces white light. Large sized dust particles are also responsible for non-selective scattering which is the sum of the contributions from the three processes involved in the interaction of the radiation with the particle, i.e. reflection from the surface of the particle with no penetration, passage of the radiation through the particle with or without internal reflections and refraction at the edge of the particle. Non-selective scattering usually results when the atmosphere is heavily dust and moisture ladden and results in a severe attenuation of the received data. However, the occurrence of this scattering mechanism is frequently a clue to the existence of large particulate matter in the atmosphere above the scene of interest, and sometimes this in itself becomes useful data.

We thus find that scattering of electromagnetic radiation by the atmospheric constituents degrades the quality of the directed signals.

3.4 TARGET-SIGNAL INTERACTION MECHANISMS

The radiation-matter interaction mechanisms across the electromagnetic spectrum are summarized in Table 3.4. Examples of remote sensing applications of each spectral regions are also given in this table. Needless to say that a good understanding on the interaction of radiation with terrestrial matter is essential for extracting correct information from remote sensing data analysis.

Table 3.4 : Radiation-matter interaction mechanisms across the electromagnetic spectrum

Spectral region	Main interaction mechanisms	Fields of remote sensing application
γ-rays, X-rays	Atomic processes	Mapping of radio-active materials
UV	Electronic processes	Presence of H and He in atmosphere
VIS, NIR	Electronic and vibrational molecular processes	Surface chemical composition, vegetation cover, biological properties
MIR	Vibrational, vibration-rotational molecular processes	Surface chemical composition, atmospheric chemical composition
Thermal IR	Thermal emission, vibrational and rotational processes	Surface heat capacity, surface temperature, atmospheric temperature, atmospheric and surface constituents
Microwave	Rotational processes, thermal emission, scattering, conduction	Atmospheric constituents, surface temperature, surface physical properties, atmospheric precipitation
Radio Frequency	Scattering, conduction, ionospheric effects	Surface physical properties, surface sounding, ionospheric sounding

3.5 ELECTROMAGNETIC SIGNALS USEFUL FOR SATELLITE REMOTE SENSING (THE ATMOSPHERIC WINDOWS)

The signal transmission pattern of the earth's atmosphere at different wavelength regions of the electromagnetic spectrum is shown in Fig.3.3. The wave bands, narrow or broad, for which the transmission percentage is very high are the spectral regions that are least attenuated by the atmospheric constituents. These spectral regions are called the atmospheric windows, and these channels play important role in remote sensing of the earth resources from satellite platforms. Some of these identified atmospheric windows are given below in Table 3.5. From Fig.3.2 and Fig.3.3 it is seen that the spectral bands of the solar spectrum which are attenuated due to absorption by the water vapor and other atmospheric gases, though can not be used in satellite remote sensing of terrestrial surface features yet they are considered extremely useful for monitoring these absorbing atmospheric gases.

Table 3.5 : Atmospheric windows for terrestrial surface monitoring from space-borne observation platforms

Spectral regions	Spectral bands
VIS (Visible)	0.4 – 0.7 μm
NIR (Near infrared)	0.7 – 1.1 μm
SWIR (Short wave infrared)	1.1 – 1.35 μm

		1.4 – 1.8 μm	
		2.0 – 2.5 μm	
	MWIR (Mid wave infrared)	3.0 – 4.0 μm	
		4.5 – 5.0 μm	
	TIR (Thermal infrared)	8.0 – 9.5 μm	
		10 – 14 μm	
	Microwave	0.1 – 100 cm	

Spectral bands for remote sensing are chosen from the entire spectrum of the electromagnetic spectrum so that they can be used for monitoring thematic earth surface features and phenomena. Table-3.6 presents such spectral bands which are already used by some important sensor systems from different remote sensing observation platforms.

Table 3.6 : Spectral bands used by some important sensor systems from different observation platforms

Sensors	Platforms	Spectral bands (μm)	Spatial resolution (m)
RBV	Landsat 1, 2	B_1 0.475 – 0.575	80
		B_2 0.580 – 0.680	80
		B_3 0.690 – 0.830	80
	Landsat 3	0.505 – 0.750	30
		(Panchromatic into NIR)	
MSS	Landsat 1,2,3,4,5	B_4 0.50 – 0.60	80
		B_5 0.60 – 0.70	80
		B_6 0.70 – 0.80	80
		B_7 0.80 – 1.10	80
MSS (TIR)	Landsat 3 only	B_8 10.4 – 12.5	240
TM	Landsat 4, 5	B_1 0.45 – 0.52	30
		B_2 0.52 – 0.60	30
		B_3 0.63 – 0.69	30
		B_4 0.76 – 0.90	30
		B_5 1.55 – 1.75	30
		B_6 10.4 – 12.5	120
		B_7 2.08 – 2.35	30
HRV (MS-TM)	SPOT 1, 2	B_1 0.50 – 0.59	20
		B_2 0.61 – 0.68	20
		B_3 0.79 – 0.89	20
PAN	SPOT 1, 2	0.51 – 0.73	10

MS-TM	SPOT 3, 4	B_1	0.50 – 0.59	20
		B_2	0.61 – 0.68	20
		B_3	0.79 – 0.89	20
		B_4	1.58 – 1.75	20
		B_5	0.43 – 0.47	20
TV Camera	Bhaskara 1, 2	B_1	0.54 – 0.66	1000
		B_2	0.75 – 0.85	1000
LISS-I	IRS- 1A, 1B	B_1	0.45 – 0.52	72.5
		B_2	0.52 – 0.59	72.5
		B_3	0.62 – 0.69	72.5
		B_4	0.77 – 0.89	72.5
LISS-II	IRS- 1A, 1B, P-2	B_1	0.45 – 0.52	36.25
		B_2	0.52 – 0.59	36.25
		B_3	0.62 – 0.69	36.25
		B_4	0.77 – 0.89	36.25
LISS-III	IRS- 1C, 1D	B_2	0.52 – 0.59	23.5
		B_3	0.62 – 0.69	23.5
		B_4	0.77 – 0.89	23.5
LISS-III (SWIR)	IRS- 1C, 1D	B_5	1.55 – 1.75	70.5
PAN	IRS- 1C, 1D		0.52 – 0.75	5.8
WiFS	IRS- 1C, 1D, P-3	B_3	0.62 – 0.69	188
		B_4	0.77 – 0.89	188
MESSR	IRS : P-3	B_1	0.51 – 0.59	50
		B_2	0.61 – 0.69	50
		B_3	0.72 – 0.80	50
		B_4	0.80 – 1.10	50
VTIRR	IRS : P-3	B_1	0.50 – 0.70	900
		B_2	6.00 – 7.00	2700
		B_3	10.5 – 11.5	2700
		B_4	11.5 – 12.5	2700
MOS-A	IRS : P-3	B_1	0.7570	5810 × 1334
		B_2	0.7606	5810 × 1334
		B_3	0.7635	5810 × 1334
		B_4	0.7664	5810 × 1334
		(Spectral half-width : 0.0014)		
MOS-B	IRS : P-3	B_1	0.408	1450 × 1570

Target — Signal Interaction in Optical Region

		B_2 0.443	1450×1570
		B_3 0.485	1450×1570
		B_4 0.520	1450×1570
		B_5 0.570	1450×1570
		B_6 0.615	1450×1570
		B_7 0.650	1450×1570
		B_8 0.685	1450×1570
		B_9 0.750	1450×1570
		B_{10} 0.815	1450×1570
		B_{11} 0.870	1450×1570
		B_{12} 0.945	1450×1570
		B_{13} 1.010	1450×1570
		(Spectral half-width : 0.010)	
MOS-C	IRS : P-3	B_1 1.600	1500
		B_2 2.200 (SWIR)	1500
OCM	IRS : P-4	B_1 0.402 – 0.422	360×236
		B_2 0.433 – 0.453	360×236
		B_3 0.480 – 0.500	360×236
		B_4 0.500 – 0.520	360×236
		B_5 0.545 – 0.565	360×236
		B_6 0.660 – 0.680	360×236
		B_7 0.745 – 0.785	360×236
		B_8 0.845 – 0.885	360×236
HRIR	Nimbus - 1, 2, 3	B_1 3.40 – 4.20	8000
	NOAA - 1, 2, 3		
HIRS/2	Nimbus-6	B_{1-5} 14.95 – 13.97	17400
		B_{6-7} 13.64 – 13.35	17400
		B_8 11.11	17400
		B_9 9.71	17400
		B_{10-12} 8.16 – 6.72	17400
		B_{13-17} 4.57 – 4.24	17400
		B_{18-20} 4.00 – 0.69	17400
VHRR	NOAA 2,3,4,5	B_1 0.60 – 0.70	1000
		B_2 10.5 – 12.5	1000
AVHRR	NOAA 6, 7, 8, 9, 10	B_1 0.58 – 0.68	1100
		B_2 0.72 – 1.10	1100
		B_3 10.3 – 11.3	4000
		B_4 11.5 – 12.5	4000

HCMR	HCMM	B_1	0.55 – 1.10	500
		B_2	10.5 – 12.5	600
CZCS	Nimbus-7	B_1	0.43 – 0.45	825
		B_2	0.51 – 0.53	825
		B_3	0.54 – 0.56	825
		B_4	0.66 – 0.68	825
		B_5	0.70 – 0.80	825
		B_6	10.5 – 12.5	825
Daedalus MSS	Airborne Observation Platform	B_1	0.38 – 0.42	24
		B_2	0.42 – 0.45	24
		B_3	0.45 – 0.50	24
		B_4	0.50 – 0.55	24
		B_5	0.55 – 0.60	24
		B_6	0.60 – 0.65	24
		B_7	0.65 – 0.70	24
		B_8	0.70 – 0.80	24
		B_9	0.80 – 0.90	24
		B_{10}	0.90 – 1.10	24
MS Camera (S-190A Experiment)	Skylab (Manned Satellite)	B_1	0.70 – 0.80	145
		B_2	0.80 – 0.90	145
		B_3	0.50 – 0.90	145
		B_4	0.40 – 0.70	85
		B_5	0.60 – 0.70	60
		B_6	0.50 – 0.60	60
MSS	Skylab	B_1	0.41 – 0.46	
		B_2	0.46 – 0.51	
		B_3	0.52 – 0.56	
		B_4	0.56 – 0.61	
		B_5	0.62 – 0.67	
		B_6	0.68 – 0.76	
		B_7	0.78 – 0.88	
		B_8	0.98 – 1.08	
		B_9	1.09 – 1.19	
		B_{10}	1.20 – 1.30	
		B_{11}	1.55 – 1.75	
		B_{12}	2.10 – 2.35	
		B_{13}	10.2 – 12.5	
AVIRIS (Hyper-spectral Scanner)	Airborne Observation Platform (U-2 aircraft, NASA)	Spectral Range 0.40 – 2.40 (224 spectral channels of band width 0.01)		20

Target — Signal Interaction in Optical Region

MODIS	Space-borne Observation Platform (EOS, NASA)	B_1	0.620 – 0.670	250
		B_2	0.841 – 0.876	250
		B_3	0.459 – 0.479	500
		B_4	0.545 – 0.565	500
		B_5	1.230 – 1.250	500
		B_6	1.628 – 1.652	500
		B_7	2.105 – 2.155	500
		B_8	0.405 – 0.420	500
		B_9	0.438 – 0.448	500
		B_{10}	0.483 – 0.493	500
		B_{11}	0.526 – 0.536	500
		B_{12}	0.546 – 0.556	500
		B_{13}	0.662 – 0.672	500
		B_{14}	0.673 – 0.683	500
		B_{15}	0.743 – 0.753	500
		B_{16}	0.862 – 0.877	500
		B_{17}	0.890 – 0.920	1000
		B_{18}	0.931 – 0.941	1000
		B_{19}	0.915 – 0.965	1000
		B_{20}	3.660 – 3.840	1000
		B_{21}	3.929 – 3.989	1000
		B_{22}	3.989 – 4.020	1000
		B_{23}	4.020 – 4.080	1000
		B_{24}	4.433 – 4.498	1000
		B_{25}	4.482 – 4.549	1000
		B_{26}	1.360 – 1.390	1000
		B_{27}	6.535 – 6.895	1000
		B_{28}	7.175 – 7.475	1000
		B_{29}	8.400 – 8.700	1000
		B_{30}	9.580 – 9.880	1000
		B_{31}	10.780 – 11.280	1000
		B_{32}	11.770 – 12.270	1000
		B_{33}	13.185 – 13.485	1000
		B_{34}	13.485 – 13.785	1000
		B_{35}	13.785 – 14.085	1000
		B_{36}	14.085 – 14.385	1000
ESMR	Nimbus-6	K_a	0.8 cm (37.5 GHz)	20 Km
SMMR	Seasat Nimbus	K_a	0.8 cm (37.5 GHz)	21 Km
		K	1.4 cm (21.4 GHz)	38 Km
		K_u	1.7 cm (17.6 GHz)	44 Km
		X	3.0 cm (10.0 GHz)	74 Km
		C	4.5 cm (6.6 GHz)	121 Km

MSMR	IRS : P-4	C 4.5 cm (6.6 GHz)	120 Km
		X 2.8 cm (10.65 GHz)	75 Km
		K 1.6 cm (18.7 GHz)	45 Km
		K 1.4 cm (21.3 GHz)	40 Km
		Polarization V and H for all frequencies	
Imaging Radar	Seasat	K 23.5 cm (1.3 GHz)	25 Km
Imagimg Radar	SIR-A	K 23.5 cm (1.3 GHz)	38 Km
Imaging Radar	SIR-B	K 23.5 cm (1.3 GHz)	25 Km
Microwave Radiometer (Dick-type)	Bhaskara-1, 2	K 1.58 cm (19 GHz)	125 Km
		K 1.36 cm (22 GHz)	125 Km
		K_a 0.97 cm (31 GHz)	125 Km

Suggestions for Supplementary Reading

These suggestions are not exhaustive but are limited to the works that will be found especially useful for developing interest and creative imagination in the field of Remote Sensing.

Books

American Society of Photogrammetry, 1983. Manual of Remote Sensing, Vol. I, R. N. Colwell (Ed.), Falls Church, Va.

Bowker, D.E. et.al., 1985. Spectral Reflectances of Natural targets for Use in Remote Sensing Studies, NASA Reference Publication – 1139, National Technical Information Service, Spring-field, Va.

Elachi, Charles, 1987. Introduction to the Physics and Techniques of Remote Sensing, John Wiley & Sons, New York.

Lillesand, Thomas M. and Ralph W. Kiefer, 1987. Remote Sensing and Image Interpretation, 2nd ed., John Wiley & Sons, New York.

Rees, W. G. 1990. Physical Principles of Remote Sensing, Cambridge University Press, Cambridge.

Sabins, Floyd, F., Jr., 1987. Remote Sensing – Principles and Interpretations, 2nd ed., W. H. Freeman and Company, New York.

Swain, P. H. and S. M. Davis, 1978. Remote Sensing : The Quantitative Approach, MacGraw-Hill, New York.

Toselli, F. (Ed.), 1989. Applications of Remote Sensing to Agrometeorology, Kluwer Academic Publishers, Dordrecht, The Netherlands.

Twomey, S. 1977. Introduction to the Mathematics of Inversion in Remote Sensing and Indirect Measurements, Elsevier.

Van de Hulst, H. C. 1957. Light Scattering by Small Particles, Wiley, New York.

Research Papers

Abrams, M. J., R. P. Ashley, L. C. Rowan, A. F. H. Goetz and A. B. Kahle, 1977. Mapping of Hydrothermal Alteration in the Cuprite Mining District, Nevada, using Aircraft Scanner Imagery for the 0.46 – 2.36 µm Spectral Region. Geology, 5 : 713

Brennan, B. and W. R. Bandeen, 1970. Anisotropic Reflectance Characteristics of Natural Earth Surface Materials, Appl. Optics, 9 : 405-412.

Buettner, K. J. K. and C. D. Kern, 1965. Determination of Infrared Emissivities of Terrestrial Surfaces. J. Geophys. Res., 70 : 1329 – 1337.

Chahine, M. T., 1968. Determination of Temperature Profile in an Atmosphere from its Outgoing Radiance. J. Opt. Soc. Am. 58 : 1634.

Chahine, Moustafa T., 1983. Interaction Mechanisms within the Atmosphere, In : Manual of Remote Sensing, Vol. I, 2nd ed., R. N. Colwell, (Ed.), American Society of Photogrammetry, Falls Church, Va., p. 165-230.

Colwell, R. N., W. Brewer, G. Landis, P. Langley, J. Morgan, J. Rinker, J. M. Robinson and A. L. Sorem, 1963. Basic Matter and Energy Relationships involved in Remote Reconnaissance. Photogramm. Eng. 29 : 761 – 799.

Conrath, B. J. 1969. On the Estimation of Relative Humidity Profiles from Medium Resolution Infrared Spectra obtained from a Satellite. J. Geophys. Res., 74 : 3347.

Conrath, B. J. 1972. Vertical Resolution of Temperature Profiles obtained from Remote Radiation Measurements. J. Atmos. Sci., 29 : 1262 – 1271.

Colwell, J. E. 1974. Vegetation Canopy Reflectance, Remote Sensing Environ. 3 : 175-185.

Herman, B. M., S. R. Bowning and R. J. Curran,1971. The Effect of Atmospheric Aerosol on Scattered Sunlight. J. Atmos.Sci., 28 : 419 – 428.

Hovis, W. W., Jr., 1966. Infrared Spectral Reflectance of Some Common Minerals, Appl. Optics, 5 : 245-248.

Hunt, G. R. and J. W. Salisbury, 1970. Visible and Near-infrared Spectra of Minerals and Rocks – I : Silicate Minerals, Modern Geology, 1 : 283-300.

Hunt, G. R. and J. W. Salisbury, 1971 a. Visible and Near-infrared Spectra of Minerals and Rocks – II : Carbonates, Modern Geology, 2 : 23-30.

Hunt, G.R., J. W. Salisbury and C. J. Lenhoff, 1971 b. Visible and Near-infrared Spectra of Minerals and Rocks – III : Oxides and Hydroxides, Modern Geology, 2 : 195-205.

Hunt, G. R., J. W. Salisbury and C. J. Lenhoff, 1971 c. Visible and Near-infrared Spectra of Minerals and Rocks – IV : Sulphides and Sulphates, Modern Geology, 3 : 1-14.

Hunt, G. R., J. W. Salisbury and C. J. Lenhoff, 1972. Visible and Near-infrared Spectra of Minerals and Rocks – V : Halides, Phosphates, Arsenates, Vanadates and Borates, Modern Geology, 3 : 121-132.

Hunt, G. R., J. W. Salisbury and C. J. Lenhoff, 1973 a. Visible and Near-infrared Spectra of Minerals and Rocks – VI : Additional Silicates, Modern Geology, 4 : 85-106.

Hunt, G. R., J. W. Salisbury and C. J. Lenhoff, 1973 b. Visible and Near-infrared Spectra of Minerals and Rocks – VII : Acidic Igneous Rocks, Modren Geology, 4 : 217-224.

Hunt, G. R. and R. P. Ashley, 1974. Spectra of Altered Rocks in the Visible and Near-infrared. Economic Geology, 74 : 1613.

Hunt, G. R., J. W. Salisbury and C. J. Lenhoff, 1974 a. Visible and Near-infrared Spectra of Minerals and Rocks – VIII : Intermediate Igneous Rocks, Modern Geology, 4 : 237-244.

Hunt, G. R., J. W. Salisbury and C. J. Lenhoff, 1974 b. Visible and Near-infrared Spectra of Minerals and Rocks – IX : Basic and Ultrabasic Igneous Rocks, Modern Geology, 5 : 15-22.

Hunt, G. R. and J. W. Salisbury, 1976 a. Visible and Near-infrared Spectra of Minerals and Rocks – XI : Sedimentary Rocks, Modern Geology, 5 : 211-217.

Hunt, G. R. and J. W. Salisbury, 1976 b. Visible and Near-infrared Spectra of Minerals and Rocks – XII : Metamorphic Rocks, Modern Geology, 5 : 219-228.

Hunt, G. R. 1977. Spectral Signatures of Particulate Minerals in the Visible and Near-infrared. Geophysics, 42 : 501 – 513.

Leibacher, J. W., R. W. Noyes, J. Toomre and R. K. Ulrich, 1985. Helioseismology. Scientific American, 253 : 48 – 59.

Lyon, R. J. P. 1965. Analysis of Rocks of Spectral Infrared Emission (8 to 25 µm), Economic Geol., 78 : 618 - 632.

Nassen, K. 1980. The Cause of Color. Scientific American, December Issue.

Radhakrishnan, K., Geeta Varadan and P. G. Diwakar, 1992. Digital Image Processing Techniques – An Overview. In : Natural Resources Management – A New Persoective. R. L. Karale (Ed.), NNRMS, Publications and Public Relations Unit, ISRO HQ, Bangalore.

Rodgers, C. D. 1976. Retrieval of Atmospheric Temperature and Composition from Remote Measurements of Thermal Radiation. Rev. Geophys. Space Phys., 14 : 609 – 624.

Rowan, L. C., A. F. H. Goetz and R. P. Ashley, 1977. Discrimination of Hydrothermally Altered and Unaltered Rocks in Visible and Near-infrared Multispectral Images, Geophysics, 42 : 522-535.

Shaw, J. H. 1970. Determination of Earth's Surface Temperature from Remote Spectral Radiance Observation Near 2600 cm^{-1}. J. Atmos. Sci., 27 : 950.

Smith, James A. 1983. Matter–Energy Interaction in the Optical Region, In : Manual of Remote Sensing, Vol. I, 2nd ed., R. N. Colwell, (Ed.), American Society of Photogrammetry, Falls Church, Va., p. 61-113.

Srivastava, S. K. and B. C. Panda, 1991. On Dynamics of Bidirectional Radiation Scattering Characteristics across and within the Canopies of Wheat, Mustard and Gram. Ann. agric. Res. 12 : 307 – 314.

Srivastava, S. K. and B. C. Panda, 1993. On Dynamics of Radiation Distribution Field and its Impact on Hemispherical Canopy Reflectance. Photonirvachak : J. Indian Soc. Remote Sensing, 21 : 49 – 58.

Tucker, C. J. and L. D. Miller, 1977. Soil Spectra Contributions to Grass Canopy Spectral Reflectance, Photogramm. Eng. Remote Sensing, 43 : 721-726.

Vlcek, J. 1982. A Field Method for Determination of Emissivity with Imaging Radiometers. Photogramm. Eng. Remote Sensing, 48 : 609 – 614.

Westwater, E. R. and O. N. Strand, 1968. Statistical Information Content of Radiation Measurements used in Indirect Sensing. J. Atmos. Sci., 25 : 750 – 758.

Yamamoto, G., M. Tanaka and S. Asano, 1970. Radiative Transfer in Water Clouds in the Infrared Region. J. Atmos. Sci. 27 : 282.

Target–Signal Interaction in Microwave Region

4

Operational microwave remote sensing is both passive and active. Passive microwave remote sensing depends on the emission characteristics of various target surfaces or the media of interest, very much like infrared (emissive part of the electromagnetic spectrum) remote sensing. In active microwave remote sensing which we call radar (radio detection and ranging), pulses of microwave radiation from an artificial source are allowed to illuminate a remote target surface, interact with it and get reflected / scattered. A part of this modified radar return carrying information about the target is received by the sensor antenna for further analysis on the identification and characterization of the remote target.

Radar systems can be of non-imaging and imaging types. Microwave scatterometers belong to the non-imaging radar system. Imaging radar systems include the side looking airborne radar (SLAR) which are further categorized as the real aperture radar (RAR) and the synthetic aperture radar (SAR). These are presented in Fig.4.1, 4.2, 4.3 and 4.4. The major advantages of radar remote sensing are that it is unaffected by the adverse weather conditions (clouds) and it is capable of both day and night operations.

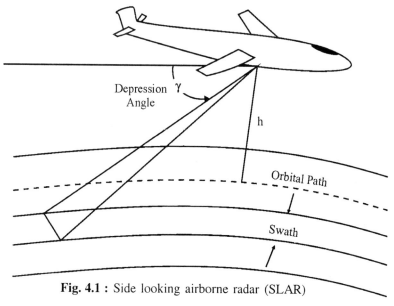

Fig. 4.1 : Side looking airborne radar (SLAR)

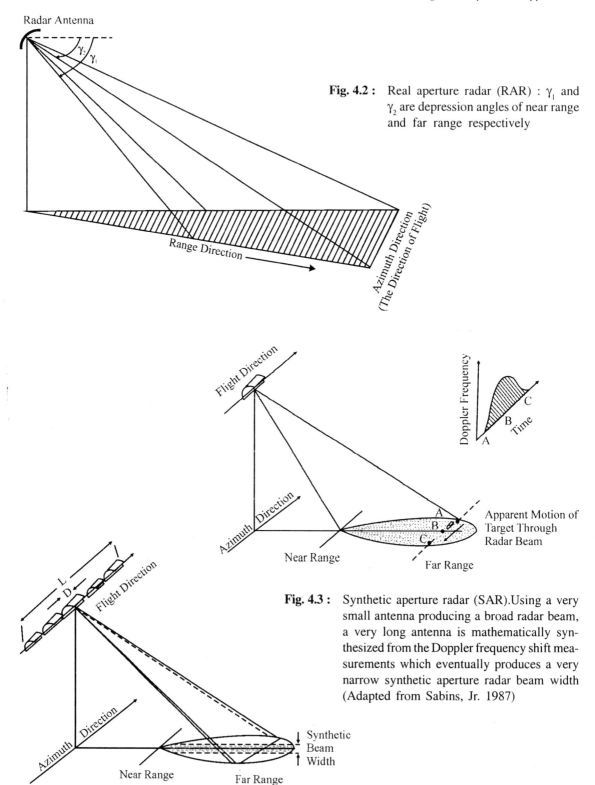

Fig. 4.2 : Real aperture radar (RAR) : γ_1 and γ_2 are depression angles of near range and far range respectively

Fig. 4.3 : Synthetic aperture radar (SAR). Using a very small antenna producing a broad radar beam, a very long antenna is mathematically synthesized from the Doppler frequency shift measurements which eventually produces a very narrow synthetic aperture radar beam width (Adapted from Sabins, Jr. 1987)

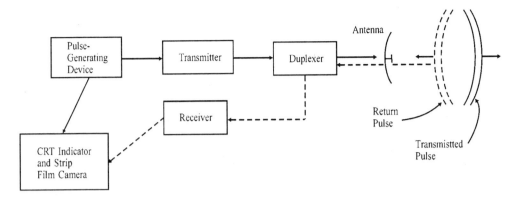

Fig. 4.4 : Block diagram of imaging radar system (from Sabins, Jr. 1987)

4.1 RADAR RETURN FROM CHARACTERISTIC SURFACES

Radar back scattering cross-sections (σ) of targets, which are responsible for displaying different tones on the imageries can be expressed as

$$\sigma = f\,(\lambda,\, \varphi,\, \rho,\, \theta,\, \varepsilon,\, R_1,\, R_2,\, V) \tag{4.1}$$

where φ = incidence angle
ρ = polarization
θ = aspect angle
ε = complex dielectric constant
R_1 = surface roughness (micro scale)
R_2 = subsurface roughness, and
V = complex volume scattering

As many of the radar targets of complicated shapes can be described in terms of facets, spheres, cylinders and corner reflectors, the scattering patterns of these simpler targets are briefly presented below :

4.1.1 Facets

For a conducting rectangular facet with sides l and b parallel to the x- and y-axes, an incident plane polarized wave with either horizontal or vertical polarization in the x-z plane, produces a back scattering cross-section

$$\sigma\,(\theta) = (4\pi l^2 b^2/\lambda^2)[\{\sin(\kappa l \sin\theta)\}/(\kappa l\,\sin\theta)]^2\,\cos^2\theta \tag{4.2}$$

where $\sigma(\theta)$ = back scattering cross-section of the incident wave
λ = wavelength of the incident radiation, and
κ = wave number $(2\pi/\lambda)$

This formula yields good results for θ in the range of 0–45 degrees, beyond which polarization effects becomes significant (Fig.4.5).

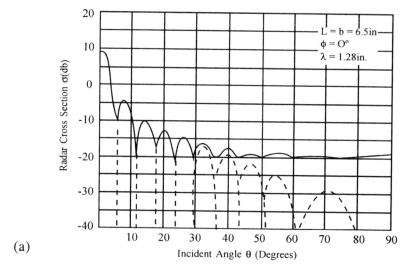

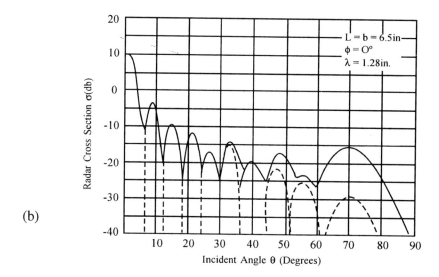

Fig. 4.5 : Radar cross-sections for a plane facet : (a) Effects of vertical polarization, (b) Effects of horizontal polarization (from Ross, 1966)

If the incident wave is defined by the direction (θ, φ), then the corresponding bidirectional back scattering cross-section is given by

$$\sigma(\theta, \varphi) = \frac{4\pi l^2 b^2}{\lambda^2} \left[\frac{\sin(kl \sin\theta \cos\varphi)\sin(kb \sin\theta \sin\varphi)}{(kl \sin\theta \cos\varphi)(kb \sin\theta \sin\varphi)} \right]^2 \cos^2\theta \qquad (4.3)$$

Equation (4.3) shows the variation of σ versus θ with constant φ, to be similar to that of $\sigma(\theta)$ in equation (4.2), except that the widths of the main lobes are now controlled by l and b.

The maximum back scattering cross-section is given by

$$\sigma_{max} = (4\pi l^2 b^2)/\lambda^2 \qquad (4.4)$$

Application : A curved surfaces like the sea surface can be modeled by a collection of facets.

4.1.2 Spheres

Scattering by a sphere is isotropic and is same for both horizontal and vertical polarizations. If the radius of the sphere

$$r \leq \lambda/10$$

then Rayleigh scattering predominates and the back scattering cross-section is given by

$$\sigma = 64\pi^5 \left[(\varepsilon - 1)/(\varepsilon + 2) \right]^2 (r^6/\lambda^4) \qquad (4.5)$$

$$\varepsilon = \varepsilon'_r - (jg/\omega \varepsilon_o) \qquad (4.6)$$

where ε'_r = relative dielectric constant of the sphere
 g = conductivity of the sphere
 ω = angular frequency of the radiation = $2\pi \nu$
 ε_o = dielectric constant of vacuum

For $r > 1.6 \lambda$

The radar back scattering cross-section is given by

$$\sigma = |R|^2 \pi r^2 \qquad (4.7)$$

where $R = (\varepsilon^{1/2} - 1)/(\varepsilon^{1/2} + 1) \qquad (4.8)$

and ε = complex dielectric constant.

For $0.1\lambda < r < 1.6 \lambda$, the radar back scattering cross-section for a conducting sphere is shown in (Fig.4.6).

Applications : Rain and clouds can be modeled by a collection of small spheres.

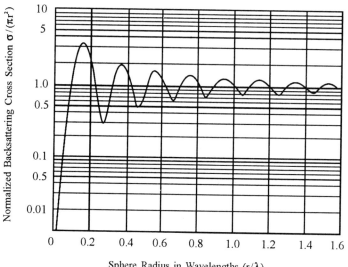

Fig. 4.6 : Radar cross-section for perfectly conducting spheres as functions of radius and wavelength (after Fung and Ulaby, 1983)

4.1.3 Cylinders

When for a conducting circular cylinder (length $l \gg r$), the radius of the cylinder (r) becomes greater than or almost equal to the wavelength of microwave radiation, i.e.

$$r \geq \lambda$$

then the back scattering cross-section for both the transverse magnetic mode and the transverse electric mode is given by

$$\sigma(\theta) = \{(2\pi r l^2 \cos\theta)/\lambda\} \, [\{\sin(\kappa l) \sin\theta\}/(\kappa l \sin\theta)]^2 \tag{4.9}$$

If the radius of the cylinder becomes

$$r \leq \lambda/10$$

then the back scattering cross-section for the transverse magnetic mode is given by

$$\sigma_m(\theta) = (\pi l^2) \, [\{\sin(\kappa l \sin\theta)\}/(\kappa l \sin\theta)]^2 / [\log_e \{(2\pi r \cos\theta)/\lambda\}]^2 \tag{4.10}$$

For the transverse electric mode, the back scattering cross-section is approximately given by

$$\sigma_e(\theta) = (9\pi/4) \, l^2 \, (2\pi r/\lambda)^4 \tag{4.11}$$

For conducting thin circular cylinders and for circular cylinders of average radius which is neither very small nor very large compared to the wavelength λ, the back scattering cross-sections are shown in Fig. 4.7 and 4.8 respectively.

Applications : Scattering from grass and reeds can be described in terms of a collection of cylinders. Similarly, volume distribution of randomly oriented cylinders of various sizes may satisfactorily model a defoliated forest stand.

4.1.4 Corner Reflectors

A dihedral corner reflector consists of two plane surfaces intersecting each other perpendicularly. For such reflectors the maximum scattering cross-section is given by

$$\sigma_{max} = (8\pi \, b^2 \, l^2)/\lambda^2 \tag{4.12}$$

at $\gamma = 90°$, $0°$ and $45°$ for vertical, horizontal and parallel polarizations. Many of the man made structures like a vertical wall and the horizontal surface intersecting each other perpendicularly can be modeled through dihedral corner reflectors. Fig.4.9, 4.10 and 4.11 show the radar cross-sections of a dihedral corner reflector for vertical, horizontal and parallel polarizations.

While monitoring back scattered signals from an extended target such as a cropped field, it is a normal practice to use the scattered signals from an isolated trihedral corner reflector as reference because it has a broad scattering pattern as shown in Fig. 4.12. The radar cross-section becomes maximum along the axis of symmetry of the trihedral corner reflector ($\theta = 0°$, $\varphi = 0°$)

$$\sigma_{max} = (\pi/3) \, (l^4/\lambda^2) \tag{4.13}$$

where l is the length of each edge of the reflector aperture.

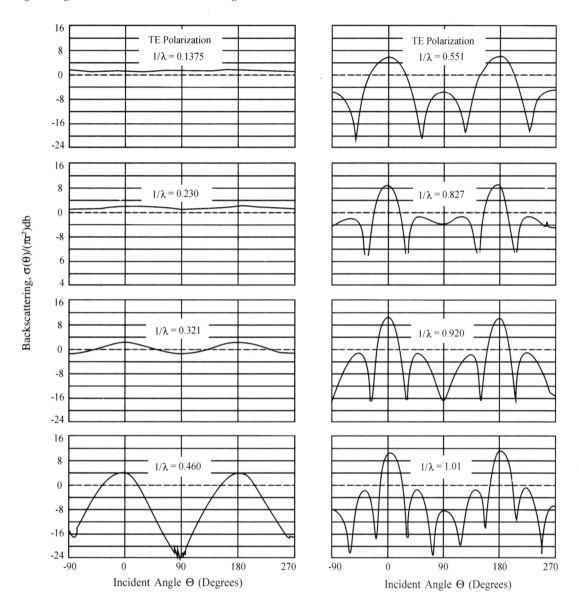

Fig. 4.7 : Backscattering cross-sections of perfectly conducting thin circular cylinders of different lengths. Radius of each cylinder is 0.115λ (after Carswell, 1965)

4.2 EFFECT OF DIELECTRIC PROPERTIES OF TARGETS ON RADAR RETURN

Electromagnetic spectrum in the microwave region interacts with the dielectric properties of natural materials. The complex dielectric constant (ε) of a material consists of a real part (ε') and an imaginary part (ε'') given by

$$\varepsilon = \varepsilon' + j\varepsilon'' \tag{4.14}$$

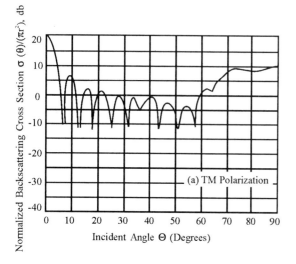
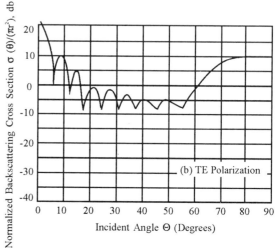

Fig. 4.8 : Backscattering cross-sections of perfectly conducting circular cylinders of average radius for (a) transverse magnetic (TM) polarization and (b) transverse electric (TE) polarization (from Fung and Ulaby, 1983)

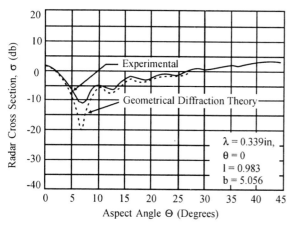

Fig. 4.9 : Radar cross-section of a dihedral corner reflector : For vertical Polarization, $\gamma = 90°$ (Skolnik, 1970)

Fig. 4.10 : Radar cross-section of a dihedral corner reflector : For horizontal polarization, $\gamma = 0°$ (Skolnik, 1970)

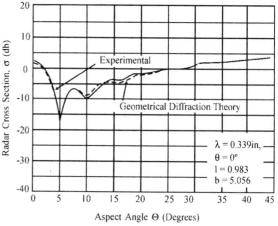

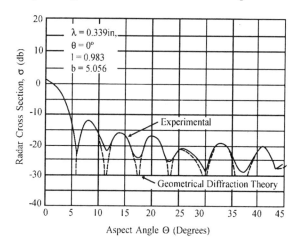

Fig. 4.11 : Radar cross-section of a dihedral corner reflector : For parallel polarization, $\gamma = 45°$ (Skolnik, 1970)

Sometimes ε' is called the relative permittivity and ε'' the loss factor. Higher loss tangent ($\tan\delta$) of a material indicates its higher conductivity status, whereas lower loss tangent shows the dominance of the real part of the dielectric property of the material.

Both real and imaginary parts of the complex dielectric constant of materials vary with temperature and frequency of radiation in the microwave region. Fig. 4.13 and 4.14 show the nature of variation of the dielectric constant of water with temperature and microwave frequency. The variation of dielectric constant (real and imaginary parts) of soil and vegetation as a function of their moisture content are shown in Fig. 4.15 and 4.16.

Basically it is due to this type of interactions of microwave radiation with earth surface materials that the radar return is affected giving rise to identifiable spectral signatures for the surface feature identification and delineation.

4.2.1 Surface and Volume Scattering

The radar return from a natural terrain consists of the contributions of surface back scattering and volume scattering. Since the strength of microwave interaction with material media changes from one type of medium to another type of medium having different dielectric properties, the interface between any two such adjacent media behaves as a surface producing microwave back scatter, usually recognized as surface scattering. The material medium in between two interfaces also interacts with the microwave radiation giving rise to what is called the volume scattering. Thus for example, the radar return from a cropped field can be taken as a combination of the surface back scattering and the volume back scattering. The surface back scattering arises from the crop canopy surface (air-canopy interface), soil surface (crop-soil interface), and from the soil horizons (interface between two soil profiles) depending on the depth of penetration of microwave in the soil. The depth of penetration of microwave into a medium, say soil, is called the skin depth which is given by the relation : $1/\alpha$ where

$$\alpha = [2\pi/\lambda] \, [\varepsilon'/2]^{1/2} \, [(1 + \tan^2\delta)^{1/2}]^{1/2} \qquad (4.15)$$

It is thus found that the smaller the loss tangent ($\tan\delta$), the greater is the penetration. Penetration also increases with the increase in the relative permittivity of the medium and the wavelength of microwave (λ) used for irradiation. Volumetric properties such as density, voids, subsurface layers, gradients etc. also affect the depth of penetration in a significant way. The volume scattering contribution comes from the crop stand and the subsurface soil layers under the present circumstance.

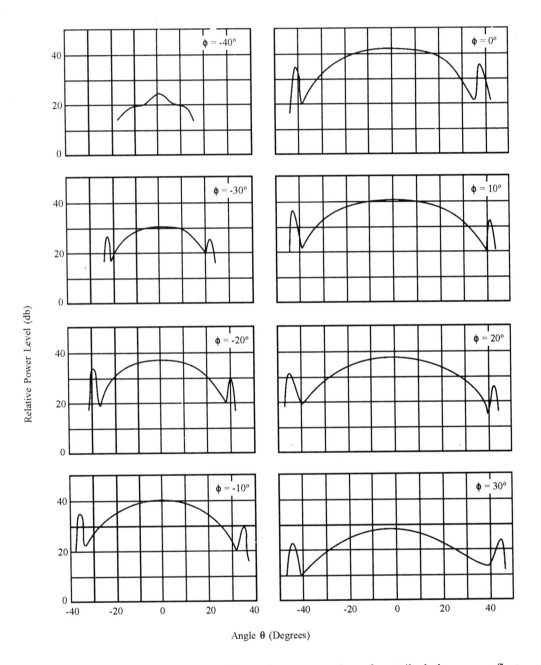

Fig. 4.12 : Angular distribution of the relative rader cross-section of a trihedral corner reflector (after Fung and Ulaby, 1983)

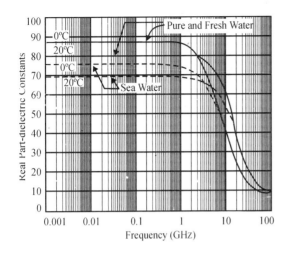

Fig. 4.13 : Dielectric constant of water (real part) as a function of frequency at different temperatures (after Fung and Ulaby, 1983)

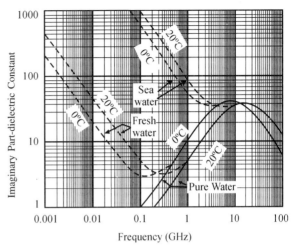

Fig. 4.14 : Dielectric constant of water (imaginary part) as a function of frequency at different temperatures (after Fung and Ulaby, 1983)

4.2.2 Influence of System and Target Parameters on Imagery Gray Tones

The system and target parameters that significantly influence the gray tones on the radar imagery are the following :

System parameters : 1. Wavelength (λ)
                              2. Polarization (HH, VV, HV, VH)
                              3. Look angle/Incidence angle
                              4. Ground resolution cell

Target parameters : 1. Complex dielectric constant/moisture
                              2. Surface/subsurface roughness
                              3. Volume/multiple volume scatterers
                              4. Contrast in neighboring surface features

Larger radar wavelength (λ) not only increases the penetration depth, as discussed earlier, and thereby increasing the scattering contributions of the volume/multiple volume scatterers, but also it defines the roughness/smoothness of surfaces through Rayleigh's criterion :

$$h < [\ \lambda/(8\cos\theta)]\ \rightarrow \text{Smooth surface} \qquad (4.16)$$

where θ is the angle of incidence and h is the measure of the surface roughness or volume inhomogeneities.

In imaging radar, the strength of like polarized returns from targets are stronger than the cross polarized returns from the same targets. The processes responsible for higher like polarized radar returns are quasi-specular surface reflection and surface/volume scattering. The depolarization process that produces cross-polarized radar return are found to be due to

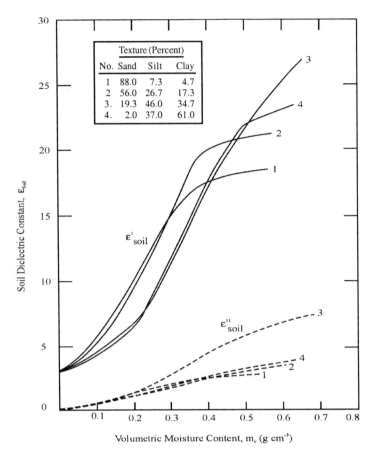

Fig. 4.15 : Relative dielectric constant as a function of volumetric moisture content for four soils measured at 5GHz (from Wang and Schmugge, 1980)

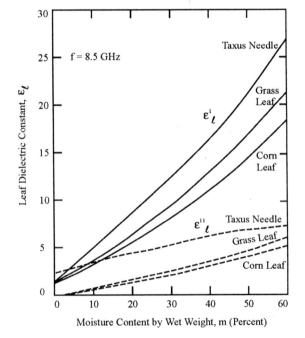

Fig. 4.16 : Relative dielectric constant as a function of gravimetric moisture content of leaves (after Fung and Ulaby, 1983)

Target — Signal Interaction in Microwave Region

(1) quasi-specular reflection for a homogeneous, two-dimensional smoothly undulating surface, (2) multiple scattering by target surface roughness, (3) multiple volume scattering arising out of inhomogeneities, and (4) anisotropic properties of the targets. Fig.4.17 shows the radar return power of different targets as a function of incident angle / depression angle and beam polarization conditions.

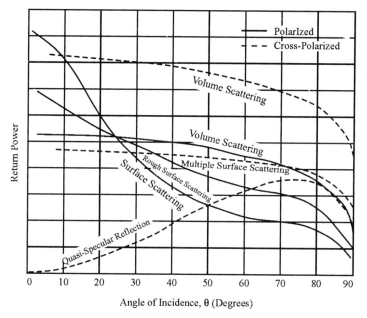

Fig. 4.17 : Radar return power in surface and volume scattering as a function of the angle of incidence, under different polarization conditions (after Fung and Ulaby, 1983)

4.3 MICROWAVE RADIOMETRY

Mechanisms of passive microwave emission from target surfaces is the same as those of passive thermal emission described earlier. Likewise, passive microwave radiometry is similar to the thermal infrared radiometry. However, because of the wavelength differences between the thermal infrared and microwaves, their interaction characteristics with the terrestrial materials and atmospheric constituents become different. The atmospheric transmission windows of microwaves for monitoring earth surface features are found in 1-20 GHz band and around 37 GHz. For atmospheric observations, microwave absorption bands around 22 and 183 GHz for water vapor and around 22 and 118 GHz for oxygen are normally used.

According to Rayleigh- Jeans formula, the brightness (B_b) of a black body in the microwave region is approximately given by

$$B_b = (2kT)/\lambda^2 \qquad (4.20)$$

where k is Boltzmann's constant, T is the temperature of the black body in Kelvin scale, and λ is the wavelength of radiation. For a loss-less narrow beam antenna, the power received from a black body with temperature T over a beam-width Δv is given by Nyquist formula

$$P_b = kT \, \Delta v \qquad (4.21)$$

Since the real terrestrial bodies (gray bodies) emit less energy at a given temperature compared to a black body at the same temperature, the brightness $B(\theta, \varphi)$ of a gray body may be thought of as the brightness of an equivalent black body at a lesser temperature $T_b(\theta, \varphi)$.

$$B(\theta, \varphi) = (2k/\lambda^2) T_B(\theta, \varphi) \quad (4.22)$$

Under the circumstance, a loss-less narrow beam antenna receiving power from a gray body with a brightness temperature $T_B(\theta, \varphi)$ along the direction (θ, φ), over a bandwidth Δv is given by an equivalent formula :

$$P_g(\theta, \varphi) = kT_B(\theta, \varphi) \Delta v \quad (4.23)$$

Thus in principle, the brightness temperature of a body $T_B(\theta, \varphi)$ can be determined from the power received at the antenna, $P_g(\theta, \varphi)$.

4.3.1 Passive Microwave Emission from Target Surfaces

The emissivity of a thermal/microwave radiator is given by

$$e(\theta, \varphi) = B(\theta, \varphi, T)/B_b(T) \quad (4.24)$$

which is the ratio of the brightness of a unit surface area of a gray body at temperature T to the brightness of the same area of a black body at the same temperature T. Since brightness temperature of the body is proportional to the brightness itself, the above equation can also be written as

$$e(\theta, \varphi) = T_B(\theta, \varphi)/T \quad (4.25)$$

where T is the brightness temperature of a black body and $T_B(\theta, \varphi)$ is the equivalent brightness temperature of a gray body emitting in (θ, φ) direction from a unit surface area.

It is found that the observed brightness is dependent on the polarization of the radiometer antenna. Hence it is logical to say that the observed emissivity and brightness temperature of the surface of a body are also polarization dependent. We can thus write

$$e_i(\theta, \varphi) = T_{Bi}(\theta, \varphi)/T \quad (4.26)$$

where $i = h$ for horizontal polarization, and

$i = v$ for vertical polarization of the emitted radiation.

Fig. 4.18 shows the computed emissivity from a plane surface as a function of incidence angle.

The emissivity of a terrestrial thermal radiator is mainly dependent on its shape or surface roughness and its material composition. For example, an infinite, perfectly flat (smooth) surface of homogeneous medium has a polarized emissivity

$$e_{V/H}(\theta, \varphi) = 1 - R_{V/H}(\theta, \varphi) \quad (4.27)$$

where R_V is the Fresnel power reflection coefficient for vertical polarization and R_H for the horizontal polarization.

The primary objective of passive microwave remote sensing is the determination of emissivity $e(\theta, \varphi)$ of the terrain, sea surface and the atmosphere, from which secondary information on

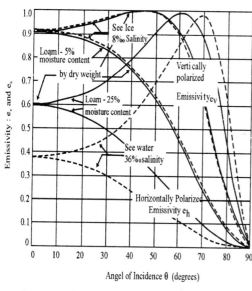

Fig. 4.18 : Computed emissivity from a plane surface as a function of incidence angle at 10GHz (after Fung and Ulaby, 1983)

surface roughness and material composition can be deduced. This can be achieved by independently measuring the surface temperature using infrared radiometer / thermometer, provided the brightness temperature $T_B(\theta, \varphi)$ can also be measured. But it is not possible to measure $T_B(\theta, \varphi)$ with spatial precision because of the smoothing effect of the antenna gain pattern on the brightness temperature distribution pattern as shown in Fig. 4.19. However, the observable antenna temperature T_A which is a gain-weighted sum of the individual brightness temperatures from each direction plays the central role for the evaluation of the emissivity of the radiator surface. Taking the main beam efficiency ε_M (which may be of the order of 0.95) and the side-lobe efficiency $(1 - \varepsilon_M)$, the observable antenna temperature is given by

$$T_A = \varepsilon_M T_{BM} + (1 - \varepsilon_M) T_{BS} \qquad (4.28)$$

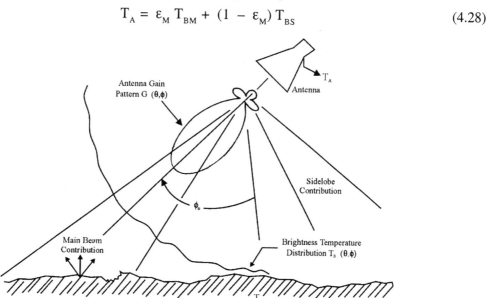

Fig. 4.19 : Main-beam and sidelobe contributions to antenna temperature T_A (after Ulaby and Carver, 1983)

where T_{BM} is the brightness temperature with respect to the main beam and T_{BS} is the brightness temperature with respect to the side-lobe of the antenna gain pattern.

Since
$$(1 - \varepsilon_M) T_{BS} \ll \varepsilon_M T_{BM}$$
one can consider
$$T_A \cong \varepsilon_M T_{BM} \tag{4.29}$$
Thus we may finally have from equations 4.25 and 4.29,
$$e(\theta, \varphi) = T_{BM}/T \cong T_A/T \tag{4.30}$$

While Fig.4.18 shows the variation of microwave emissivity of a plane surface as a function of incident angle, the effect of surface roughness on emission from an inhomogeneous irregular layer is shown in Fig.4.20 to 4.22.

Normalized antenna temperature as a function of incidence angle is presented in Fig. 4.23 for a bare field with smooth soil surface of different soil moisture conditions. In Fig.4.24 the response of normalized antenna temperature to soil moisture content of a smooth bare field under nadir view is shown. The effect of surface roughness on microwave emission from a bare soil at 1.4 GHz is given in Fig. 4.25.

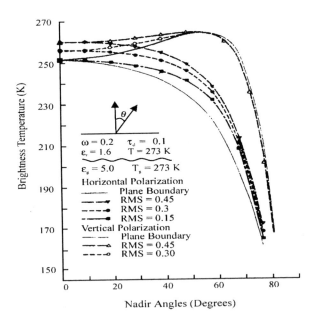

Fig. 4.20 : Effect of surface roughness on emission from an inhomogeneous layer with an irregular bottom interface (from Fung and Chen, 1981)

Target — Signal Interaction in Microwave Region

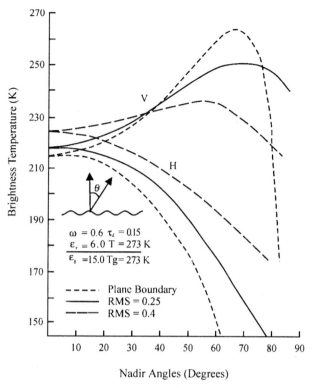

Fig. 4.21 : Effect of surface roughness on emission from an inhomogeneous layer with an irregular top interface (from Fung and Chen, 1981)

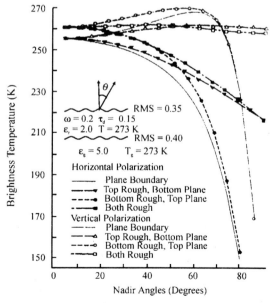

Fig. 4.22 : Effect of surface roughness on emission from an inhomogeneous irregular layer (from Fung and Chen, 1981)

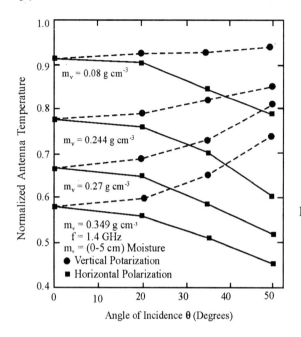

Fig. 4.23 : Normalized antenna temperature as a function of incidence angle for a bare field with smooth soil surface, for each of four different soil moisture conditions (from Newton and Rouse, 1980)

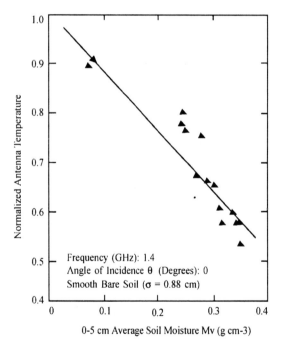

Fig. 4.24 : Response of normalized antenna temperature to soil moisture content of a smooth, bare field under nadir view (from Newton and Rouse, 1980)

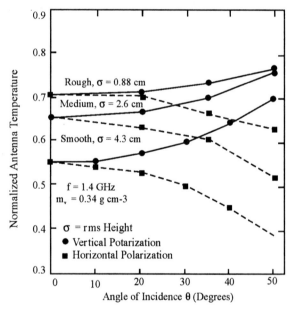

Fig. 4.25 : Effect fo surface roughness on microwave emission from bare soil at 1.4GHz (from Newton and Rouse, 1980)

Suggestions for Supplementary Reading

These suggestions are not exhaustive but are limited to the works that will be found especially useful for developing interest and creative imagination in the field of Remote Sensing.

Books

American Society of Photogrammetry, 1983. Manual of Remote Sensing, Vol. I, R. N. Colwell (Ed.), Falls Church, Va.

Beckmann, P. and A. Spizzichino, 1963. The Scattering of Electromagnetic Waves from Rough Surfaces. The Macmillan Co., New York.

Brekhovskikh, L. M., 1960. Waves in Layered Media. Academic Press, New York.

Elachi, Charles, 1987. Introduction to the Physics and Techniques of Remote Sensing, John Wiley & Sons, New York.

Kerr, D. E. 1951. Propagation of Short Radio Waves. McGraw-Hill Book Co., New York.

Sabins, Floyd, F., Jr., 1987. Remote Sensing – Principles and Interpretations, 2nd ed., W. H. Freeman and Company, New York.

Swain, P. H. and S. M. Davis, 1978. Remote Sensing : The Quantitative Approach, MacGraw-Hill, New York.

Toselli, F. (Ed.), 1989. Applications of Remote Sensing to Agrometeorology, Kluwer Academic Publishers, Dordrecht, The Netherlands.

Tsang, L., J. A. Kong and R. T. Swin, 1985. Theory of Microwave Remote Sensing. Wiley-Interscience Publication.

Research Papers

Carswell, A. I. 1965. Microwave Scattering Measurements in the Rayleigh region using a Focussed-beam Syatem. Canad. J. Phys. 43 : 962 - 977.

Cumming, W. 1952. The Dielectric Properties of Ice and Snow at 3.2 cm. J. Appl. Phys., 23 : 768 – 773.

Elachi, C. and N. Engheta, 1982. Radar Scattering from a Diffuse Vegetation Layer. IEEE Trans. Geosci. Remote Sensing, GE-20 : 212.

Fung, A. K. and M. F. Chen, 1981. Emission from an Inhomogeneous Layer with Irregular Interfaces. Radio Sci. 16 : 289 – 298.

Fung, Adrian K. and Fawwaz T. Ulaby, 1983. Matter – Energy Interaction in the Microwave Region. In : Manual of Remote Sensing, Vol.I, 2nd ed., R. N. Colwell (Ed.), American Society of Photogrammetry, Falls Church, Va. p. 115 – 164.

Hasselmann, K. 1966. Feynman Diagrams and Interaction Rules of Wave-Wave Scattering Processes. Rev. Geophys. 4 : 1 – 32.

Katzin, M. 1957. On the Mechanisms of Radar Sea Clutter. Proc. IEEE, 45 : 44 – 54.

Newton, R. W. and J. W. Rouse, Jr. 1980. Microwave Radiometer Measurements of Soil Moisture Content. IEEE Trans. Antennas Propag. AP-28 : 680 – 686.

Ross, R. A. 1966. Radar Cross-section of Rectangular Flat Plates as a Function of Aspect Angle. Trans. IEEE Antennas Propag., AP-14 : 329 – 335.

Skolnik, M. I. 1970. Radar Handbook. McGraw-Hill Book Co., p. 27 – 36.

Ulaby, Fawwaz T. and Keith R. Carver, 1983. Passive Microwave Radiometry. In : Manual of Remote Sensing, Vol.I, 2nd ed., R. N. Colwell (Ed.), American Society of Photogrammetry, Falls Church, Va. p. 475 – 516.

Wang, J. R. and T. J. Schmugge, 1980. An Empirical Model for the Complex Dielectric Permittivity of Soils as a Function of Water Content. IEEE Trans. Geosci. Remote Sensing, GE-18 : 288 – 295.

The Sensor and Sensor Platforms $\boxed{5}$

5.1 SENSOR MATERIALS

The sensor or detector transforms the energy of the incoming radiation into a form of recordable information. It is found that no single sensor material is equally sensitive to the entire range of electromagnetic spectrum. Therefore, different sensor materials are used for the construction of detectors in different wavelength ranges. In general, there are two types of electromagnetic signal detectors, namely optical film detectors and opto-electronic detectors.

Sensor materials in the film detectors are silver bromide grains. Usually black and white, true color and infrared false color films are in use. Black and white and true color films are sensitive to the visible band of the electromagnetic spectrum (0.4 – 0.7 μm). Spectral sensitivity curves of these photographic films are found to attain a maximum around 0.69 μm and then falls to a very small value at 0.7μm. It is because of this that all panchromatic films operate within 0.4 – 0.7 μm spectral band. However, spectral sensitizing procedures can be applied to the emulsions of black and white films so that they can be made sensitive to near infrared radiation up to wavelength of approximately 0.9 μm. The sensitivity of the false color film is increased up to 0.9 μm including a short but highly reflective part of the near infrared spectrum (0.7 – 0.9 μm).

Opto-electronic detectors are classified into two types on the basis of the physical processes by which radiant energy is converted to electrical outputs. They are thermal detectors (sensitive to temperature changes) and quantum detectors (sensitive to changes in the incident photon flux).

Typical thermal detectors are : thermocouples/thermopiles and thermister bolometers. In thermocouples and thermopiles, change in voltage takes place due to the change in temperature between thermoelectric junctions. In thermister bolometers change in resistance takes place due to the change of temperature of the sensor materials. Thermister bolometers usually use carbon or germanium resistors with resistance change of 4% per degree. Thermal detectors are slow, have low sensitivity and their response is independent of the wavelength of the electromagnetic radiation.

Quantum detectors are of three types. They are:

1. Photo-emissive detectors (photocells and photomultipliers use alkali metal surface coatings such as cesium, silver-oxygen-cesium composite)
2. Photo-conductive detectors (photosensitive semiconductors whose conductivity increases with incident photon flux), and
3. Photo-voltaic detectors (here modification takes place in the electrical properties of a semi-conductor p-n junction such as backward bias current on being irradiated with light)

In the wavelength region < 1 μm, quantum detectors like photo-multipliers and Si-photodiodes are found quite efficient and do not need cooling devices. However, for satellite platforms the photo-multipliers are found less suitable because of its accompanying high voltage power supplies that increases the weight of the payload substantially. In 1 to 3 μm wavelength region, Ge-photodiode, Ge-photo-conductor, InSb photodiode, InAs photodiode and HgCdTe photodiode can be used for sensor materials. However, these sensors are to be kept cooled as their electrical characteristics change with increase in their temperature. In the thermal wavelength region (10 to 14 μm), HgCdTe photo-conductors and PbSnTe photodiodes can be used as sensor materials with effective cooling systems. For the microwave region of the electromagnetic spectrum, the sensor or detector invariably used is the microwave antenna.

5.2 SENSOR SYSTEMS

In a sensor system the sensor material is integrated into its appropriate circuitry and housing to detect and process the input signals and give out the corresponding outputs for further analysis to generate information on the target surface from which the signals are received. Sensor systems are of two types: non-imaging sensor system and imaging sensor system.

Non-imaging sensor system include sounders and altimeters for measurement of high accuracy locations and topographic profiles, spectrometer and spectroradiometer for measurement of high spectral resolution along track lines or swath, and radiometers, scatterometers and polarimeters for high accuracy intensity measurements and polarization changes measurements along track lines or wide swath.

Imaging sensor systems are again of two types: framing systems and scanning systems (Fig.5.1). In framing systems images of the targets are taken frame by frame. These include imagers like photographic film cameras and return beam videcon. The scanning systems include across track scanners and along track (push broom) scanners. Imagers and scanning altimeters/sounders are used for three dimensional topographic mapping. Multispectral scanners/thematic mappers are used for limited spectral resolution with high spatial resolution mapping. Imaging spectrometers are meant for high spectral and spatial resolutions. Imaging radiometers and imaging scatterometers (microwave) are used for high accuracy intensity measurement with moderate imaging resolution and wide coverage. Brief descriptions of some of the important framing and scanning systems mentioned above are presented in the following paragraphs.

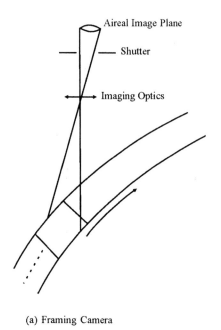
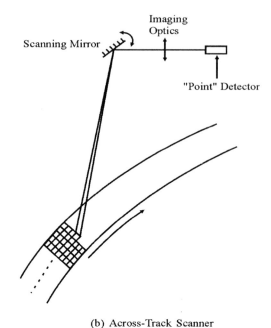

Fig. 5.1 : Schematic presentation of different types of Imaging System :
    (a) The Framing System (Camera)
    (b) The Scanning System : Across–Track Scanner / Whisk Broom Scanner
    (c) The Scanning System : Along–Track Scanner / Push Broom Scanner

5.2.1 Framing Systems

Large Format Camera (LFC)

This high performance photographic film camera was used for the first time in Space Shuttle flight mission of October,1984. Large format camera (Fig.5.2) has an achromatic lens (in the wavelength range of 0.4 to 0.9 mm) of focal length 305 mm. Its format is 230 × 460 mm and exposure range is 3 to 24 ms. The camera is flown with the long dimension of its format in the flight direction to obtain the necessary stereo coverage for the topographic map compilation. The magazine capacity of the large format camera is 1200 m of film for 2400 frames.

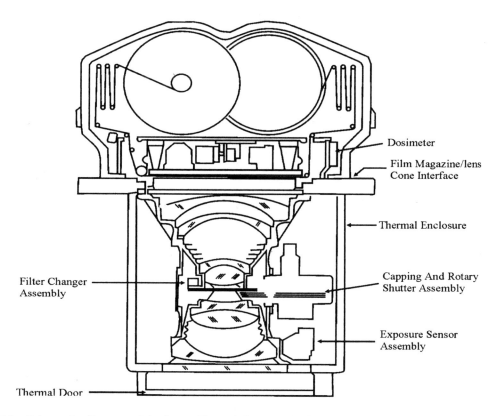

Fig. 5.2 : Schematic diagram of the Large Formal Camera (LFC) – A Framing System (from Slater, 1983)

Hasselblad Camera

This was the multi-band camera flown on Apollo-9 flight mission of March,1969 (Fig.5.3) to obtain the first multi-band pictures of earth from space in the NASA program. It had in the assembly four Hasselblad cameras with filters in the green, yellow, red and deep red regions of the electromagnetic spectrum. It was this Hasselblad camera with infrared ektachrom film which was used to obtain the first historic aerial photograph of the coconut plantation of Kerala in India in late nineteen sixties. The analysis of the coconut crown intensity showed its strong relationship with the coconut production by the plants. Investigating the cause of low crown intensity it was found that such plants were affected by a viral disease attributed to the coconut root wilt virus which was confirmed from electron microscopic studies. The clear practical demonstration of the fact that the onset of a disease could be identified by the near-infrared aerial photography, much before its visual symptoms become apparent, gave a strong boost to the development of remote sensing program of the Indian Space Research Organization, Department of Space, Government of India.

Fig. 5.3 : A hasselblad-camera array belonging to the framing system (from Slater, 1983)

Return Beam Vidicon (RBV)

Return beam vidicon is a framing system which is an electron imager and works like a television camera (Fig.5.4). When the shutter is opened and closed, its optical system frames an image on a photoconducting plate and retained as a varying charge distribution just like a photographic camera frames the image on a photochemical film. The pixels of the image receiving higher intensity of light becomes more conducting, meaning thereby that more of the electrons from their corresponding backside pixels move to the front side. Thus the backside pixels of the photoconducting plate becomes positively charged to different amounts. Now a fine electron beam from the electron gun is allowed to scan the pixels line by line. Thus while the electron beam neutralizes the pixels of the photoconductor from the backside, a part of the beam, now weaker than the forward scanning beam, returns back to a detector carrying the image describing signals. The strength of these feeble signals is increased in an electron multiplier and final signals are digitized and recorded as the image output. These electronically processed image data are amenable to rapid transmission from sensor platform to a ground receiving station. Being a framing device, the return beam vidicon collects a full frame data practically instantaneously.

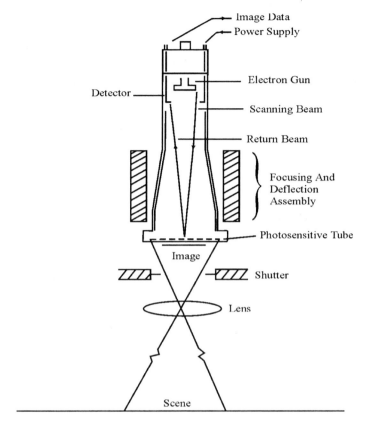

Fig. 5.4 : Schematic diagram of a return beam vidicon (RBV), essentially a TV camera belonging to the framing system (After Sabins, Jr. 1987)

Alternately, the images may be recorded on a magnetic tape for replay at a later time when the sensor platform is within the range of a ground receiving station.

Landsat-1 and-2 carried three RBVs with spectral filters to record green (0.475 – 0.575 μm), red (0.580 – 0.680 μm), and photographic infrared (0.698 – 0.830 μm) images of the same area on the ground. During image processing these data are taken as due to blue, green and red bands to develop the false color image of the scene. Landsat-3 carried two RBVs operating in the panchromatic band (0.505 – 0.750 μm) to obtain higher resolution (24m) compared to its MSS ground resolution (79m). Moreover, each of these two RBVs has a covered adjacent ground area of 99 × 99Km with 15Km sidelap. Successive pairs of RBV images also have a 17Km forward overlap. Thus two pairs of RBV images (Fig.5.5) cover an area of 181 × 183 Km which is equivalent to the area corresponding to the MSS scene (185 × 185 Km). The time sequence of RBV scene acquisition was kept at 25 seconds, because of the driver for the framing sequence of the multi-spectral scanner of the Landsat series of remote sensing satellites.

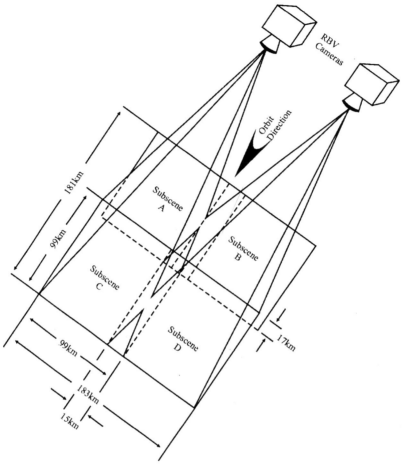

Fig. 5.5 : Schematic diagram of the assembly of two Return Beam Vidicons showing sidelap in their swaths. Essentially this framing system increases the effective swath-width. (after Sabins, Jr. 1987)

5.2.2 Scanning Systems

Across Track Multispectral Scanner (MSS)

This form of imaging is used in Landsat series of satellites. The scanning system (Fig.5.6) employs a single detector per band of the multispectral signal. It has an electrical motor, to the axel of which is attached a solid metal cylinder whose free end is cut at 45 degrees to the axis of its rotation and highly polished to act as a scanning mirror. The field of view (FOV) is restricted by an aperture so that the mirror will receive signals in almost nadir view from 2000 ground resolution cells that makes one scan line. The signal received by the rotating scanning mirror from a ground resolution cell (corresponding to GIFOV) is a white one and contains spectral information in different bands. This white beam is reflected by the mirror in the flight direction (parallel to the ground) and is allowed to pass through a monochromator/spectroscope which splits the composite beam into its color cmponents. The detectors with their designed apertures are so placed that they now receive the spectral information from the ground resolution cells in the specified bandwidths of various color components of the white signal. The current produced in the sensor material (Si- photodiodes etc.) by the different bands of the electromagnetic spectrum are digitized and transmitted to the ground receiving stations or stored in magnetic tapes till a ground receiving station is within its range for transmission/reception. Since the rotating mirror scans the swath line by line perpendicular to the track (flight direction), the scanner is called across-track scanner.

The dwell time of the scanner is computed by the formula :
Dwell Time = (Scan rate per line)/(Number of ground resolution cells per line)
The Spatial Resolution of the Scanner
= Ground resolution Cell
= GIFOV × Attitude of the sensor

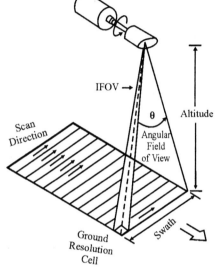

Fig. 5.6 : Schematic diagram of Across-Track Scanner. IFOV expressed in radian measure defines the angular resolution of the ground resolution cell. IFOV multiplied by the altitude of the sensor yields the spatial resolution of the Across-Track Scanner.

Along Track Multispectral Scanner/Push Broom Scanner

In this type of scanner, the scan direction is along the track (direction of flight) and hence the name along track scanner (Fig.5.7). It is also called push broom scanner because the detectors are analogous to the bristles of a push broom sweeping a path on the floor.

Development of charge-coupled device (CCD) has contributed to the successful design of the along track scanner. In this the sensor elements consist of an array of silicon photodiodes arranged in a line. There are as many silicon photodiodes as there are ground resolution cells (corresponding to IFOV) accommodated within the restricted FOV of the sensor optics. Each silicon photodiode, in turn, is coupled to a tiny charge storage cell in an array of integrated circuit MOS (metal oxide semiconductor) device forming a charge coupled device (CCD) (Fig.5.8). When light from a ground resolution cell strikes a photodiode in the array, it generates a small current proportional to the intensity of light falling on it and the current charges the storage cell placed behind the diode. The charged cells form a part of an electronic shift register which can be activated to read out the charge stored in the cells in a sequential fashion. The output signals are correlated with the shift pulses, and digitized to reconstitute the image.

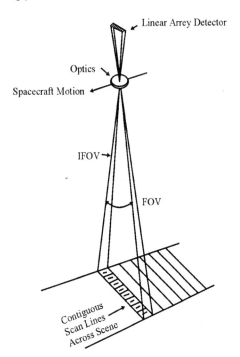

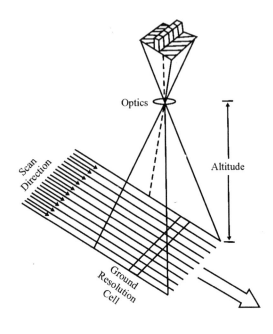

Fig. 5.7(a) : Schematic diagram of Along-Track Scanner. IFOV is the angle subtended by a sensor element or by a ground resolution cell at the sensor optics. IFOV expressed in radian measure multiplied by the altitude of the sensor optics yields the spatial resolution of the along-track scanner

Fig. 5.7(b) : Schematic diagram of Along-Track Scanner showing the Along-Track scan lines corresponding to each of the linear array detector elements (in CCD configuration).

Fig. 5.8 : Schematic Diagram illustrating the function of a CCD Linear Array Imager.

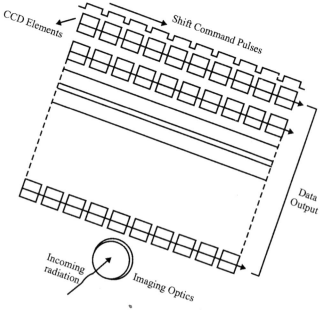

Because of the linear array of the sensor elements in this type of imaging / scanning system, it is also called linear imaging self scanning (LISS) system. SPOT satellites and IRS series of satellites extensively use this type of payloads for scanning purposes.

For multispectral scanning, the along track scanner accommodates one line of detector array for each spectral band. However, IRS satellites use as many cameras as there are spectral bands each with a characteristic filter and a linear array of CCD detectors.

The spatial resolution of the sensor = ground resolution cell
$$= \text{GIFOV} \times \text{altitude of the scanner}$$

The dwell time for the along track scanner is given by

Dwell time = (ground resolution cell dimension)/(velocity of sensor platform)

Under identical setup, it is found that the dwell time of along track scanner is far greater than the dwell time of the across track scanner. The increased dwell time of the along track scanner which provides higher signal intensity from ground resolution cells provides two important improvements for incorporation in the scanner system, namely to minimize the detector size corresponding to a smaller GIFOV producing higher spatial resolution and to narrow down the spectral bandwidth, thereby increasing the spectral resolution.

Side Viewing / Side Looking Scanner

The across track and along track scanners described above are used in passive remote sensing in visible, infrared and microwave regions of the electromagnetic spectrum. These scanners always receive signals in the nadir view. However, if the user demands (on payment basis) to observe a dynamic scene frequently, then there is provision in SPOT and IRS satellites to steer the cameras to look off-nadir at the required scene, some days before to some days after the normal nadir viewing date (Fig.5.9).

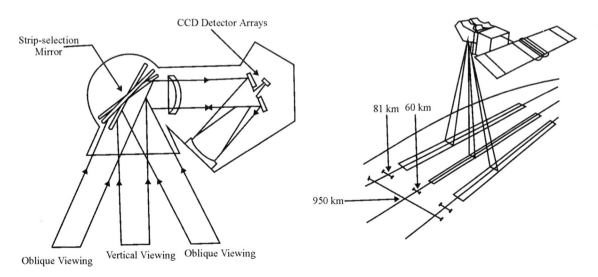

Fig. 5.9 : Schematic diagram showing the oblique viewing capability of SPOT

But scanners of active remote sensing, like the scanners used for radar remote sensing, in their normal mode of scanning look to the sides and not to the nadir for technical reasons which will be described later. Therefore such scanners are called side looking airborne radar (SLAR). The most sought after, sophisticatedly designed synthetic aperture radar (SAR) belongs to the side looking scanner system.

5.3 SCALE, RESOLUTION AND MAPPING UNITS

5.3.1 Scale

Remote sensing information is scale dependent. Images can be described in terms of scale which is determined by the effective focal length of the optics of the remote sensing device, altitude from which image is acquired and the magnification factor employed in reproducing the image.

Normally we talk about small scale, intermediate scale and large scale while describing an image. These scales are quantitatively specified in the following manner :

Small scale	(> 1 : 500,000)	1cm = 5 Km or more
Intermediate scale	(1 : 50,000 to 1 : 500,000)	1cm = 0.5 to 5 Km
Large scale	(< 1 : 50,000)	1cm = 0.5 Km or less

Thus large scale images provide more detailed information than the small scale images. For achieving optimum land use planning at different levels the following scales are often recommended :

National level	1 : 1,000,000
State level	1 : 250,000
District level	1 : 50,000
Farm/Micro-watershed/ Village/Command area level	1 : 4,000 to 1 : 8,000

These scales are equally applicable for images of visible, infrared and microwave remote sensing.

With regard to photo scale (S) of the aerial or satellite imageries, it is computed as the ratio of the photo distance (d) to the ground distance (D) between any two known points, i.e.

$$S = d/D = (\text{photo distance})/(\text{ground distance})$$

For a photograph taken in the nadir view, the image scale is a function of the focal length of the camera used to acquire the image, the flying height of the sensor platform above the ground and the magnification factor (M) employed in reproducing the image, i.e.

Image Scale = Mf/H

= (magnification factor) (camera focal length)/(flying height above terrain)

The Sensor and Sensor Platforms

However, it is observed that the image scale of the photograph of a landscape may vary from point to point because of displacements caused by the topography.

5.3.2 Resolution

An image can be described not only in terms of its scale, as mentioned earlier, but also in terms of its resolution. In remote sensing we basically need three different types of information to be acquired, such as spatial information, spectral information and radiometric (intensity) information. Accordingly the sensor systems or the instruments vary in principles of detection and construction. For obtaining spatial information we use imagers, altimeters and sounders. Spectral information can be acquired by spectrometers and intensity (radiometric) information by radiometers and scatterometers. A suitable modification of the remote sensing equipment helps to acquire two of the three types of information at a time. For example, with an imaging spectrometer one can obtain spatial as well as spectral information from a target. Spectral and radiometric information can be gathered by a spectroradiometer. An imaging radiometer collects not only spatial information but also the radiometric information. Acquisition of all the three types of information from a target is possible by the multispectral scanner and the multispectral push broom imager. Fig.5.10 summarizes the information needed for remote sensing analysis and the types of sensor systems or instruments used to acquire such information.

As already mentioned, remote sensing information is not only scale dependent but also resolution dependent. Four types of resolutions are considered in remote sensing work. They are :

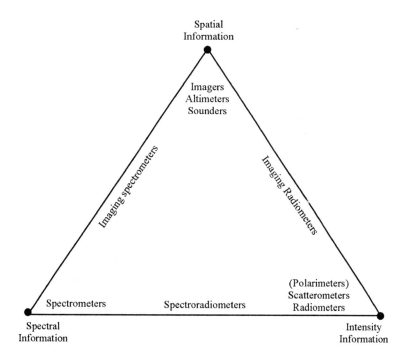

Fig. 5.10 : Diagram summarizing the basic information needed for remote sensing and the sensor systems used to acquire such information

1. Spatial Resolution
2. Spectral Resolution
3. Radiometric Resolution, and
4. Temporal Resolution

Spatial Resolution

Spatial resolution is the minimum distance between two objects that a sensor can record distinctly. Description of spatial resolution may be placed into one of the following four categories :

1. The geometric properties of the imaging system
2. The ability to distinguish between point targets
3. The ability to measure the periodicity of repetitive targets, and
4. The ability to measure spectral properties of small finite objects

The geometric properties of the imaging system is usually described by the ground projected instantaneous field of view (GIFOV) (Fig.5.6 and 5.7). However, a GIFOV value is not in all cases a true indication of the size of the smallest object that can be detected. An object of sufficient contrast with respect to its background can change the overall radiance of a given pixel so that the object becomes detectable. Because of this, features smaller than 80 m, such as railway lines and low-order drainage patterns are detectable on Landsat MSS images. Conversely, on Landsat imagery, objects of medium to low contrast may only be detectable if they are of the size 250 m or more.

For measurements based on the distinguishability between two point targets, Rayleigh criterion is used (Fig.5.11). According to Rayleigh, the two objects are just resolvable if the maximum (center) of the diffraction pattern of one falls on the first minimum of the diffraction pattern of the other. Consequently the smallest angle resolvable is

$$\theta_m = 1.22 \ (\lambda/D) \text{ radians} \tag{5.1}$$

where λ and D are the wavelength of light and diameter of the lens of a telescopic system respectively. In remote sensing terminology, θ_m is called the instantaneous field of view (IFOV).

Measures of resolution using periodicity of repetitive targets are expressed in line pairs/cm (Fig. 5.12 and 5.13).

Resolution employing spectral properties of the target is the effective resolution element (ERE). This measure of spectral resolution is of interest because of the increasing importance of automated classification procedures which are highly dependent upon the fidelity of the spectral measurements recorded by the sensor system.

Effective resolution element (ERE) is defined as the size of an area for which a single radiance value can be assigned with reasonable assurance that the response is within 5% of the value representing the actual relative radiance.

Spatial resolution of a system must be appropriate if one is to discern and analyze the phenomena of interest, namely the detection and identification of objects and their analysis. To move from detection to identification, the spatial resolution must improve by about 3 times. To pass from identification to analysis, a further improvement in spatial resolution of 10 or more times may be needed.

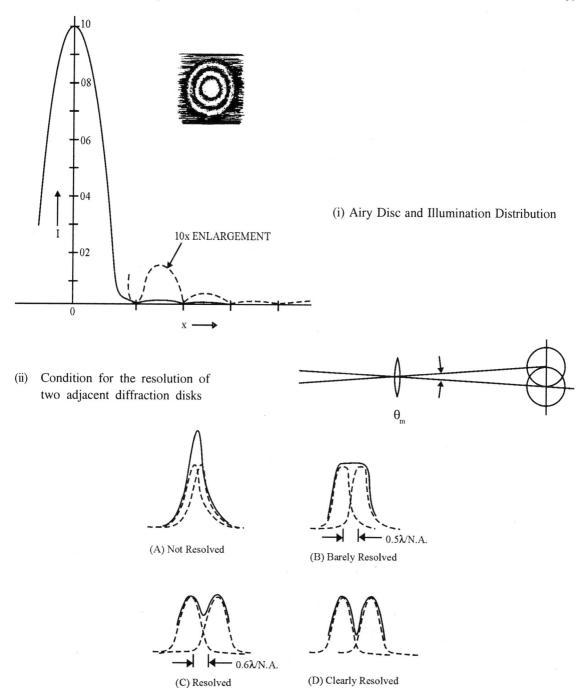

(i) Airy Disc and Illumination Distribution

(ii) Condition for the resolution of two adjacent diffraction disks

(iii) Dotted lines indicating diffraction patterns of two point objects at various distances from each other. Case C is just resolved following Rayleigh's criterion.

Fig. 5.11 : Rayleigh's criterion for resolution of two distant point objects from their adjacent diffraction patterns.

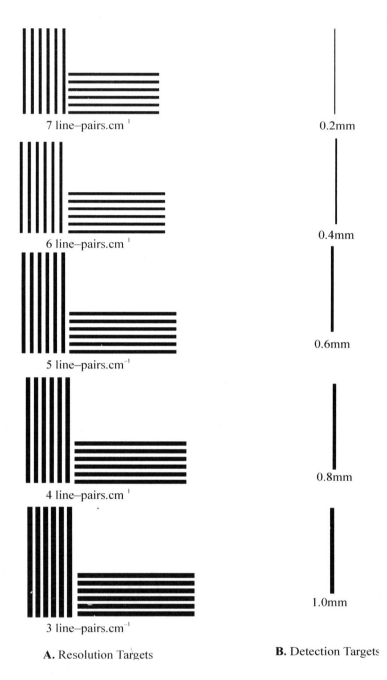

Fig. 5.12 : Chart for resolution and detection of targets with high contrast ratio. View this chart from a distance of 5m. For resolution of targets (in A), determine the most closely spread set of bars you can resolve. For detection of targets (in B), determine the narrowest bar you can detect.

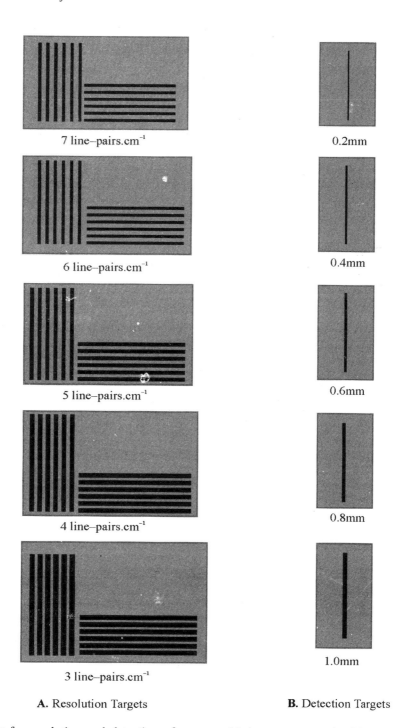

Fig. 5.13 : Chart for resolution and detection of targets with low contrast ratio. View this chart from a distance of 5m. For resolution of Targets (in A), determine the most closely spread set of bars you can resolve. For detection of targets (in B), determine the narrowest bar you can detect.

Spectral Resolution

For a remote sensing instrument, spectral resolution is determined by the bandwidth of the channels used. High spectral resolution is achieved by narrow bandwidths which are collectively likely to provide more accurate spectral signature for discrete objects than by broad bandwidths. However, narrow band instruments tend to acquire data with low signal-to-noise ratio lowering the system's radiometric resolution. This problem may be eliminated if relatively long dwell times are used during scanning/imaging.

Radiometric Resolution

Radiometric resolution is determined by the number of discrete levels into which a signal strength maybe divided (quantization). However, the maximum number of practically possible quantizing levels for a sensor system depends on the signal-to-noise ratio and the confidence level that can be assigned when discriminating between the levels.

With a given spectral resolution, increasing the number of quantizing levels or improving the radiometric resolution will improve discrimination between scene objects. It is found that the number of quantizing levels has a deciding effect upon the ability to resolve spectral radiance that is related to plant-canopy status. Interdependency between spatial, spectral and radiometric resolutions for each remote sensing instrument affects the various compromises and tradeoffs.

Temporal Resolution

Temporal resolution is related to the time interval between two successive visits of a particular scene by the remote sensing satellite. Smaller the revisit time the better is the temporal resolution of the sensor system of the remote sensing satellite. High temporal resolution satellites are more suitable for monitoring dynamic surface features or processes like the growth and development of agricultural crops, floods and so on. To monitor changes with time, temporal resolution is also an important consideration when determining the resolution characteristics of a sensor system. For agricultural applications, the use of time as a discriminant parameter may allow crops over large areas to be identified with sensors possessing spatial and spectral resolutions that are too coarse to identify them on the basis of the spectral and morphological characteristics alone.

5.3.3 Mapping Units

Minimum size of an area which can be recognized over a map is taken as 3mm × 3mm. Thus this area is taken as the minimum mapping unit. If the scale of the map is 1: 50,000, then the length of the mapping unit can be calculated as

        1 mm on the map = 50,000 mm on the ground
        3 mm on the map = 150,000 mm on the ground = 150 m on the ground

Therefore the minimum mapping unit is 150m × 150m.

Similarly when we process the remote sensing digital data, the smallest area of a land surface feature which can be visually recognized on the screen is 3 × 3 pixels. Thus if the ground resolution cell of the satellite sensor is 80m (as in case of Landsat MSS), then the smallest classifiable area becomes 240m × 240m. In this case, as the smallest classification distance (namely 240m) is more (coarser) than the minimum mapping unit, therefore, we can not use Landsat MSS data to undertake mapping of land surface features in the scale of 1: 50,000. For mapping on a specific scale, it is advisable to use remote sensing data that provides a smaller value of the smallest classification distance (3 pixels) than the minimum mapping unit. In the present example, Landsat TM data (ground resolution 30m × 3 = 90m) or IRS LISS-II data (ground resolution 36.25m × 3 = 108.75m) or IRS LISS-III data (ground resolution 23.5m × 3 = 70.5m) can be used for mapping land surface features on a map with 1: 50,000 scale.

5.4 REMOTE SENSOR PLATFORMS

Essentially three different types of platforms are used to mount the remote sensors wherefrom they collect information on earth's surface features and record / transmit the information to an earth receiving station for their further analysis and interpretation. These sensor platforms are :

1. Ground Observation Platform
2. Airborne Observation Platform, and
3. Space-borne Observation Platform

A diagrammatic presentation of the concept of multi-level sensor platforms for remote sensing data acquisition is given in Fig.5.14.

5.4.1 Ground Observation Platform

Ground observation platforms are necessary to develop the scientific understanding on the signal-object and signal-sensor interactions. These studies, both at laboratory and field levels, help in the design and development of sensors for the identification and characterization of the characteristic land surface features. Important ground observation platforms include – handheld platform, cherry picker, towers and portable masts. Portable handheld photographic cameras and spectroradiometers are used for laboratory and field experiments to collect the ground truth. In cherry pickers automatic recording sensors can be placed at a height of about 15m from the ground. Towers can be raised for placing the sensors at a greater height for observation. Towers can be dismantled and moved from place to place. Portable masts mounted on vehicles can be used to support cameras and other sensors for testing and collection of reflectance data from field sites.

5.4.2 Airborne Observation Platform

Airborne observation platforms are important to test the stability performance of the sensors before they are flown in the space-borne observation platforms. Important airborne observation platforms include – balloons, drones (short sky spy), aircraft and high altitude sounding rockets.

Balloon Platform

Balloons are used for remote sensing observations (aerial photography) and nature conservation studies. Balloons developed by the Scociété Européanne de Propulsion can carry the payloads upto altitudes of 30Km. It consists of a rigid circular base plate for supporting the entire sensor

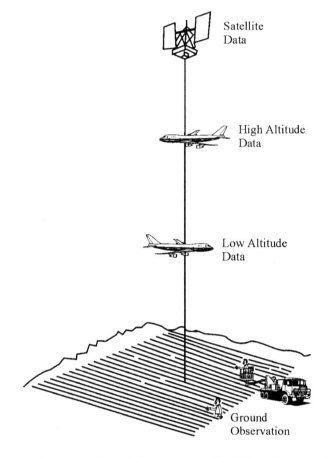

Fig. 5.14 : Diagrammatic presentation of the concept of multilevel sensor platforms for remote sensing data acquisition

system which is protected by an insulating and shockproof light casing. It is roll stabilized and temperature controlled. Essential equipments carried by the balloon includes – camera, multispectral photometer, power supply units and remote control system. The sensor system is brought back to earth by tearing the carrying balloon through remote control.

Drone

Drone is a remotely piloted vehicle which looks like a miniature aircraft. This sensor platform essentially consists of a central body in the shape of a circular tube for carrying the engine, propelling fan, fuel tank and the sensor system. The tail of the drone has small wing structures and a tail plane with control mechanisms. Thus drone is a pilotless vehicle, remotely controlled from the ground station. It is capable of a climb rate of 4m/s with an operating altitude of about 0.5 Km, a forward speed of about 100 Km/h and can also exhibit hovering flight. The servomotor systems, operating the aerodynamic controls, receive signals related to the altitude and position of the aerial vehicle from sensors within the drone and from the ground. The drone sensors can provide information to maintain the drone in the altitude demanded by the

The Sensor and Sensor Platforms 75

ground control or by a self-contained (programmed) navigation system. Drone's payload include equipment of photography, infrared detection, radar observation and TV surveillance.

An example of a drone developed in Britain during World War second is the short sky-spy which was originally conceived as a military reconnaissance. But it is found to play potentially useful role in the testing and operation of remote sensing devices. The unique advantage of such a device is that it could be accurately located above the area for which data was required. It is also an all-weather type of platform capable of both night and day observations. The most sophisticated unmanned aerial vehicle – The Predator Drone, recently developed by the United States Military and Intelligence Wing is capable of searching even specific individual enemy supported by the necessary accumulated human intelligence in its database. A modern Drone empowered with human intelligence input is shown in Fig. 5.15.

Fig. 5.15 : A modern Drone empowered with human intelligence input

Airborne High Altitude Photography

Special aircraft carrying large format cameras on vibrationless platforms are traditionally used to acquire aerial photographs of land surface features. While low altitude aerial photography results in large scale images providing detailed information on the terrain, the high altitude, smaller scale images offer advantage to cover a larger study area with a fewer photographs.

NASA acquired aerial photographs of the United States from U-2 and RB-57 reconnaissance aircraft at altitudes of 18 Km above the terrain on film of 23 × 23 cm format. Each photograph covered 839 square kilometers at a scale of 1: 120,000. Black and white, normal color and infrared color films were used in these missions.

The National High Altitude Photography (NHAP) program (1978), coordinated by the US Geological Survey, started to acquire coverage of the United States with a uniform scale and format. From aircraft at an altitude of 12 Km, two cameras (23 × 23 cm format) acquired black and white and infrared color photographs. The black and white photographs were acquired with the camera of 152 mm focal length to produce photographs at 1: 80,000 scale which cover 338 square kilometers per frame. The infrared color photographs were acquired with a camera of 210 mm focal length to produce photographs at 1: 58,000 scale covering 178 square kilometers per frame.

Airborne Mulitspectral Scanners

Aircraft platforms offer an economical method of testing the remote sensors under development. Thus photographic cameras, electronic imagers, across-track and along-track scanners, and radar and microwave scanners have been tested over ground truth sites from aircraft platforms in many countries, especially in the NASA programmes.

NASA U-2 aircraft acquired images using a Daedalus across-track multispectral (10 bands between 0.38 – 1.10 mm) scanner. Similarly the M-7 airborne scanner having 12 spectral bands covering ultraviolet, visible and infrared regions, was also tested. Signals from the scanners were monitored and controlled in flight at the operator console and were recorded in analog form by a wide-band magnetic tape recorder. The recorded signals were later digitized and reformatted on the ground for digital image processing and information extraction.

High Altitude Sounding Rockets

High altitude sounding rocket platforms are useful in assessing the reliability of the remote sensing techniques as regards their dependence on the distance from the target is concerned. Synoptic imageries can be obtained from such rockets for areas of some 500,000 square kilometers per frame. The apogee of the European rockets is approximately 300 Km which can be extended to 400 Km with 200 Kg payload. The high altitude sounding rocket is fired from a mobile launcher. During the flight its sensors are held in a stable attitude by an automatic control system. Once the desired scanning work is over from a stable altitude, the payload and the spent motor are returned to the ground gently by parachute enabling the recovery of the data/ photographic records.

The Skylark Earth Resource Rocket is also a platform of this type. The sensor payload in this case consists of two cameras of the Hasselblad type. This rocket system has been used in surveys over Australia and Argentina.

5.4.3 Space-borne Observation Platforms

Essentially these are satellite platforms. Two types of satellite platforms are well recognized, they are : manned satellite platform and unmanned satellite platform.

Manned Satellite Platforms

Manned satellite platforms are used as the last step, for rigorous testing of the remote sensors on board so that they can be finally incorporated in the unmanned satellites. This multi-level remote sensing concept is already presented in Fig.5.14. Crew in the manned satellites operate the sensors as per the program schedule. Information on a series of NASA's manned satellite programs are given in Table-5.1.

The Sensor and Sensor Platforms

Table 5.1 : Some manned satellite programs of NASA

Program	Year	Crew	Sensors used
Mercury	1962 – 1963	One	Hand-held Camera
Gemini	1964 – 1965	Two	Hand-held Camera
Apollo	1968 – 1972	Three	Multispectral Camera
Skylab	1973 – 1974	Three	Hand-held Camera, Multispectral Scanner
Space Shuttle	1981 –	Three to Seven	Hand-held Camera, LFC, SIR, MOMS
International Space Station	2000 Nov. 2, 2000 Mar. 10, 2001 Aug. 12, 2001 Dec. 7, 2001 June 7, 2002 Nov. 25, 2002 Feb. 1, 2003 April 28, 2003 Oct. 20, 2003 April 17, 2004	Variable 1st Station Crew Arrived 2nd Station Crew Arrived 3rd Station Crew Arrived 4th Station Crew Arrived 5th Station Crew Arrived 6th Station Crew Arrived Space Shuttle Columbia Disaster 7th Station Crew Arrived 8th Station Crew Arrived 9th Station Crew Arrived	Multiple sensors for Remote Sensing and a range of laboratory equipments for conducting physico-chemical and biological experiments and synthesis of high-tech engineering materials. It is planned to serve as the base for launching smaller unmanned satellites into polar orbits from which remote sensing data can be relayed to earth stations. Crew from the Space Station can also go to these polar satellites to repair and refuel them. Space Shuttle to provide transportation of astronauts and necessary cargo between Earth and the Space Station.

Unmanned Satellite Platforms

Landsat series (Fig.5.16, 5.17), SPOT series (Fig.5.18, 5.19) and IRS series (Fig. 5.20, 5.21, 5.22) of remote sensing satellites, the NOAA series of meteorological satellites (Fig.5.23), the entire constellation of the GPS satellites and the GOES and INSAT series of geostationary environmental, communication, television broadcast, weather and earth observation satellites all belong to this unmanned satellite category. We may add to this list the unmanned satellites launched by Russia, Canada, European Space Agency, China and Japan to indicate the current state of development in space technology to tackle our problems from global perspective.

These satellites are space observatories which provide suitable environment in which the payload can operate, the power to permit it to perform, the means of communicating the sensor acquired data and spacecraft status to the ground stations, and a capability of receiving and acting upon commands related to the spacecraft control and operation. The environment of the space-borne observation platform is considered as both structural and physical and includes such factors as the framing mount for the payload and functional support systems, the means of maintaining the thermal levels of the payload within allowable limits, and the ability to maintain the orbital location of the spacecraft so that the sensors look at their targets from an acceptable and known perspective. The torso of the observatory is the sensory ring which mounts most of the observatory subsystems that functionally support the payload. The satellite mainframe subsystems designed to meet these support functions include : the structure subsystem, orbit control subsystem, attitude control subsystem, attitude measurement subsystem, power subsystem, thermal control subsystem, and the telemetry, storage and telecommand subsystems.

5.5 GROUND SYSTEMS

Ground communication stations are the radio telemetry links between satellites and the earth. They fall into two general categories : (1) those having the sole function of receiving the sensor and attitude data from the satellites, and (2) those which, in addition to receiving the sensor and attitude data, can receive satellite house keeping data and transmit commands.

5.6 SATELLITE LAUNCH VEHICLE

Rocket-vehicles are used for launching the satellites. The type of rocket used for Landsat launches was of Mc Donnell-Douglas Delta-900 series. This was a two-stage rocket that is thrust-augmented with a modified Thorbooster, using nine Thiokol solid propellant strap-on engines. The French launch vehicle is an Ariane Rocket. Russian launch vehicles use cryo-engines. The Indian satellite launch vehicles are of PSLV and GSLV types.

5.7 REMOTE SENSING SATELLITE ORBITS

A space-borne remote sensing platform is placed and stabilized (by special orbit maneuvers) in an orbit in which it moves. From geometrical characteristics point of view, the orbits of the space-borne platform can be circular, elliptic, parabolic or hyperbolic. Although the operational orbits for terrestrial remote sensing are supposed to be circular, it is difficult in practice to establish and maintain an exactly circular orbit. Therefore, the so-called nominally circular orbits are slightly elliptical in form. Parabolic and hyperbolic orbits are not used for terrestrial remote sensing. However, they are used primarily in extraterrestrial flights for sending us information on the extraterrestrial objects.

From the point of view of periodicity of satellite movement, orbits can be classified as geo-synchronous (geo-stationary) and sun-synchronous.

5.7.1 Geosynchronous Orbit

It is an important special case of the circular orbit class which is achieved by placing the satellite at an altitude (35,786, 103 m) such that it revolves in synchrony with the earth, namely from west to east at an angular velocity equal to the earth's rotation rate. The geosynchronous orbit maintains the satellite over a narrow longitude band over the equator. When this band shrinks to a line the orbit is called geostationary. These orbits are most frequently used for communication / television broadcast satellites. They are also used for meteorological and other applications. Insat series of satellites launched by Indian Space Research Organization, Department of Space, Government of India belong to this class of satellites.

The Sensor and Sensor Platforms 79

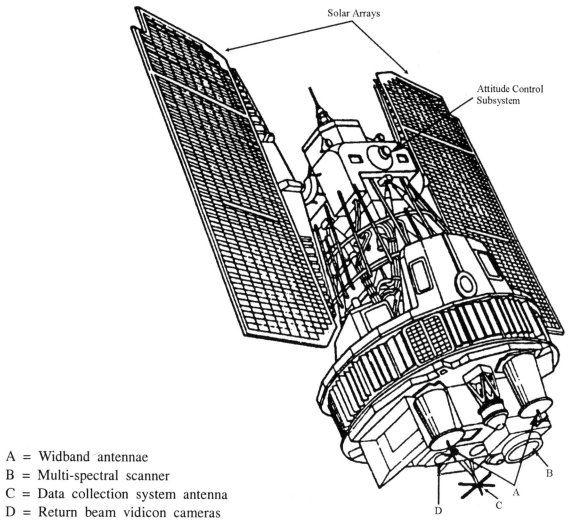

A = Widband antennae
B = Multi-spectral scanner
C = Data collection system antenna
D = Return beam vidicon cameras

Launch		Orbital Parameters	
Landsat 1:	23 July 1972 Operation ended: 6 January 1978	Orbit:	near polar sun-synchronous
		Altitude:	919 km
Landsat 2:	22 January 1975 Operation ended: 25 February 1982	Inclination:	99.09°
		Coverage:	82° N to 82° S
		Period:	103 minutes, crossing the equator at 09:30 hrs local time
Landsat 3:	5 March 1978 Standby mode: 31 March 1983	Repeat Cycle:	18 days

Fig. 5.16 : Schematic diagram of Landsat 1, 2 and 3.

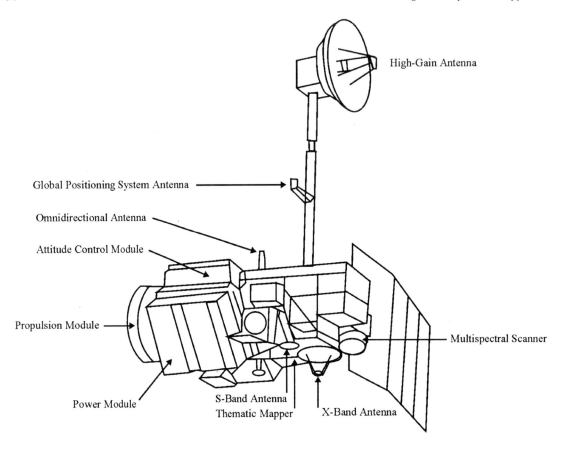

Launch	Orbital Parameters
Landsat 4 : 16 July 1982 Landsat 5 : 1 March 1984	Orbit : Near polar sun-synchronous Altitude : 705 km Inclination : 98.2° Coverage : 81° N to 81° S Period : 99 minutes, crossing the equator at 09:45 hrs local time Repeat cycle : 16 days

Fig. 5.17 : Schematic diagram of Landsat 4 and 5.

The Sensor and Sensor Platforms

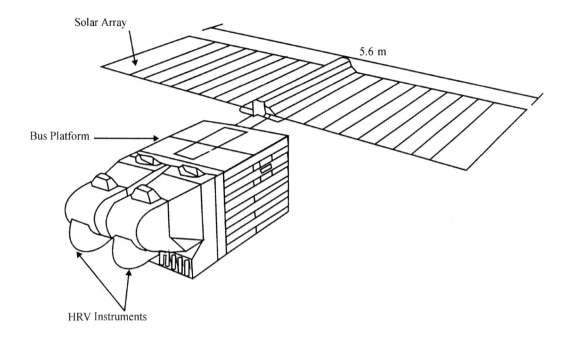

Launch		Orbit	
SPOT 1 :	22 February 1986 Operational	Orbit	: Near polar sun-synchronous
		Altitude	: 830 km
		Inclination	: 98.7°
SPOT 2 :	22 January 1990 Operational	Coverage	: 81° N to 81° S
		Period	: 101.4 minutes, crossing the equator at 10:30 a.m. local time
		Repeat Cycle :	26 days

Fig. 5.18 : Schematic diagram of SPOT 1 and 2.

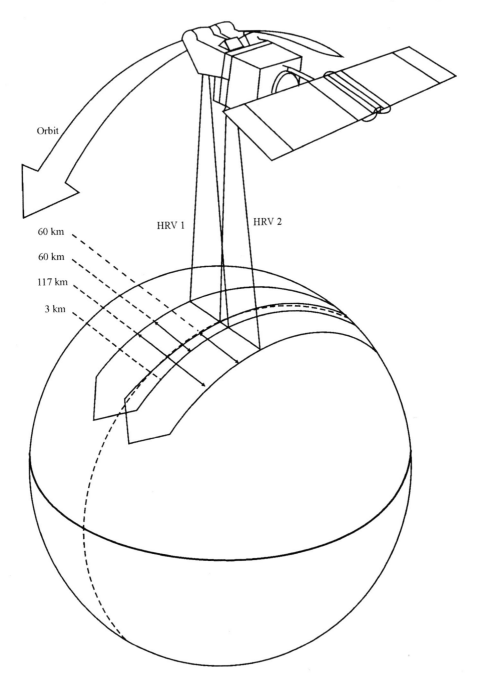

Fig. 5.19 : Schematic diagram showing the combined swath of SPOT HRV sensors.

The Sensor and Sensor Platforms

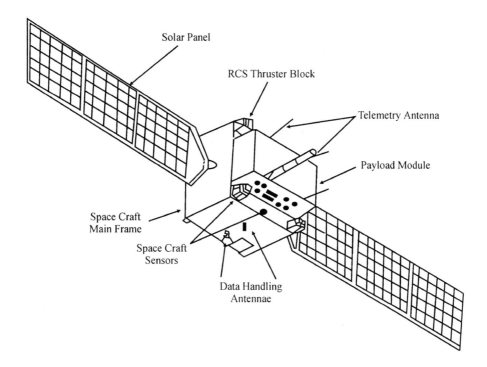

Launch		Orbital Parameters	
IRS-1A :	17 March 1988 (lifetime 3 years)	Orbit	: Near polar sun-synchronous
		Altitude	: 904.1 km
		Inclination	: 99°
IRS-1B :	29 August 1991	Period	: 103.192 minutes
		Repeat Cycle	: 22 days
		Orbits/day	: 14

Fig. 5.20 : Schematic diagram of Indian remote sensing satellite (IRS–1A/1B).

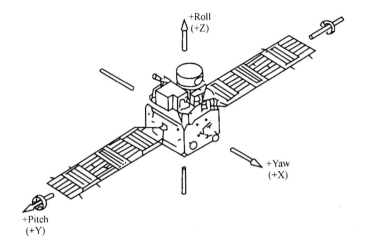

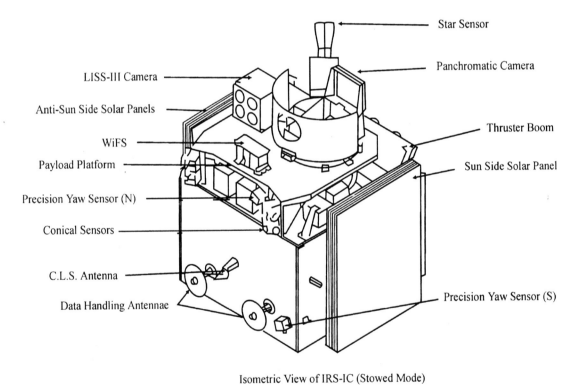

Isometric View of IRS-IC (Stowed Mode)

Fig. 5.21 : Schematic diagram of Indian remote sensing satellite (IRS–1C/1D).

The Sensor and Sensor Platforms

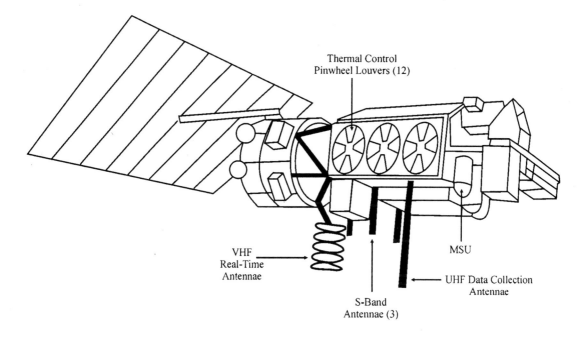

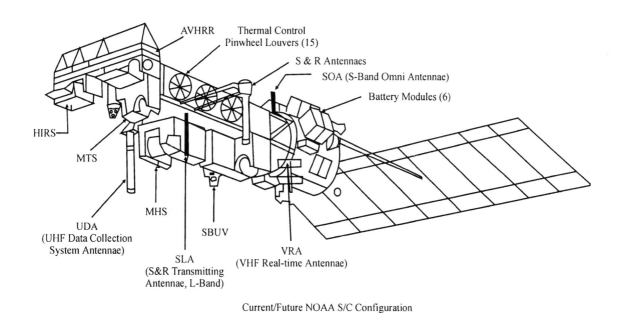

Fig. 5.22 : Schematic diagrams showing the early and the current NOAA spacecraft configurations.

5.7.2 Sunsynchronous Orbit

If the orbit precession exactly compensates for the earth's revolution around the sun, the orbit is called sunsynchronous. It is an important case of elliptical orbit class in which the orbital plane is near polar (> 85 degrees from the equatorial plane) and the altitude is such that the satellite passes over all places on earth having the same latitude twice daily revolving in the same mode (ascending or descending) at the same local sun time. Here solar incidence angle which is held almost constant over the samelatitude finds potential applications in earth resource survey and management. All remote sensing satellites like Landsat, SPOT and IRS belong to this class of satellites. With sunsynchronous satellites, remote sensing observations of a particular scene (location) can only be made at one fixed time in nadir view during a predetermined date which eliminates multitemporal observations within its revisit period.

5.7.3 Shuttle Orbit

Shuttle orbits range from 200 to 300 kilometers above the surface of the earth at an inclination of 30° to 60° to the equatorial plane. Payloads are mounted on the cargo bay. Space shuttles are the manned space observation platforms. Space stations launched by the United States of America and Russia to carry out important physico-chemical and biological experiments and to develop some hightech materials are also placed in these orbits so that the crew can carry food items and ferry the scientists to-and-fro between the earth and the space station.

5.7.4 Coverage

Coverage represents the areas of the earth which are observed by the satellite in one repeat cycle. For polar (near polar for technical reasons) orbits, it is given as the maximum north and south latitudes. An important aspect of the remote sensing mission design of satellite is to provide coverage of a particular geographic area on some schedule. Coverage has two elements : (1) the nadir trace or ground track of the satellite, and (2) the sensor view area or swath. The ground track is determined by the satellite orbit while the swath is determined not only by the orbit but also by the field of view and the look direction of the sensor relative to the nadir. Landsat, SPOT and IRS series of satellites essentially provide coverage through nadir traces. However, SPOT and some recent IRS series of satellites have been provided with tiltable mirrors to sweep off-nadir traces involving programs pointing sequences.

5.7.5 Passes

If one observes the pass sequence of the satellite nadir trace then one finds a daily shift through N days when on the Nth day the nadir trace over a particular location exactly coincides. Thus the temporal resolution of the satellite is N days. The characteristic number of passes and the temporal resolution of different satellite systems are presented in Table 5.2.

Table 5.2 : Characteristics of some remote sensing satellites

Satellite	Altitude (Km)	Orbits/day	Repetivity (days)	Ground Resolution	Radiometric Resolution
Landsat 1, 2, 3	918	14	18	79 m	128 MSS 4,5,6 64 MSS 7
Landsat 4, 5	705	14.5	16	30 m TM 120 m TM 6	256 TM
SPOT 1, 2	832	13.8	26	20 m MS 10 m PAN	256
IRS-1A,1B	904	14	22	72.5 m LISS-I 36.25m LISS-II	128 LISS-I 128 LISS-II
IRS-1C	817		24	5.8m PAN 188 m WiFS	64 128
IRS-P3	817		24	188 m WiFS 580 – 1000m MOS	128
IRS-1D	821		24	5.8m PAN 23.5m LISS-III 70.5m LISS-III MIR 188m WiFS	64 128 128 128

5.7.6 Pointing Accuracy

The attitude of the sensor platform is affected by the rotational drifts about the roll, pitch and yaw axes which in turn affect the location accuracy of the sensor. It is for this reason that for every satellite configuration, the minimum tolerable roll, pitch and yaw drift angles are decided earlier. The satellite is regularly monitored for these parameters by the ground tracking stations during its passes and corrections are applied accordingly.

Suggestions for Supplementary Reading

These suggestions are not exhaustive but are limited to the works that will be found especially useful for developing interest and creative imagination in the field of Remote Sensing.

Books

American Society of Photogrammetry, 1983. Manual of Remote Sensing, Vol. I, R. N. Colwell (Ed.), Falls Church, Va.

American Society of Photogrammetry, 1968. Manual of Color Aerial Photigraphy, Falls Church, Va.

American Society of Photogrammetry, 1980. Manual of Photogrammetry, 4th ed., Falls Church, Va.

Barbe, D. F. 1980. Charge – Coupled Devices. Topics in Applied Physics, Vol.38. Springer- Verlag, Berlin.

Cheng, Cheng Ji, (Ed.). 1989. Manual for the Optical-Chemical Processing of Remotely Sensed Imagery. ESCAP / UNDP Regional Remote Sensing Programme (RAS/86/141), Beijing, China.

Curran, P. J. 1985. Principles of Remote Sensing, Longman, London.

Elachi, Charles, 1987. Introduction to the Physics and Techniques of Remote Sensing, John Wiley & Sons, New York.

Gregory, R. L. 1966. Eye and Brain – The Psychology of Seeing. World University Library, McGraw-Hill Book Co., New York.

Lillesand, Thomas M. and Ralph W. Kiefer, 1987. Remote Sensing and Image Interpretation, 2nd ed., John Wiley & Sons, New York.

Sabins, Floyd, F., Jr., 1987. Remote Sensing – Principles and Interpretations, 2nd ed., W. H. Freeman and Company, New York.

Slater, P. N. 1980. Remote Sensing : Optics and Optical Systems. Addition-Wesley, Reading, Mass.

Toselli, F. (Ed.), 1989. Applications of Remote Sensing to Agrometeorology, Kluwer Academic Publishers, Dordrecht, The Netherlands.

Research Papers

Allison, Lewis J. and Abraham Schnapf, 1983. Meteorological Satellites. In : Manual of Remote Sensing, Vol. I, 2nd ed., R. N. Colwell, (Ed.). American Society of Photogrammetry, Falls Church, Va. p. 651 – 679.

Barzegar, F. 1983. Earth Resources Remote Sensing Platforms. Photogramm. Eng. Remote Sensing, 49 : 1669.

Doyle, F. J. 1979. A Large Format Camera for Shuttle. Photogramm. Eng. Remote Sensing, 45 : 73 – 78.

Doyle, F. J. 1985. The Large Format Camera on Shuttle Mission 41-G. Photogramm. Eng. Remote Sensing, 5! : 200 – 203.

Elachi, Charles, 1983. Microwave and Infrared Remote Sensors. In : Manual of Remote Sensing, Vol. I, 2nd ed., R. N. Colwell, (Ed.). American Society of Photogrammetry, Falls Church, Va. p. 571 – 650.

Fraysse, G., 1989. Platforms. In : Applications of Remote Sensing to Agrometeorology, F. Toselli, (Ed.). Kluwer Academic Publishers, Norwell, Ma. p. 35 – 55.

Freden, Stanley, C. and Frederick Gordon, Jr., 1983. Landsat Satellites. In : Manual of Remote Sensing, Vol. I, 2nd ed., R. N. Colwell, (Ed.). American Society of Photogrammetry, Falls Church, Va. p. 517 – 570.

Hartl, Ph. 1989. Sensors. In : Applications of Remote Sensing to Agrometeorology, F. Toselli, (Ed.). Kluwer Academic Publishers, Norwell, Ma. P. 19 – 34.

Jones, R. C. 1968. How Images are Detected. Scientific American, 219 : 111 – 117.

Joseph George, V. S. Ayengar, Ram Rattan, K. Nagachenchaiah, A. S. Kiran Kumar, B. V. Aradhye, K. K. Gupta and D. R. M. Samudraiah, 1996. Cameras for Indian Remote Sensing Satellite IRS-1C. Current Science (Special Issue), 70 (7) : 510 – 515.

Kalyanaraman, S. and . K. Rajangam, 1996. Advanced Features and Specifications of IRS-1C Spacecraft. Current Science (Special Issue), 70 (7) : 501 – 509.

Meisner, D. E. 1986. Fundamentals of Airborne Video Remote Sensing. Remote Sensing Environ. 19 : 63 – 80.

Meisner, D. E. and O. M. Lindstrom, 1985. Design and Operation of a Color Infrared Aerial Video System. Photogramm. Eng. Remote Sensing, 51 : 555 – 560.

Norwood, Virginia T. and Jack C. Lansing, Jr. 1983. Electro-Optical Imaging Sensors. In : Manual of Remote Sensing, Vol. I, 2nd ed., R. N. Colwell, (Ed.). Americal Society of Photogrammetry, Falls Church, Va. p. 335 – 367.

Robinson, B. F. and D. P. DeWitt, 1983. Electro-Optical Non-Imaging Sensors. In : Manual of Remote Sensing, Vol. I, 2nd ed., R. N. Colwell, (Ed.). American Society of Photogrammetry, Falls Church, Va. p. 293 – 333.

Rosenberg, P. 1971. Resolution, Detectability and Recognizability. Photogramm. Eng. Remote Sensing, 37 : 1244 – 1258.

Shivakumar, S. K., K. S. Sarma, N. Nagarajan, M. G. Raykar, H. R. RaoK. V. S. R. Prabhu and N. Ranjan Paramananthan, 1996. IRS-1C Mission Planning, Analysis and Operations. Current Science (Special Issue), 70 (7) : 516 – 523.

Specht, M. R., D. Needler and N. L. Fritz, 1973. New Color Film for Water Penetration Photography. Photogramm. Eng. 40 : 359 – 369.

Slater, Philip N., 1983. Photographic Systems for Remote sensing. In : Manual of Remote Sensing Vol. I, 2nd ed., R. N. Colwell, (Ed.). American Society of Photogrammetry, Falls Church, Va. p. 231 – 291.

Venkatachary, K. V., P. Soma, V. Jalaramaiah, A. Karuppaiyan, N. Sengupta, O. Chiranjeevi and Y. K. Singhal, 1996. TTC Network and Spacecraft Control Centre for IRS-1C. Current Science (Special Issue), 70 (7) : 524 – 533

Remote Sensing Data Acquisition And Dissemination | 6 |

6.1 RADIO COMMUNICATION AND DATA TRANSMISSION

Remote sensors mounted on a suitable platform (ground observation platform, air-borne observation platform, manned space-borne platform or unmanned satellite platform) receive spectral signatures from earth surface features, pixel wise, in a number of spectral bands. These data are stored as latent image on photographic films or are digitized and stored on magnetic tapes. Once this mission is completed, these data are physically brought to the laboratory where images are processed from the photographic films or digital data and then information about the targets are extracted. However, this method of mechanical data transmission is not possible from an unmanned space-borne observation platform in which case the satellites continuously remain in space for a number of years. Under the circumstance, radio/microwave communication provides a link between the remote sensing systems on satellites and the ground data receiving stations. Radio/microwave transmission of remote sensing data may be made either in real time or it may be delayed until the ground receiving station is well within the range of data transmission and reception. A delay in transmission from the real time requires an onboard data storage facility. In such cases data is transmitted from the space-borne observation platform to the ground receiving station upon a command from the ground control station.

Before transmitting the sensor received signals to the ground receiving station, in the satellite platform itself the imagery is scanned by various processes to organize the information in a series of analogue or its equivalent digital form. These signals then modulate an RF career through appropriate amplitude modulation, frequency modulation or phase modulation. After the data modulation, the communication process accomplishes power amplification in which the power amplifier raises the modulator output to the power level required to meet the system requirement for transmission from an antenna.

When remote sensing data are conveyed by radio link, from the transmitting to the ground receiving station, the reception begins with a large high-gain antenna, thereby maximizing the signal power level from the remote satellite. Fig. 6.1 summarizes the Landsat Data Collection System (DCS).

Since there are now only a limited number of ground receiving stations, they are inadequate to handle the very high volume of image data transmitted from a large number of satellites launched by USA, Russia, France, Canada, India, China, Japan, the European Agency and so

Remote Sensing Data Acquisition and Dissemination 91

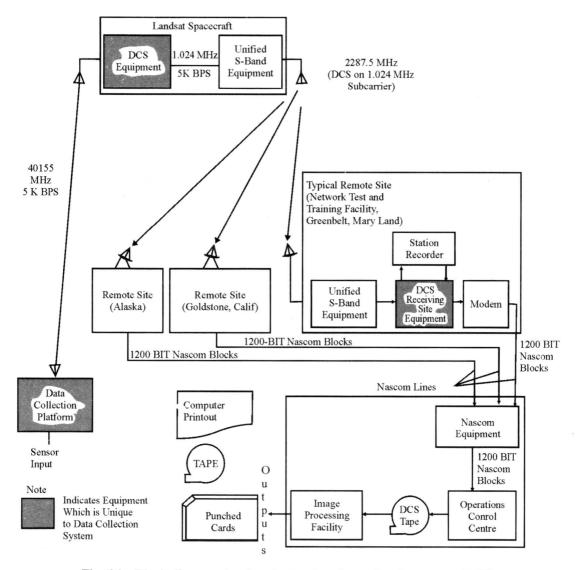

Fig. 6.1 : Block diagram showing the Landsat data collection system (DCS)

on. Thus to avoid large on-board data storage facilities, the USA has launched the tracking and data relay satellites (TDRS) which are placed in geo-stationary orbits to receive data directly from the remote sensing satellites and transmit them to the TDRS ground receiving stations. These TDRS-linked ground receiving stations, in turn, relay the image data via a domestic communication satellite (DOMSAT) to different centers where facilities are available to convert the satellite data into master film negatives, computer compatible tapes or other data products for distribution to users. Although presently the number of TDRS platforms are also not adequate for the acquisition of all the remote sensing satellite data, yet when the requisite number of TDRS platforms are fully deployed, the image data of all remote sensing satellites can be acquired on a worldwide basis. The practical aspects of Landsat remote sensing data communication and ground processing system after the initiation of Domsat relay are presented in Fig. 6.2. A diagram showing the Landsat total communication and processing system is given in Fig. 6.3. The block diagrams of the Ground Segment Organization of IRS-1A /1B and IRS-1C are shown in Fig.6.4 and 6.5 respectively.

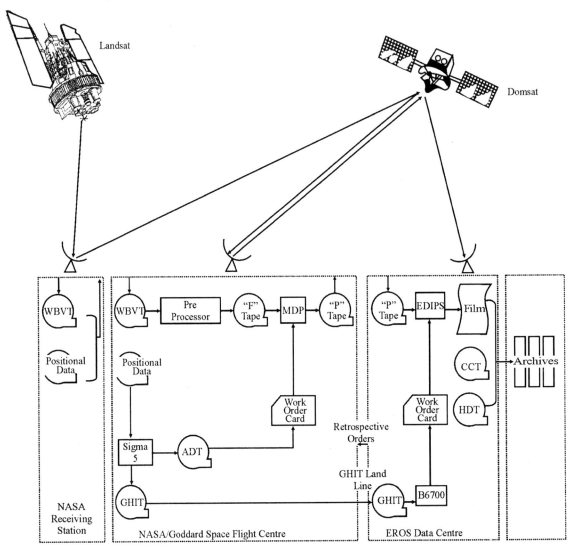

Fig. 6.2 : Schematic diagram showing Landsat data communication and ground processing system after initiation of Domsat relay.

6.2 PREPROCESSING OF SATELLITE DIGITAL DATA

Preprocessing of satellite digital data aims at improving the quality of data so that the users can restore an image which on further enhancement will yield acceptable and reproducible results. This is achieved by the following operations :

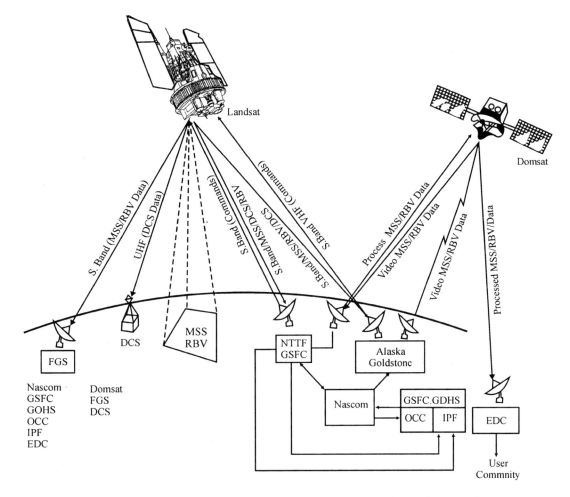

Fig. 6.3 : Schematic diagram showing landsat total communication and processing system.

1. Radiometric correction of remote sensor data
2. Geometric correction of remote sensor data, and
3. Noise removal from remote sensor data

6.2.1 Radiometric Correction of Remote Sensor Data

Radiometric errors may be internal or external or both by origin. The internal radiometric errors are introduced by the sensor design. Therefore these errors are systematic and constant, and can be removed by internal calibration measurements. For multispectral scanners the data collected in each band are radiometrically calibrated using an internal reference source. Black body sources are internally used to calibrate the thermal channel data whereas non-thermal channels are calibrated by internal calibration lamps.

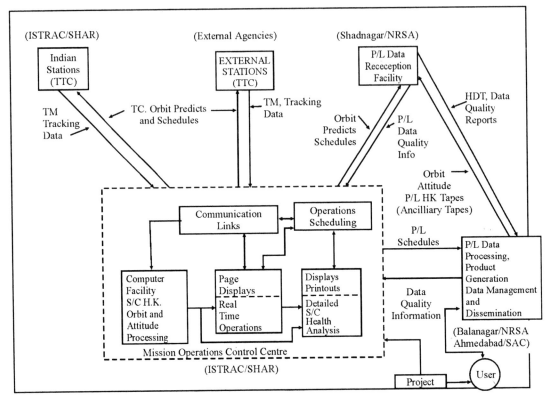

Fig. 6.4 : Block Diagram showing IRS-1A/1B Ground Segment Organization

Some radiometric errors originated from external factors include variation in solar illumination and viewing geometry, and variation in the atmospheric transmittance due to Rayleigh scattering by aerosols. These errors can be corrected during preprocessing of the raw data. Between band ratio techniques can be used profitably to reduce these errors.

6.2.2 Geometric Correction of Remote Sensor Data

Geometric errors in remote sensor data are of external origin. We find both systematic and nonsystematic geometric errors in remotely sensed data. Systematic geometric errors originating from space-borne sensor platform velocity and simultaneous earth's rotation transforms the square shaped ground resolution cells to skewed ones, namely parallelograms. This error is corrected mathematically by deskewing transformations.

Nonsystematic errors introduced into the remote sensor data by the fluctuations in the altitude and attitude of the sensor platform can not be corrected by analytical means. However, these can be corrected up to an acceptable accuracy with the help of a sufficient number of well distributed ground control points present in the image.

Two more geometric corrections which are not undertaken at the preprocessing stage in the remote sensing data center before the data are suitably formatted and archieved are the image rectification and image registration. In addition, noise correction is also not done in the data center. These latter three corrections are left to the users to take up during image restoration

Remote Sensing Data Acquisition and Dissemination

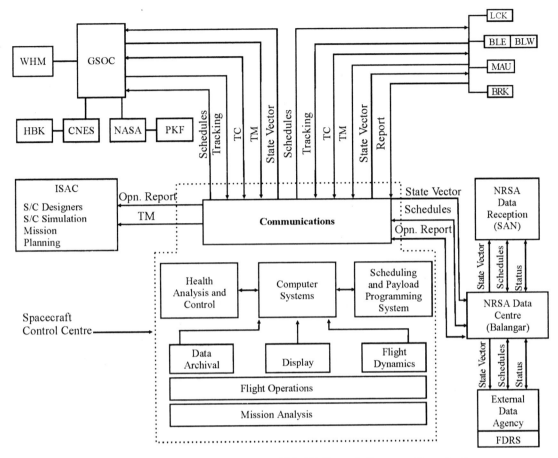

Fig. 6.5 : Block Diagram showing IRS-1C Ground Segment Organization

and image enhancement stages. Thus it is decided to appropriately describe them in the next chapter.

6.3 DATA FORMATTING AND ARCHIVAL

After applying the above mentioned radiometric and geometric corrections, the multispectral digital data from remote sensors are suitably formatted before archival. Data Centers normally choose some of the most prevalent formats for storing these digital data. These formats are :

(1) Band Sequential (BSQ), (2) Band Interleaved by Line (BIL), (3) Band Interleaved by Pixel (BIP) and (4) Run-length Encoding.

More information on data formats will also be given in the next chapter.

6.4 REMOTE SENSING DATA DISSEMINATION

Remote Sensing Data Centers associated with the Earth Stations (ground receiving stations) have the sole responsibility to disseminate the data to individual users and user organizations on payment basis.

The users are to order for the satellite data/data products for full scenes or quadrants (subscenes) indicating the path and row of a specific satellite pass and use of the sensor. At the Data Centers, the technicians browse the scenes of interest to the users and if they are cloud free, the data are extracted from the archive in the required format in floppies/cartridges/CDs/CCTs and sent to the users. Now-a-days, due to easily available internet services, the browsing facilities are also extended to the user's end. In case the user wants, the data product can be supplied in the form of geocoded B/W imagery or FCC. State-wise or district-wise mosaiced products, films and slides of data products can also be supplied on request.

Earlier, data for small areas were supplied on floppies based on fixed grids which resulted in more number of floppies to cover such small areas. Now with the introduction of floating segment floppy products, the floating segment is used to generate precisely the area of interest.

About some leading Data Centers of the world, it is worthy of record to say that the United States of America had promoted the establishment of Earth Stations all over the world to acquire, process and distribute Landsat data from mid-seventies. Earth Observation Satellite (EOSAT) Company was formed in 1985 by merging several units to take over the Landsat 4/5 operations. EOSAT has now linkages with several ground stations and an extensive distributor network. Similarly, National Remote Sensing Agency (NRSA), Hyderabad under the Department of Space, Government of India has Earth Receiving Stations at Shad Nagar near Hyderabad to acquire, process and market remote sensing data products of IRS, Landsat, SPOT, ERS and NOAA series of satellites. These data products mostly cater the needs of South-East Asian countries including India, middle-east and also of the African countries.

The launch of IRS-1A and 1B operations by ISRO/DOS, Government of India proved beyond doubt that India can emerge as a reliable supplier of high quality remote sensing data products. Continuity of data products was also assured from its future missions : IRS-1C, IRS-1D, IRS-P3/P4/P6 and so on. National priorities and the need to promote application oriented projects nationwide made NRSA to concentrate on creating infrastructure to meet indigenous demands /requirements. The space segment development capabilities, production infrastructure and systems of ISRO/DOS, and applications expertise based on NRSA's own resources and know-how prompted it to look beyond India for marketing the IRS series data products. The worldwide need for remote sensing data is quite huge and there is a positive demand for a wider choice of data products.

Keeping in view the above mentioned background, NRSA, on behalf of the DOS, Government of India and its corporate arm – Antrix Corporation, entered into a series of agreement with EOSAT Company of USA for reception of IRS-1B/1C data by EOSAT Earth Station at Norman, USA and processing/marketing the data products. Mutual cooperation on marketing dataproducts from each other's missions was also assured.

The salient features of these important agreements are :
- NRSA and EOSAT are interested in promoting the use of remote sensing for improved management of earth resources to benefit mankind.

- NRSA and EOSAT will act as mutual sales representatives to market each other's products.
- EOSAT will promote sale of IRS data worldwide.
- NRSA will continue sale of Landsat data in India.
- EOSAT and NRSA will cooperate to establish an appropriate network of ground stations to receive and process IRS data, and
- NRSA and EOSAT will extend technical support to each other.

The benefits of this historic tie-up are:

- Greater choice of data products to users.
- It will enable better understanding of the user needs.
- Utilize the strengths of EOSAT in reaching the users and NRSA/DOS strength of continued availability of data from its present/future missions.
- With lower investment existing ground stations can be upgraded to receive/process IRS data.
- Value added services and geographic information from one source as the tie-up matures.
- Archived data is a valuable asset and it can now be utilized in larger quantities.
- The quality of services/products will be state-of-art due to the joint efforts of NRSA and EOSAT.
- Financial benefits to the partners of the program.

Suggestions for Supplementary Reading

These suggestions are not exhaustive but are limited to the works that will be found especially useful for developing interest and creative imagination in the field of Remote Sensing.

Books

American Society of Photogrammetry, 1983. Manual of Remote Sensing, Vol. I, R. N. Colwell (Ed.), Falls Church, Va.

Bennett, W. R. and J. R. Davey, 1965. Data Transmission. McGraw-Hill, New York.

Blake, L. V., 1966. Antennas. John Wiley & Sons, New York.

Lillesand, Thomas M. and Ralph W. Kiefer, 1987. Remote Sensing and Image Interpretation, 2nd ed., John Wiley & Sons, New York.

Sabins, Floyd, F., Jr., 1987. Remote Sensing – Principles and Interpretations, 2nd ed., W. H. Freeman and Company, New York.

Taub, H. and D. L. Schilling, 1971. Principles of Communication Systems. McGraw-Hill, New York.

Research Papers

Beckman, John A., 1983. Communication and data Transmission Systems. In : Manual of Remote Sensing, Vol. I, 2nd ed., R. N. Colwell (Ed.). American Society of Photogrammetry, Falls Church, Va. p. 681 – 698.

Duck, Kenneth I. And Joseph C. King, 1983. Orbital Mechanics for Remote Sensing. In : Manual of Remote Sensing, Vol. I, 2nd ed., R. N. Colwell (Ed.). American Society of Photogrammetry, Falls Church, Va. p. 699 – 717.

Edwards, David T. and Tina Cary, 1996. International Marketing of IRS-1C Data. Current Science (Special Issue), 70 (7) : 638 – 641.

Hebbar, Jairam K., K. L. Majumdar, S. S. Palsule, K. M. M. Rao, J. D. Murthy, V. H. Patel, P. K. Srivastava, B. Lakshmi, V. Vittal Reddy, Francis Xavier, R. Joseph Arokiadas, R. Ramakrishnan, N. S. Parandhaman and D. Kaveri Devi, 1996. Data Processing System of IRS-1C and Data Quality Evaluation. Current Science (Special Issue), 70 (7) : 543 – 550.

Hebbar, Jairam K., J. Dakshina Murthy, K. M. M. Rao, B. Laxmi, V. Vittal Reddy, R. Joseph Arokiadas, D. Kaveri Devi, D. Chandrashekharan, G. Padma Rani and V. Raghu, 1996. IRS-1C Data Products Generation and Dissemination. Current Science (Special Issue), 70 (7) : 551 – 561.

Kumar Anil and T. C. Sarma, 1996. Data Reception System for IRS-1C. Current Science (Special Issue), 70 (7) : 534 – 542.

Kumar Sheelan Santosh, 1991. IRS-1A Data Dissemination. Current Science (Special Issue), 61 (3 & 4) : 288 – 291.

Radhakrishnan, K., V. Jayaraman and P. P. Nageshwara Rao, 1991. The Economics of Remote Sensing. Current Science (Special Issue), 61 (3 & 4) : 272 – 277.

Rao Mukund, V. Jayaraman, K. R. Sridhar Murthy, N. Sampath and M. G. Chandrasekhar, 1996. Commercialization of Remote Sensing – Issues and Perspectives. Current Science (Special Issue), 70 (7) : 642 – 647.

Scientific American, 1972. Communications. (Special Issue), September.

Sheelan Santosh Kumar, 1991. IRS-1A Data Dissemination. Current Science (Special Issue), 61 (3 & 4) : 288 – 291.

The Sensing 7

7.1 REQUIREMENTS OF DIGITAL IMAGE PROCESSING (HARDWARE AND SOFTWARE)

Digital image processing is carried out using : (i) mainframe-based, (ii) mini-computer-based, or (iii) microcomputer-based digital image processing systems. At this stage it is necessary to present only a general outline to the hardware and software configurations of the digital image processing systems. However, before fixing the image processing characteristics, one must be clear about the purpose for which these systems are to be used to achieve the objectives. For example, if the purpose is to allow 10 – 15 persons to access the image processing facilities simultaneously then a workstation which may accommodate maximum of 5 persons at a time is not the final solution. Similarly if the image processing systems are intended to educate the above number of trainees in an interactive environment, then a mainframe system operating in batch processing mode will serve no useful purpose. Thus to serve the purpose mentioned above, at least three digital image processing workstations networked to a minicomputer (server) is found ideal.

7.1.1 Characteristics of Digital Image Processing System

Central Processing Unit (CPU)

CPU does numerical calculations and exchanges the input/output data to peripheral storage devices, color monitor, printer etc. To handle large volumes of digital data faster, it is better to have CPUs with 16-bit registers or preferably higher, as 8-bit registers are slower for this purpose.

Maths Coprocessor

This is used to enhance the speed of the CPU further for numerical computation. An array of maths coprocessors is ideal for handling voluminous digital remote sensing data, particularly during image enhancement and image analysis operations.

Random Access Memory (RAM)

The CPU must have provision of a very high RAM for handling the operating system, image processing software and the necessary remote sensing data/image that are to be held in memory during performing calculations. A minimum of 256K or preferably higher RAM is ideal.

Operating System and Compiler

Operating system and compiler for the digital image processing are configured to satisfy the criteria that they are both easy and powerful. Thus the user can program his own algorithms to compute some parameters for specific use. The most widely used operating systems are DOS and UNIX, and the often used compilers in digital image processing software are BASIC, FORTRAN and Assembler.

Storage Disk

Digital image processing system requires very high memory of the order of several gigabytes, for storage of data and random access. In addition to the hard disc for mass storage, the provision of local disc subsystems becomes desirable to backup the hard disc, to provide data input to the system, to transfer data between workstations and for archival of processed image data.

Display Resolution

Image processing systems normally provide the display of 512×512 or 1024×1024 pixels on the screen of the CRT monitor. This helps in a terrain evaluation in regional context.

Color Resolution

The number of colors displayable at a time on the color monitor speaks about the color resolution. Higher color resolution is required for color compositing of images, black and white image display, color density slicing, pattern recognition, classification, visual image analysis and so on. 1-bit of image processor memory is required for maximum of $2^1 = 2$ displayable colors (black and white). Similarly, for 4-bit, 8-bit, 12-bit and 24-bit memory of the image processor, the displayable colors are 16 ($= 2^4$), 256 ($= 2^8$), 4096 ($= 2^{12}$) and 16,700,000 ($= 2^{24}$) respectively. However, as we move towards higher levels of color resolution (say 16,700,000 colors), the system becomes quite expensive because under this circumstance every pixel location is to be bit-mapped and the computer memory has to be substantially increased for the image processor. Practically 4096 ($= 2^{12}$) displayable colors are taken as a minimum set for color compositing of images and for the other jobs mentioned above. This scheme of 12-bit memory of the image processor is shared by three color bands (green, red, near-infrared) each having 4-bit of it.

Software

Image processing software resides as an executable command module in the hard disk. The software is menu-driven and interactive. Some of the largely used commercial image processing software packages are : ERDAS, EASI/PACE, GRASS, IDIMS, ELAS, GYPSY, ERIPS, SMIPS and so on.

7.2 DATA LOADING AND IMAGE RESTORATION

7.2.1 Data Loading

Users, on demand, receive remote sensing digital multispectral data from the archives of the Data Centre in CCTs or CDs written in Band Sequential (BSQ) format which is easier for extracting the useful data during image processing. In BSQ format the data (DN values) of all pixels covering the area of interest for a single band are kept in one file with the necessary header/trailer information. Thus there will be as many files of data as there are bands in the multispectral scanner. These data are loaded to the storage disk of the computer through the appropriate devices, namely streamer tape drive (for CCTs) and CD drive.

7.2.2 Image Restoration/Display

The pixel DN values/gray levels of each band are provided with a 8-bit system (2^8 gray levels). That is to say that the DN values of the pixels in the area of interest are recorded in the gray scale ranging from 0 to 255. To proceed with image restoration/display, the digital multi-band data of the study area is transferred from the mass storage disk to the local disk. To get a real color imagery we need color compositing of blue, green and red band data. But while processing the imagery from MSS data of Landsat, or LISS data of IRS satellites, the blue band data are usually not used to avoid its high degree of noise content due to Rayleigh scattering from atmospheric air molecules. Thus when we use band 2 (green), band 3 (red) and band 4 (near-infrared) data, we give a false instruction to the image processor that these bands are to be treated as blue, green and red respectively. Thus the resulting color imagery is called a false color composite (FCC). To achieve this we first draw the band 2 data from the local disk. The display controller and image processor are then instructed to read the DN values of the pixels and on the basis of color look-up table convert the DN values to analogue gray values of which 512×512 pixels (imagery of a subscene) are displayed at one time on the monitor screen. Next, following this procedure, band 3 and band 4 imageries of the same subscene are displayed one after another on the monitor. Finally these three imageries are superposed for color compositing. As explained earlier, instead of using a costly 24-bit digital image processing system, we use a moderate priced one of 9-bit or 12-bit image processing system which will allow us to use either 512 or 4096 color elements for the final FCC display on the RGB-display monitor. In this way a false color imagery of the area of interest is restored from its digital data. But this restored color image may not be of a very good quality to start with. It, therefore, needs further improvement in quality before one tries to extract reliable information from it. The procedures to achieve this are undertaken under interactive digital image processing which is schematically shown in Fig.7.1, and will be described in the next sub-sections. However, before doing that we would like to discuss how to change the scale for presenting the image of the entire study area and the way to study the spectral features lying on a transect of the image.

7.2.3 Image Reduction

This is needed to display the entire image of the study area for the satisfaction of the user and to identify as many identifiable locations, the so-called ground control points, which are useful to orient the image to make it to a geo-coded one following the procedure of image registration. The top of such an image points to the geographical north and the coordinate scheme assigned faithfully reproduces the latitude-longitude of different surface features found in the image. However, due to the reduced scale the information available in this image is much reduced.

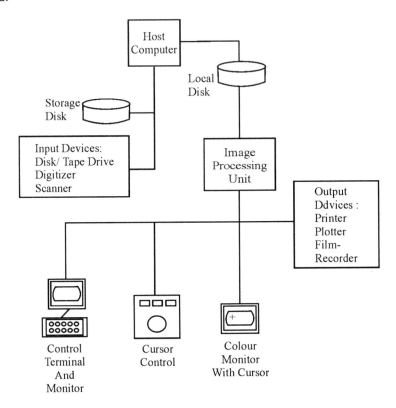

Fig. 7.1 : Schematic diagram of an interactive digital image processing system

For example, let the study area be covered by 1024×1024 pixels whose multispectral digital data are to be processed to get the image and that our image processing system can display only 512×512 pixels or less at any one time, then to display the image of the entire study area we have to reduce the pixel numbers of the rows and columns by a factor of half and this is done following the rule of periodic row and column stripping. This is done by choosing every alternate rows and columns of pixel data (starting from the first row and first column) for image processing. Consequently this will strip out the alternate rows and columns of pixels (starting from the second row and second column) from the study area. Thus we will be left with data of 512×512 pixels for image processing which can be displayed by the monitor screen at one go. In the illustration presented in Fig. 7.2, you can easily find that the image area is reduced to ¼ the original area in the process of alternate periodic stripping.

7.2.4 Roam Mechanism

By this mechanism provided by the image processor, one can roam about a very large area of the image viewing only 512 × 512 pixels at a time on the monitor. This is done by scrolling

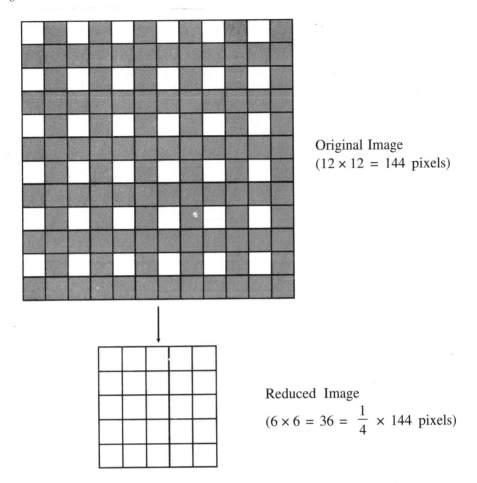

Fig. 7.2 : Image reduction by periodic stripping. Here the image is reduced to ¼th by stripping the alternate rows and columns.

the image about any x-y direction of the image. Thus one can inspect the entire image without any loss of original information, not at one go, but subscene by subscene.

7.2.5 Image Magnification

Digital image magnification or digital zooming is essential for presenting the image with a large scale to help visual image interpretation. This is done by replication of the rows and columns. For example, if the scale of the image is to be magnified by a factor of 2, then each row and column of pixels are replicated twice (Fig.7.3). It will be seen in this case that each pixel is replaced by a 2×2 block of pixels and the area of the final image is increased to 4 times the original area.

7.2.6 Transects

When one is interested to extract information along a few user-specified transects [say from pixel $P(i,j)$ to pixel $Q(k,l)$], then the image processing system helps in reading the DN values

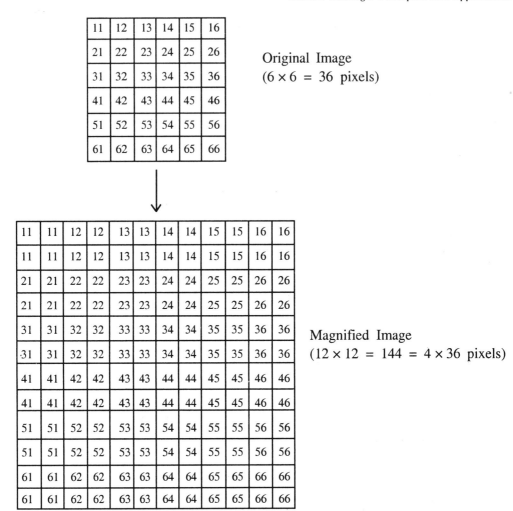

Fig. 7.3 : Image magnification by periodic replication. Here the original image is magnified 4 times by replicating (once) every row and column of the original image. (Here pixels are designated by their row and column numbers.)

of all pixels from P to Q once the image is geometrically rotated and translated so that the end points of each transect fall on the same scan line (Fig.7.4). The histograms of the transects in each band may then be formed and stored which will find use for the transect image quality enhancement leading to better transect feature classification and interpretation.

7.3 IMAGE RECTIFICATION AND REGISTRATION

7.3.1 Image Rectification

To start with, an image of a scene restored from raw digital data obtained from the aerial or satellite platform looks quite distorted and hence needs rectification. The degraded quality of this image is mainly due to the systematic errors introduced during the data acquisition processes.

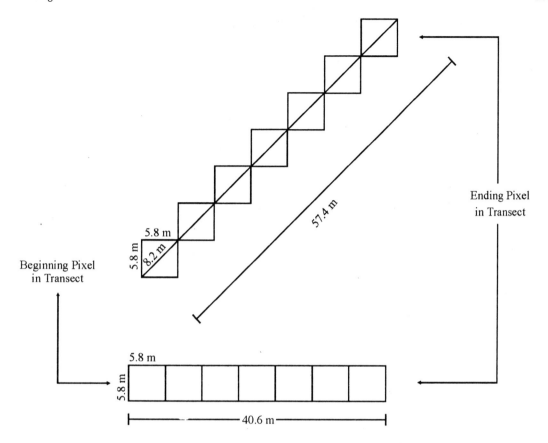

Fig. 7.4 : Linear measurements and pixel intensities (DN values) along a transect are easily obtained when the beginning and ending pixels of the transect fall on the same scan line. If this is not the case, the image has to be geometrically rotated till the end points of the transect fall on the same scan line, otherwise as shown here, the hypotenuse of the stair stepped pixels must be considered.

Thus the geometric and radiometric corrections applied to the raw satellite data at the Satellite Data Center, in fact, forms the first step towards image rectification. However, the image restored from such corrected digital data still does not correspond to the geocentric lat-long coordinate system. Thus this work is undertaken further by the user through the process of image rectification in which the image is aligned to a map (through translation and rotation operations) so that the image becomes planimetric just like its map. Practically this is done by the following procedures :

Let there be only one image. To rectify this image, obtain the topographic map of this area from the Survey Department. Digitize the map along with lat-long coordinate system and store the map in a file with the north of the map pointing to the earth's geographical north. Project the image and the map on the image processor screen. Find out as many ground control points (GCPs) in the map with their known latitudes and longitudes. Identify in the image scene as many spatial features corresponding to these GCPs to the nearest precision. Finally establish the coordinate transformation by pairing the GCPs and their corresponding spatial features by correspondence matching and the image is thus rectified. Rectification is also known as georeferencing. Geocoding is a special case of rectification which includes scaling to a uniform, standard pixel. The use of standard pixel sizes and coordinates permits convenient layering of images in a geographic information system(GIS).

7.3.2 Image Registration

Image registration can be of two types : map-to-image registration which is otherwise known as the primary registration, and image-to-image registration In fact, in the image rectification process described above, we have really presented the primary registration technique. While undertaking primary image registration for a study area, it is understood that we do not have a rectified image of the study area already stored in our image processing system. However, once we have already stored a rectified image of the study area in our image processing system, the subsequent images of the study area are subjected to image–to-image registration. Here the process is the alignment of one image (unrectified) to another image (already rectified) of the same area. And when it is done, any two pixels at the same location in both the images are said to be 'in register' and represent two samples at the same point on earth. Pixel-by-pixel registration requires extracting characteristic identical spatial features (points, lines, regions etc.) in both the images and subjecting them to pixel-by-pixel correspondence matching. Single-pixel GCPs can be manually extracted by zooming both the images at crossings of rails and roads and their bends, runway of the aerodromes, bridges over rivers, characteristic geological structures, edge points, special structures in the built up areas etc.which serve as useful conventional GCPs. But for quick and automated image-to-image registration, small areas (m x m pixels) in each image called 'chips' can serve as the characteristic spatial features for automated GCP location identification.

7.4 IMAGE STATISTICS EXTRACTION USING RADIOMETRIC DATA

Before proceeding to the digital image processing from the preprocessed/rectified multispectral digital data, it is advisable to compute a few univariate and multivariate statistical parameters that provide information on the data quality, data redundancy and the directions to be followed to digitally process the data to get an acceptable quality image.

7.4.1 Univariate Image Statistics

This statistics is on digital data set of an imagery pertaining to a single spectral band. These data are digital numbers (DN) which are nothing but the brightness values recorded on each pixel of the imagery. If the imagery consists of m rows and n columns of pixels (Fig.7.5), then the total number of pixels become equal to N = mn. This (N) then is the total number of DN values to be subjected to statistical calculation.

The univariate statistical parameters to be computed are :

DN_r = Range of DN values [If the scale in which the brightness (DN) values are recorded is 2^8, then the DN_r may range from 0 to 255, otherwise it is from $DN_{minimum}$ to $DN_{maximum}$] (7.1)

The Sensing

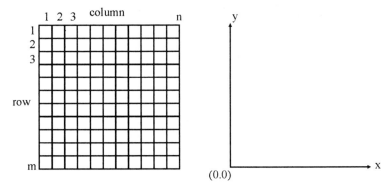

Fig. 7.5 : Diagramatic presentation of a single band digital data of an imagry. Mark the preferred discrete (row, column) notation as compared to the continuous spatial (x, y) notation.

μ_k = Mean of DN values for a spectral band k of the imagery

$\quad = (1/N) \sum_{i=1}^{m} \sum_{j=1}^{n} DN_{ijk}$ (7.2)

At this stage, for simplicity, the pixel designation (ij) may be replaced by the pixel counting number (p) which will take values p = 1 to N. Thus μ_k will be given by

$\mu_k \quad = (1/N) \sum_{p=1}^{N} DN_{pk}$ (7.3)

$\sigma_k \quad$ = Standard deviation of DN values for a spectral band k of the imagery

$\quad = [\sum_{i=1}^{m} \sum_{j=1}^{n} (DN_{ijk} - \mu_k)^2/(N-1)]^{1/2}$ (7.4)

$\sigma_k^2 \quad$ = Variance of DN values for a spectral band k of the imagery

$\quad = [\sum_{i=1}^{m} \sum_{j=1}^{n} (DN_{ijk} - \mu_k)^2/(N-1)]$ (7.5)

Skewness = $(1/N) \sum_{i=1}^{m} \sum_{j=1}^{n} [(DN_{ijk} - \mu_k)/\sigma_k]^3$ (7.6)

Kurtosis = $[(1/N) \sum_{i=1}^{m} \sum_{j=1}^{n} \{(DN_{ijk} - \mu_k)/\sigma_k\}^4] - 3$ (7.7)

Both skewness and kurtosis are dimensionless numbers.

HistDN$_k$ = histogram of DN values for a spectral band k of the imagery

$\quad$ = (count number of pixels in each DN 'bin')/N (7.8)

A single band image histogram can be produced by plotting 'fraction of total pixels' against the DN values (Fig.7.6).

Chist$_k$ = cumulative histogram of DN values for a spectral band k of the imagery

$\quad = \sum_{DN_{min}}^{DN_{max}} hist\ DN_k$ (7.9)

Since the area under the histogram is unity, the cumulative histogram curve asymptotically approaches one (Fig.7.7).

Mode : This useful statistical parameter indicates the DN value at which the histogram is maximum.

Median : This is the DN value which divides the histogram area in half, 50% pixels being below the median and 50% above it.

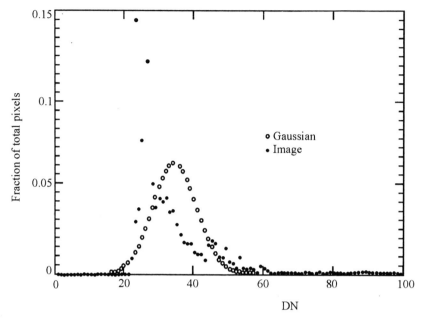

Fig. 7.6 : A typical image histogram and its Gaussian distribution.

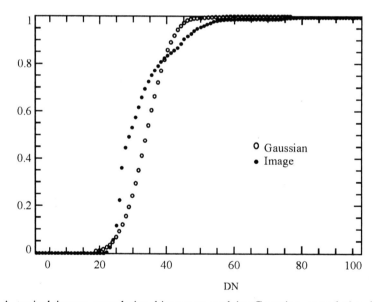

Fig. 7.7 : A typical image cumulative histogram and its Gaussian cumulative histogram.

7.4.2 Information Obtained from Univariate Image Statistics

The mean, standard deviation and variance of DN values for a specific spectral band of the imagery are treated as useful measures of central tendency. For example, for a thematic band like the red band around 0.6 mm, if these statistical parameters work out to be all small, then one can infer that the imagery may be of a vegetation canopy which strongly absorbs solar radiation in this band and efficiently produces biomass through the process of photosynthesis. Similarly a high mean value with small standard deviation for the DN values in the NIR band indicates a luxuriantly and uniformly growing crop due to the spongy mesophyll cell expansion.

The mean, standard deviation and variance are sufficient to specify a normal or Gaussian distribution. In case the histogram is found unimodal and symmetric, then a Gaussian distribution may satisfactorily describe the actual data. However, if the histogram is asymmetric and multimodal, then the existence of a number of classes of surface features equal to the number of modes are indicated. In such cases, statistical parameters indicated above may be worked out for each class to draw preliminary information about the feature classes. Image standard deviation can be used as a measure of image contrast, because it is a measure of the histogram width, namely the spread in the DN values.

Image histogram is a useful tool for image contrast enhancement. The contrast enhancement algorithm stretches the range of DN values and thresholds it at one or both ends which results in the saturation of a certain percentage of pixels. The required DN thresholds are obtained from the histogram already drawn. If the histogram is confined within a small range of DN values, then it indicates an imagery of low contrast, requiring contrast enhancement to produce a quality image.

Skewness is the measure of asymmetry of the histogram.

Skewness = 0   for any symmetric histogram

= + ve for an asymmetric histogram with a long tail towards the larger DN values

Similarly, kurtosis indicates sharpness of the peak relative to normal distribution.

Kurtosis = 0   for a normal distribution

= + ve indicates that the peak of the histogram is sharper than that of a Gaussian distribution

= − ve indicates that the peak of the histogram is less sharp than that of a Gaussian distribution

In general, the mean, standard deviation, skewness and kurtosis indicate the degree of asymmetry and sharpness in the peak of the histogram. Skewness and kurtosis are also found to be quite sensitive to outliers, which are pixels with DN values far removed from the majority distribution.

7.4.3 Multivariate Image Statistics

This statistics is on digital data set of an imagery pertaining to more than one spectral bands, say spectral bands k and l, and so on. To begin with, the parameters of multivariate image statistics provide a general information about the measure of the mutual interaction between the DN values in two spectral bands on individual pixels of the imagery. The parameters of the multivariate image statistics to be computed are:

Cov_{kl} = covariance between pixel DN values in two bands k and l

It is a measure of the joint variation of two variables about their respective means. Just as we computed $Variance_k$ (i.e., σ_k^2) using corrected sum of squares (SS_k), namely

$$Var_k = \sigma_k^2 = \sum_{p=1}^{N} (DN_{pk} - \mu_k)^2/(N-1) = SS_k/(N-1) \qquad (7.10)$$

Similarly, covariance can be computed using the corrected sum of products (SP_{kl}):

$$Cov_{kl} = [\sum_{p=1}^{N} (DN_{pk} - \mu_k)(DN_{pl} - \mu_l)]/(N-1) = SP_{kl}/(N-1) \qquad (7.11)$$

For multispectral data sets like IRS: LISS-I / LISS-II and Landsat MSS, which have 4 band data sets, the variance-covariance matrix is given in the following format (Table 7.1).

Table 7.1 : Format of variance-covariance matrix

Bands→ Bands ↓	Band 1	Band 2	Band 3	Band 4
Band 1	SS_1	Cov_{12}	Cov_{13}	Cov_{14}
Band 2	Cov_{21}	SS_2	Cov_{23}	Cov_{24}
Band 3	Cov_{31}	Cov_{32}	SS_3	Cov_{34}
Band 4	Cov_{41}	Cov_{42}	Cov_{43}	SS_4

Here variances occupy the principal diagonal elements and the covariances occupy the off-diagonal elements of the variance-covariance matrix. It is to be noted here that since the covariance matrix is symmetric, i.e. $Cov_{kl} = Cov_{lk}$, in the above matrix format, the covariance matrix elements above the principal diagonal have their duplicates below the principal diagonal, and hence only one set of covariances need to be computed. Also note that the principal diagonal elements of the matrix being variances, they are all positive. However, the off-diagonal elements (covariances between any two different bands) may be positive or negative.

7.4.4 Correlation Between Pixel Values in Two Bands : k and l

To get the degree of interrelation between pixel DN values of two bands k and l, we compute the correlation coefficient (r_{kl}) given by

$$r_{kl} = Cov_{kl}/(\sigma_k \cdot \sigma_l) = Cov_{kl}/(Var_k \cdot Var_l)^{1/2} \qquad (7.12)$$

The value of r_{kl} which is a dimensionless number varies between

$$-1 \text{ and } +1, \text{ that is } -1 \leq r_{kl} \leq +1 \qquad (7.13)$$

This condition dictates that

$$|Cov_{kl}| \leq |\sigma_k \cdot \sigma_l| \qquad (7.14)$$

A correlation coefficient of +1 indicates a monotonically perfect relationship between the DN values of the two bands making one of the bands completely redundant. But a correlation coefficient of −1 indicates that the pixel DN values of one band is perfectly but inversely related to the DN values of the other band. A high positive correlation coefficient shows the existence of high degree of redundancy between the bands. Zero or near zero value of correlation coefficient indicates that the pixel DN values of the two bands are uncorrelared or independent of each other. Correlation coefficient lying between 0 and +1 shows moderate correlation indicating that these bands provide some type of information not found in other bands. The correlation matrix is presented in the following format (Table 7.2):

Table 7.2 : Format of correlation matrix for 4 band data

Band→ Band↓	Band 1	Band 2	Band 3	Band 4
Band 1	1	r_{12}	r_{13}	r_{14}
Band 2	r_{21}	1	r_{23}	r_{24}
Band 3	r_{31}	r_{32}	1	r_{34}
Band 4	r_{41}	r_{42}	r_{43}	1

Here also one has to remember that

$$r_{kl} = r_{lk} \tag{7.15}$$

The correlation coefficient indicates the shape of the scatterogram of a multispectral image as shown in Fig.7.8.

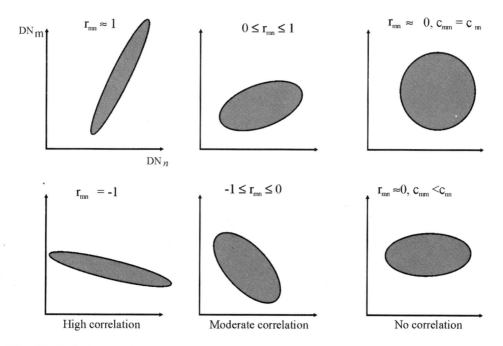

Fig. 7.8 : Typical shapes of Scattrograms of multipectral image data indicating the inter-band correlation coefficients.

Later in this chapter, we shall show that the variance-covariance and the correlation matrices are very useful in principal component analysis of remote sensing data, feature selection and classification.

7.4.5 Statistical Evaluation of Image Quality Parameters

The quality of a digital image is characterized by a number of indices. We present below some of the commonly used ones for image quality measures.

Contrast

Contrast (C) of a digital image may be variously defined as

$$C_{ratio} = DN_{max}/DN_{min} \tag{7.16}$$

$$C_{range} = DN_{max} - DN_{min} \tag{7.17}$$

$$C_{std} = \sigma_{DN} \tag{7.18}$$

where DN_{max} and DN_{min} are the maximum and minimum DN values in the imagery. C_{ratio} and C_{range} are sensitive to the outliers.

To the above set of contrast parameters, one can add one more set on visual contrast. For this the DN values in the above equations should be changed to gray levels (GL) :

$$C_{ratio}(visual) = GL_{max}/GL_{min} \tag{7.19}$$

$$C_{range}(visual) = GL_{max} - GL_{min} \tag{7.20}$$

$$C_{std}(visual) = \sigma_{GL} \tag{7.21}$$

Since any one of the above contrast parameters does not lead to the best image contrast, the combination of a number of them should be used to take a final decision on the quality of the image.

Modulation

Modulation which is also a measure of the image quality is defined by

$$M = (DN_{max} - DN_{min})/(DN_{max} + DN_{min}) \tag{7.22}$$

Modulation is always positive and ranges from 0 to 1. It is a dimensionless parameter. Modulation is appropriate in describing the quality of spatially repetitive (periodic) signals. From equation (7.16) it is also found that modulation is related to contrast ratio :

$$M = (C_{ratio} - 1)/(C_{ratio} + 1) \tag{7.23}$$

Like C_{ratio} and C_{range}, modulation (M) is also found to be sensitive to the outliers.

Signal-to-Noise Ratio

The signal which carries information from the target of interest is essentially the noiseless part of the measurement by the sensor. But since the sensor can not completely avoid registering a number of different types of noise along with the main signal, the signal data we finally receive from the sensor is really a corrupted one. Thus some measures of the relative amount of signal and noise is needed for : (i) better engineering design of the sensor, (ii) assessment of data quality, (iii) development of noise reduction algorithms and (iv) development of some information extraction algorithms.

We take signal-to-noise ratio (SNR) as an indicator of the image quality. It is defined by

$$SNR_{amplitude} = C_{ratio} \text{ (signal)}/C_{ratio} \text{ (noise)} \qquad (7.24)$$

$$SNR_{range} = C_{range} \text{ (signal)}/C_{range} \text{ (noise)} \qquad (7.25)$$

$$SNR_{std} = C_{std} \text{ (signal)}/C_{std} \text{ (noise)} = \sigma_{DN} \text{ (signal)}/\sigma_{DN} \text{ (noise)} \qquad (7.26)$$

Because C_{ratio} and C_{range} are sensitive to outliers, the last expression (eqn. 7.26) of SNR_{std} is taken as more reliable. Alternately we can choose the most commonly used indicator with respect to variance given by

$$SNR_{var} = Var \text{ (signal)}/Var \text{ (noise)} = \sigma^2 \text{ (signal)}/\sigma^2 \text{ (noise)} \qquad (7.27)$$

We may also define SNR with respect to power as

$$SNR_{power} = (SNR_{amplitude})^2 = (C_{signal}/C_{noise})^2 \qquad (7.28)$$

SNR is a dimensionless number. However, we can express it in decibels (dB) using the following relationship

$$SNR_{dB} = 10 \log(SNR) \qquad (7.29)$$

where SNR may be given by any of the above mentioned definitions.

Normally the random uniform noise levels can be estimated from uniform areas of the imagery which are assumed to have no useful information content. When the noise level is not high, the slightly noisy imagery may be taken for granted as due to the signal itself. It is found that the existence of periodic noise is more visible in the imagery than the random noise. However, it is also easier to remove periodic noise than the random noise for which suitable software is not readily available.

7.5 IMAGE ENHANCEMENT USING SPECTRAL TRANSFORMS

Satellite sensors are designed to monitor, in every band, a wide range of scenes characterized by very low radiance pixels of water bodies to very high radiance pixels of snow and sand. However, normally in the area of our study, all these surface features do not exist and the pixels of the subscene/scene are confined to a rather narrow range of DN values out of the full radiometric scale (quantization range of 0 to 255) in 8-bit system. Thus it produces an image of low contrast. Moreover, pixels occupied by all biophysical materials produce low DN values in visible to near infrared part of the spectrum giving rise to low contrast images. Thus to begin with radiometric transforms are applied to improve the visual quality of the displayed black and white image in each band.

Contrast enhancement is the mapping of the original data space to the full scale gray level (GL) display space. Thus the mapping function transforms the DN input image to the output GL image. In this contrast enhancement operation, since DN values confined within a narrow range are expanded to fill the entire available GL range, the transform is also called contrast stretch. A few useful contrast stretches are described below:

7.5.1 Linear Contrast Stretch

This is also called the min-max stretch because the actual image DN range (minimum to maximum DN values) is expanded to fill the entire dynamic range of display device (0 to 255). The mapping function of the min-max transform is

$$GL = [(DN - DN_{min})/(DN_{max} - DN_{min})] \times 255 \qquad (7.30)$$

The linear contrast stretch is found sensitive to the outliers. Linear contrast enhancement is satisfactorily applied to images of Gaussian or near-Gaussian unimodal histograms. In this case $d(GL)/d(DN)$ which is constant can be taken as the gain factor for contrast (Fig.7.9). However, when the image histogram is non-Gaussian (multimodal), it is possible to perform stepwise linear contrast stretch to the imagery (Fig.7.10).

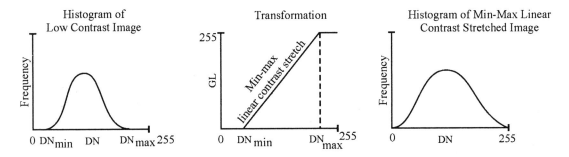

Fig. 7.9 : DN-to-GL transformation indicating the applicability of Min-max linear contrast stretch operation for Image Contrast Enhancement.

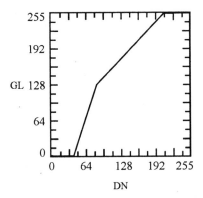

Fig. 7.10 : DN-to-GL transformations indicating stepwise linear contrast stretch operation for Image Contrast Enhancement.

The image DN range of some sensors of AVHRR (10-bits/pixel) and hyperspectral sensors (12-bits/pixel) exceeds the dynamic range of the display device. In such cases a linear transformation can similarly be used to decrease the image contrast.

7.5.2 Saturation Stretch

Saturation stretch is executed starting with a linear stretch for all the pixels within a DN range which is smaller than the min-max range to achieve higher contrast increase (Fig.7.11). Usually saturation of a few percent of image pixels is allowed. Pixels with a DN outside the range are mapped to either 0 or 255. This clipping (saturation) at the extremes is acceptable unless it destroys the image structure.

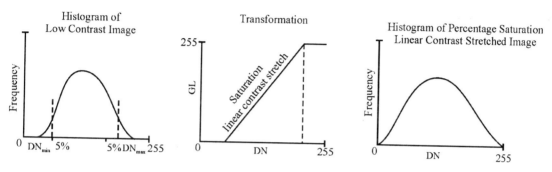

Fig. 7.11 : DN-to-GL transformation for a 5% saturation linear contrast stretch operation for Image Contrast Enhancement

7.5.3 Nonlinear Contrast Stretch

When the image histogram is asymmetric, the simple linear contrast stretch is not applicable. This can be done as a two segment min-max stretch, with the left segment having higher gain than the right segment.

7.5.4 Histogram equalization

Histogram equalization is an example of the largely used nonlinear transform. This is done using the cumulative distribution function (CDF) of the image as a transformation function

$$GL = 255 \; CDF \; (DN) \qquad (7.31)$$

to produce uniform histograms (Fig.7.12).

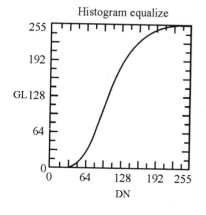

Fig. 7.12 : DN-to-GL transformation to equalize histogram for image contrast enhancement

7.5.5 Normalization Stretch

This is a powerful contrast enhancement procedure. In this algorithm the image is transformed through linear stretch such that it has a specified GL mean and standard deviation, and is clipped at the extremes of the display GL range. The procedure is somewhat like the saturation stretch. The mapping functions are given by

(a) $GL = (\sigma_{ref}/\sigma)(DN - \mu) + \mu_{ref}$ (7.32)

(b) $GL = 255$, $GL > 255$

$GL = 0$, $GL < 0$ (7.33)

This matches means and variances. As shown in (b) above, clipping becomes necessary as the resulting GLs usually extend beyond the limits of the display range.

7.5.6 Reference Stretch

This is required to compare multispectral and multitemporal imageries of the same scene or of adjacent scenes provided the histograms of the two images are similar.

Normalization stretch may be generalized to match histogram shapes which is accomplished by matching the CDF of an image to a reference CDF. This is done by forward mapping the image DNs through the image CDF, and then backward mapping through the CDF_{ref} to DN_{ref}. CDF_{ref} may be from another image or from a theoretical image with specified CDF, say a Gaussian histogram. Here the mapping function is given by

$$GL = CDF_{ref}^{-1}[CDF(DN)]$$ (7.34)

It matches histograms.

Another technique for reference stretching is the statistical pixel matching. For this it is necessary that the two images registered cover the same area on the ground. Important application area of this technique is the mosaicing of the sidelap area of two adjacent satellite paths.

7.5.7 Thresholding

This type of contrast enhancement segments the image into two categories defined by a single DN threshold (DN_T) as shown in Fig.7.13. The threshold mapping functions are :

$GL = 255$; $DN \geq DN_T$
$GL = 0$; $DN < DN_T$ (7.35)

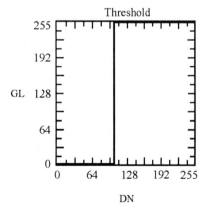

Fig. 7.13 : DN-to-GL transformation indicating the presence of a DN threshold for making appropriate decision on further Digital image processing.

The Sensing

These transformations produce binary output with sharply defined spatial boundaries which may be used for masking portions of the image. Each portion may then be separately subjected to contrast enhancement and finally recombined. Thresholding may be viewed as a simple classification algorithm.

At this point one should be clear that contrast enhancement techniques are primarily meant to improve visual image analysis. Therefore the enhanced image should not be subjected to further computer analysis for digital classification and information extraction. This is due to the fact that contrast stretch operations change the original pixel DN values, often in a nonlinear manner.

7.5.8 Color Images

So far the image contrast enhancement techniques using radiometric transforms have been described for single band image to help visual image analysis. However, color display of images and their enhancement using radiometric transforms further improves the visual image interpretation. A few such simple techniques are outlined below :

Min-Max Stretch

In this technique, linear (min-max) stretch is applied to the R, G, B band images separately and these three enhanced images are composited to get the final enhanced color image.

Normalization Stretch

For getting the enhanced color image, normalization stretch is applied to each of the three bands (designated R, G, B) independently by setting their means and standard deviations equal across the bands and finally compositing the three. The flow diagram of normalization algorithm for color images is presented in Fig.7.14.

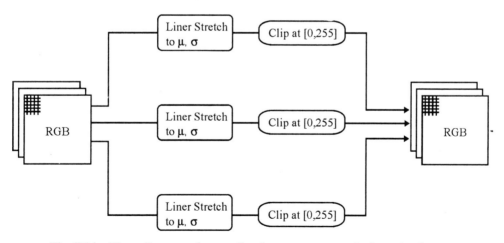

Fig. 7.14 : Flow diagram of normalization contrast stretch for color imagery.

Decorrelation Stretch

When three spectral band data of the image are found to be correlated, then color contrast enhancement becomes poor since very little color space is utilized by these three correlated spectral bands. However, if we decorrelate the bands, stretch the principal components (PCs) to fill the color space and then inverse transform to the RGB color space, we enhance the spectral information whatever is present in the data. This, in fact, is the basis of the principal component transformation (PCT) decorrelation stretch. The flow diagram of the PCT decorrelation contrast stretch algorithm is presented in Fig. 7.15.

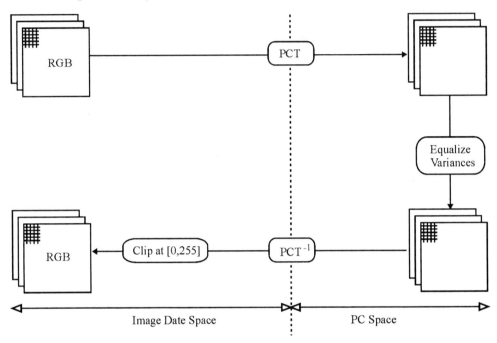

Fig. 7.15 : Flow diagram of the PCT decorrelation contrast stretch for color imagery.

Color Space Transform (CST)

To describe visually perceived color of an image, instead of using red, green and blue components, we use hue, saturation and intensity (HSI) for subjective sensations of color, color purity and brightness respectively. Thus using a color-space transform (CST), we first change RGB components to hue, saturation and intensity components. In this perceptual color-space of HIS, one can predictably modify any one or all of the components as desired. Now using the inverse CST, the processed images are converted back to RGB for display. The flow diagram for CST to modify the perceptual color characteristics of an image is presented in Fig. 7.16.

Ratioing

Radiances/reflectances of the ground resolution cells having similar materials differ due to their varying surface topography, shadows, seasonal changes in solar illumination angle and intensity of incident radiation. This creates difficulties in feature discrimination and classification. However, simple ratio (R_{mn}) transform of any two multispectral band radiances minimizes such environmental effects.

$$R_{mn} = R_m/R_n \qquad (7.36)$$

where R_m and R_n are reflectances/radiances in the multispectral bands m and n respectively.

The Sensing

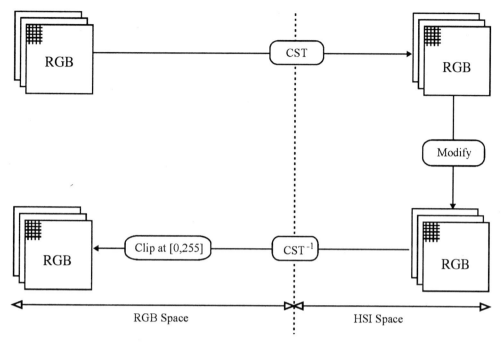

Fig. 7.16 : Flow diagram of the color-space transform (CST) for improving visual color characteristics of a multispectral imagery.

Ratio Vegetation Index (RVI)

Apart from minimizing the effects of environmental factors, ratioing also provides unique information not available in any single band that is useful for discriminating soils and vegetation. Ratio vegetation index defined by

$$RVI = R_{NIR}/R_{RED} \qquad (7.37)$$

can enhance radiance differences between soils and vegetation. While soils and geological materials exhibit similar ratio values near one, vegetation shows larger ratio value of two or more, the higher values indicating more luxuriant crop growth conditions.

Modulation Ratio (M_{mn})

This is a useful nonlinear variant of the simple ratio

$$M_{mn} = (R_m - R_n)/(R_m + R_n) = (R_{mn} - 1)/(R_{mn} + 1) \qquad (7.38)$$

Normalized Difference Vegetation Index (NDVI)

This is nothing but the modulation ratio of the near infrared and red bands

$$NDVI = (R_{NIR} - R_{RED})/(R_{NIR} + R_{RED}) \qquad (7.39)$$

It is used extensively to monitor vegetation. But it is a poor indicator of vegetation biomass if the ground cover is low, as in arid and semiarid regions.

Soil-Adjusted Vegetation Index (SAVI)

This is a better vegetation index than NDVI for low ground cover

$$\text{SAVI} = [(R_{NIR} - R_{RED})/(R_{NIR} + R_{RED} + L)] (1 + L) \quad (7.40)$$

where L is a constant which is empirically determined to minimize the vegetation index sensitivity to soil background reflectance variation.

If $L = 0$, then SAVI = NDVI

For intermediate vegetation cover $L \cong 0.5$

The factor $(1 + L)$ insures that the range of SAVI is between -1 to $+1$, like that of NDVI.

Transformed Vegetation Index (TVI)

This is found to be more linearly related to vegetation biomass than NDVI and hence used extensively for biomass estimation.

$$\text{TVI} = (\text{NDVI} + 0.5)^{1/2} \quad (7.41)$$

where 0.5 is the bias term that prevents negative values under the square root for most of the images having growing crops. To eliminate completely the possibility of the term $(\text{NDVI} + 0.5)^{1/2}$ to become imaginary, the following algorithm may be adopted

$$\text{TVI} = [(\text{NDVI} + 0.5)/|(\text{NDVI} + 0.5)|] |(\text{NDVI} + 0.5)|^{1/2} \quad (7.42)$$

Here $|(\text{NDVI} + 0.5)|$ denotes the absolute value and $0/0$ is to be taken as 1.

Perpendicular Vegetation Index (PVI)

This is taken as the perpendicular distance of a pixel vector from the soil line in a NIR versus Red reflectance-space (Fig.7.17). It is an indicator of plant development. PVI is given by the formula

$$\text{PVI} = [(R_{RED}^{soil} - R_{RED}^{veg})^2 + (R_{NIR}^{soil} - R_{NIR}^{veg})^2]^{1/2} \quad (7.43)$$

Principal Component Transformation (PCT)

As mentioned earlier, the multispectral sensor bands are often found highly correlated. Analysis of such a correlated set of multispectral data becomes inefficient because of this redundancy. The principal component transformation (PCT) is therefore designed to decorrelate these multispectral data set and thereby remove the spectral redundancy (Fig.7.18). Moreover, PCT also compresses the information content of a number of bands (e.g., the seven thematic mapper bands of Landsat) into just two or three transformed principal components. This ability to reduce the dimensionality (i.e., the number of bands in the data set) is an important economic consideration, particularly when the information recoverable from the transformed data is just as good as that from the original remote sensor data set.

To achieve this, we start with the covariance matrix (C) of the multispectral (k-bands) data. This is a $k \times k$ matrix :

$$C = \begin{bmatrix} w_{11} & w_{12} & \cdots & w_{1k} \\ w_{21} & \cdots & \cdots & w_{2k} \\ \cdots & \cdots & \cdots & \cdots \\ w_{k1} & \cdots & \cdots & w_{kk} \end{bmatrix} \quad ; \text{ with } w_{ij} = w_{ji} \qquad (7.44)$$

Then we construct an eigen vector W_{PC} (principal component transformation matrix) and its transpose (W_{PC}^T). The structure of the PC-eigen vector (W_{PC}) is given by

$$W_{PC} = \begin{bmatrix} e_{11} & e_{12} & \cdots & e_{1k} \\ e_{21} & \cdots & \cdots & e_{2k} \\ \cdots & \cdots & \cdots & \cdots \\ e_{k1} & \cdots & \cdots & e_{kk} \end{bmatrix} \quad ; \text{ with } e_{ij} = e_{ji} \qquad (7.45)$$

Components of this transformation matrix (e_{ij}) consist of weights on each of the original spectral bands and the direction cosines of the new axes relative to the original axes.

In the next step we diagonalize the covariance matrix (C) to get the principal component covariance matrix (C_{PC}) :

$$C_{PC} = W_{PC} \, C \, W_{PC}^T = \begin{bmatrix} \lambda_{11} & 0 & \cdots & 0 \\ 0 & \cdots & \cdots & 0 \\ 0 & \cdots & \cdots & 0 \\ 0 & \cdots & \cdots & \lambda_{kk} \end{bmatrix} \qquad (7.46)$$

Each eigen value (λ_{ii}, the diagonal elements of C_{PC}) is equal to the variance (self-covariance) of the respective PC image along the new coordinate axes. The sum of all eigen values equals the sum of all band variances of the original image preserving the total variance in the data. Since C_{PC} is diagonal, the principal component images are uncorrelated and, by convention, are ordered by decreasing variance. Thus PC_{11} ($= \lambda_{11}$) has the largest variance and PC_{kk} ($= \lambda_{kk}$) has the lowest variance. The result is the removal of any correlation present in the original k-dimensional data, with a simultaneous compression of most of the total image variance into a fewer dimensions. The eigen values (λ_{ii}) contain important information. It is possible to compute the percent of total variance contributed by each of the principal components using the relationship

$$\% \, (\lambda_{ii}) = (\lambda_{ii} \times 100) / (\textstyle\sum_{i=1}^{k} \lambda_{ii}) \qquad (7.47)$$

Finally we project the pixel DN values of the few chosen spectral bands to the new PC axes by the transformation matrix W_{PC}:

$$PC \, (= DN) = W_{PC} \cdot DN \qquad (7.48)$$

And these new DN values of the pixels are used to digitally process the images showing highly enhanced quality. PCT is data (scene) dependent, namely on spectral covariance matrix. Since covariance matrix changes from scene to scene and also for the same scene monitored

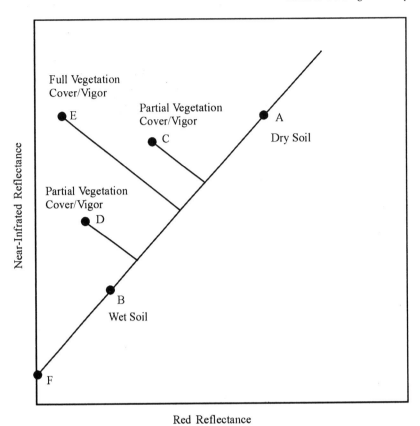

Fig. 7.17 : Schematic diagram of perpendicular vegetation index showing the relationship of vegetation cover/vigor to the soil background.

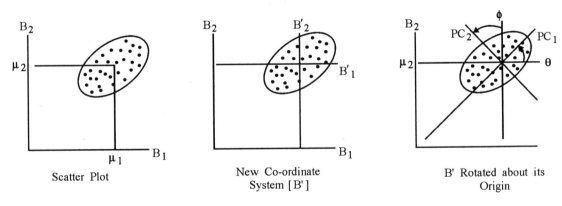

Fig. 7.18 : Scatter plot of Band 1 and Band 2 of a satellite imagry shows correlation between these bands. The principal-component transform (PCT) is then applied to the Scatterogram to generate a new coordinate system indicating the two principal components : PC_1 and PC_2.

The Sensing

from time to time, the PCT matrix has to be different for different scenes. Thus interpretation of PCs must change from scene to scene.

Standardized Principal Component Transformation (SPCT)

This transform is based on the correlation matrix instead of the covariance matrix used in PCT. In using the correlation matrix, one effectively normalizes the original bands to equal and unit variance. This transform is advantageous for combining data with different dynamic ranges in a PCT. But SPCT does not exhibit the quality of optimum information compression of the PCT.

Tasseled-Cap Transformation (TCT)

Tasseled-cap transformation is a fixed feature space (say, agricultural scene) spectral transformation using specific transformation matrix W_{TC} :

$$TC = DN = W_{TC} \cdot DN \tag{7.49}$$

Unlike in PCT, the weights in the TCT matrix are fixed and independent of the scenes. TCT is sensor dependent (primarily on spectral bands), and hence the W_{TC} must be rederived for every sensor.

Kauth and Thomas (1976) while working on agricultural monitoring using Landsat-2 MSS data (4-dimensional) observed that the DN scatterograms exhibit certain consistent properties during the crop growing season whose visualization in 4-dimension yields a shape described as a 'tasseled cap' with a base called the soil plane (Fig.7.19). Crop pixels move up the tasseled cap as the crop grows and come down to the soil plane as the crop progresses to the scenescent stage. Kauth and Thomas developed the tasseled-cap transformation and interpreted the new set of orthogonal axes in terms of the physical properties of the growing crop. The first transformed axis TC_1 is associated with soil brightness which represents pre-emergence time regime. The second axis TC_2 is associated with greenness. As the crops grow, their feature-space signature moves away from the soil axis in the direction of TC_2. The third transformed axis TC_3 is called yellowness. As the crops move towards maturity and become scenescent, their signature increases in yellowness direction while decreasing in the greenness axis. The fourth axis TC_4 which could not be characterized satisfactorily was called non-such. However, with Landsat-4 TM data, the transformed axis TC_4 was characterized as indicating haze.

In the honor of Kauth and Thomas, the proposer and interpreter of this sensor dependent, scene dependent, fixed feature-space transformation, the Tasseled-cap transformation is also called Kauth-Thomas transformation. Kauth-Thomas transformation leads to a superior separation of scene components, particularly the soil and vegetation.

The transformation matrix W_{TC} (Tasseled-cap matrix coefficients) for Landsat-1 MSS, Landsat-2 MSS and Landsat-4 TM data are shown in Table 7.3.

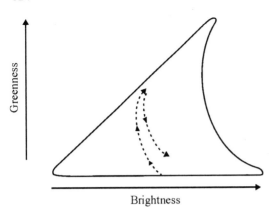

Fig. 7.19 (a): Diagrammatic presentation of Kauth-Thomas Tasseled-cap transformation. Crop development in the brightness-greenness plane of vegetation.

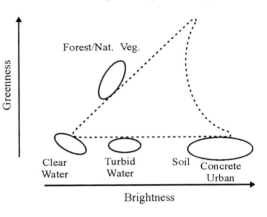

Fig. 7.19 (b): Location of different land-cover types in the brightness greeness spectral space. Brightness is highly correlated with bare soil while greeness is highly correlated with leaf area index/ percent canopy cover/green biomass.

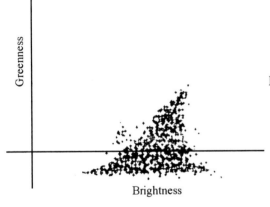

Fig. 7.19 (c): Actual scattrogram of brightness and greenness values for an agricultural area which looks like a tasseled-cap. (Adapted form Jensen, 1986)

Table 7.3: Tasseled-cap transformation coefficients for Landsat-1 MSS, Landsat-2 MSS and Landsat-4 TM bands*

Sensor	Axis designation	Bands	W_{TC} 1	2	3	4
Landsat-1 MSS	Soil brightness		+0.433	+0.632	+0.586	+0.264
	Greenness		−0.290	−0.562	+0.600	+0.491
	Yellowness		−0.829	+0.522	−0.039	+0.194
	Non-such		+0.223	+0.120	−0.543	+0.810
Landsat-2 MSS	Soil brightness		+0.332	+0.603	+0.676	+0.263
	Greenness		+0.283	−0.660	+0.577	+0.388
	Yellowness		+0.900	+0.428	+0.0759	−0.041
	Non-such		+0.016	+0.428	−0.452	+0.882

	Bands	1	2	3	4	5	7
Landsat-4 TM	Soil brightness	+0.3037	+0.2793	+0.4743	+0.5585	+0.5082	+0.1863
	Greenness	−0.2848	−0.2435	−0.5436	+0.7243	+0.0840	+0.1800
	Wetness	+0.1509	+0.1973	+0.3279	+0.3406	−0.7112	−0.4572
	Haze	−0.8242	+0.0849	+0.4392	−0.0580	+0.2012	−0.2768
	TC_5	−0.3280	+0.0549	+0.1075	+0.1855	−0.4357	+0.8085
	TC_6	+0.1084	−0.9022	+0.4120	+0.0573	−0.0251	+0.0238

* *Source:* Robert A. Schowengerdt, 1997 (adapted from Kauth and Thomas,1976; Thompson and Whemanen,1980; Crist and Cicone,1984; and Crist, Laurin and Cicone, 1986)

7.6 IMAGE ENHANCEMENT USING SPATIAL TRANSFORMS

Spatial transforms are applied to enhance the image quality through filtering remote sensing images in a desired manner. An image (x,y) is made up of spatial components, represented by roughness of tonal variation, of different sizes/scales (say 3×3, 5×5, 7×7 pixels and so on).

The filters used in the spatial transformations of the image emphasize or de-emphasize the various spatial frequencies in the image. For example, a low-pass filtered (LPF) image highlights low frequency (i.e., high wavelength or larger) spatial components and suppresses the high frequency (i.e., low wavelength or smaller) spatial components. In a similar manner, a high-pass filtered (HPF) image highlights the high frequency spatial components and suppresses the low frequency spatial components. As said earlier, these smaller or larger spatial components appear as smaller or larger grains of tonal variations in the image.

In general, we may describe the image developed from the pixel DN values in terms of its low frequency and high frequency spatial components as

$$\text{Image } (x,y) = \text{LPI } (x,y) + \text{HPI } (x,y) \tag{7.50}$$

where LPI (x,y) and HPI (x,y) are the low-pass and high-pass filtered images respectively. This relationship is valid for any size (scale) neighborhood. Thus decomposition of an image into a sum of its components at different scales forms the basis of all spatial filtering. It is obvious that addition of the components of the filtered images to synthesize the original image is the process of superposition.

Spatial filtering is a local operation in which the original image is modified on the basis of gray levels of neighboring pixels. For example, a simple low-pass filter may be implemented by passing a moving window of 3X3 pixels throughout the original image, thereby creating a second image whose central pixel's DN value corresponds to the local average value of the 9 pixels in the original image contained in the window at that point. A simple high-pass filter may be implemented by subtracting a low-pass filtered image from the original image.

7.6.1 Convolution Filters

In convolution operation, use of a moving window (kernel/operator) is made on the original image, pixel by pixel and modifying the DN values of the original pixels by the average DN value of the window. When the window finishes a row of pixels, it is moved down to the next row of pixels and the process is repeated till the entire image is transformed by this operation. Convolution operation is quite general in the sense that any function, (linear, statistical, gradient etc.) can be programmed with the moving window (Fig. 7.20).

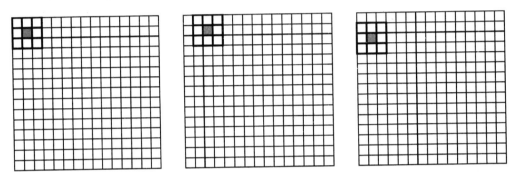

Fig. 7.20: Schematic diagram of a convolution filter : A moving window on the image for spatial filtering.

7.6.2 Linear Filters

This is a special case of convolution filters in which the operator (kernel) is chosen depending upon specific application and one can have freedom to use any weighting function suited to the problem undertaken. Examples of such filters are LPF, HPF, band-pass filter (BPF) and high-boost filter (HBF) which find applications in image enhancement and sensor simulation. As shown in eqn. 7.50, the LPF and HPF are a pair of complementary filters, the sum of which equals a unit filter. Table-7.4 shows the box filters having uniform weights in the LPF and the complementary weights in its complementary HPF. The normalizing factor is assigned to both the filters to preserve the relationship given in eqn. 7.50.

Table 7.4: Box filters with uniform weights in the low-pass filter and complementary weights in its complementary high-pass filter.

Filter size	LPF	HPF
3 × 3	$\dfrac{1}{9} \begin{bmatrix} +1 & +1 & +1 \\ +1 & +1 & +1 \\ +1 & +1 & +1 \end{bmatrix}$	$\dfrac{1}{9} \begin{bmatrix} -1 & -1 & -1 \\ -1 & +8 & -1 \\ -1 & -1 & -1 \end{bmatrix}$
5 × 5	$\dfrac{1}{25} \begin{bmatrix} +1 & +1 & +1 & +1 & +1 \\ +1 & +1 & +1 & +1 & +1 \\ +1 & +1 & +1 & +1 & +1 \\ +1 & +1 & +1 & +1 & +1 \\ +1 & +1 & +1 & +1 & +1 \end{bmatrix}$	$\dfrac{1}{25} \begin{bmatrix} -1 & -1 & -1 & -1 & -1 \\ -1 & -1 & -1 & -1 & -1 \\ -1 & -1 & +24 & -1 & -1 \\ -1 & -1 & -1 & -1 & -1 \\ -1 & -1 & -1 & -1 & -1 \end{bmatrix}$

7.6.3 High-Boost Filters (HBF)

The low-pass and high-pass filters can be combined in a variety of ways to form a set of complex filters. A high-boost filter image, for example, can be created by adding a weighted HPF image to the original image.

$$HBF = \text{original}(x,y) + K \cdot HPF(x,y) \qquad (7.51)$$

HBF with different boosting factor K are presented in Table-7.5. High-boost filtered images produce edge enhancement proportional to the K factor.

Table 7.5 : High-boost box filters with different boost factors K

K = 1	K = 2	K = 3
$1/9 \cdot \begin{bmatrix} -1 & -1 & -1 \\ -1 & +17 & -1 \\ -1 & -1 & -1 \end{bmatrix}$	$1/9 \cdot \begin{bmatrix} -2 & -2 & -2 \\ -2 & +25 & -2 \\ -2 & -2 & -2 \end{bmatrix}$	$1/9 \cdot \begin{bmatrix} -3 & -3 & -3 \\ -3 & +33 & -3 \\ -3 & -3 & -3 \end{bmatrix}$

7.6.4 Directional Filters

Directional filters are useful in preferential processing for oriented features in the image to obtain the required edge enhancement. Thus there can be directional filters to enhance features in the vertical, horizontal, diagonal or any arbitrary azimuthal directions. The operators (kernel) of these filters are given in Table-7.6.

Table 7.6 : Directional filters

Vertical	Horizontal	Diagonal	Azimuthal
$[\,-1\ +1\,]$	$\begin{bmatrix} -1 \\ +1 \end{bmatrix}$	$\begin{bmatrix} -1 & 0 \\ 0 & +1 \end{bmatrix} \quad \begin{bmatrix} 0 & -1 \\ +1 & 0 \end{bmatrix}$	$\begin{bmatrix} \sin\theta & 0 \\ -\sin\theta - \cos\theta & \cos\theta \end{bmatrix}$

The azimuthal angle θ is measured anti-clockwise from the horizontal axis.

7.6.5 Cascaded Linear Filters

When a number of linear filters operate on an image in sequence, the final result of the operation becomes equal to the convolution of the individual filters, which is nothing but a single net-filter.

$$F_c = (f\ w_1)\ w_2 = f\ (w_1\ w_2) = f\ w_{net} \qquad (7.52)$$

where $w_{net} = w_1\ w_2$. If w_1 and w_2 are each of $n \times n$ pixels in size, then w_{net} will be of $(2n-1) \times (2n-1)$ pixels in size. A few examples of cascaded filter operators are shown in Fig.7.21.

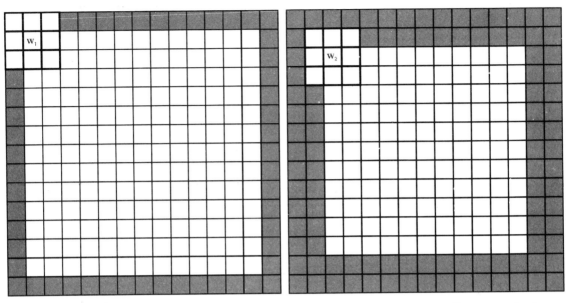

(a) Operation of a filter w_1 on the Image

(b) Subsequent operation of a filter w_2 on the Image

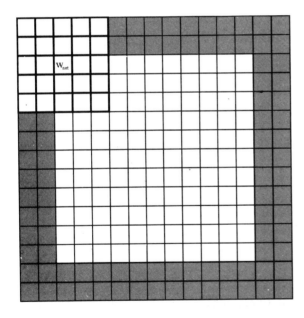

(c) Operation of a filter w_{net} on the Image

Fig. 7.21 : Schematic diagram of cascaded linear filter operations :
    (a) A filter w_1 of 3×3 pixel size operates on the image leaving out a layer of border pixels.
    (b) A filter w_2 of 3×3 pixel size operates on the above filtered image leaving out the second layer of border pixels. Thus sequential application of two filters (ie. w_1 w_2), each of 3×3 pixels size, leaves out two layers of border pixels.
    (c) The above two sequential (3×3) filter applications on the image is shown equivalent to the application of a net filter (w_{net}) of 5×5 pixel size.

The Sensing

7.6.6 Statistical Filters

Statistical filters can change the pixel DN values by a local property such as minimum, maximum, median, mode, standard deviation etc., and give us the output image depicting the pixel values of these statistical parameters. These statistical operators are useful in noise reduction, textural feature extraction, signal-to-noise ratio measurement and so on.

7.6.7 Gradient Filters

Abrupt changes in DN values from one pixel to another indicates the presence of a physical boundary in the scene, such as a coast line, the river bank, a lineament, a paved road and so on. Directional filters described earlier, in fact, calculate the directional gradient.

Under an isotropic condition, the local gradient may be computed by filtering the image in two orthogonal directions, namely horizontally and vertically, and then combining the results vectorially. Thus the magnitude of the local image gradient is given by

Gradient magnitude = $|g| = (g_x^2 + g_y^2)^{1/2}$ (7.53)

and, the direction of the local gradient is given by

Gradient angle = $\theta = \tan^{-1}(g_y/g_x)$ (7.54)

Some gradient filters are given in Table-7.7.

Table 7.7 : Gradient filters

Filter	Horizontal component	Vertical component
Roberts	$\begin{bmatrix} 0 & +1 \\ -1 & 0 \end{bmatrix}$	$\begin{bmatrix} +1 & 0 \\ 0 & -1 \end{bmatrix}$
Sobel	$\begin{bmatrix} +1 & +2 & +1 \\ 0 & 0 & 0 \\ -1 & -2 & -1 \end{bmatrix}$	$\begin{bmatrix} -1 & 0 & +1 \\ -2 & 0 & +2 \\ -1 & 0 & +1 \end{bmatrix}$
Prewitt	$\begin{bmatrix} +1 & +1 & +1 \\ +1 & -2 & +1 \\ -1 & -1 & -1 \end{bmatrix}$	$\begin{bmatrix} -1 & +1 & +1 \\ -1 & -2 & +1 \\ -1 & +1 & +1 \end{bmatrix}$

The vector diagram of computing the image gradient is shown below in Fig. 7.22

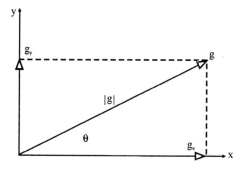

Fig. 7.22 : Vector Diagram for computing Image gradient at a pixel. Directional high-pass filters applied to an image can compute directional gradients g_x and g_y, say, in two orthogonal directions x and y. The magnitude ($|g|$) and direction (θ) of the local image gradient at a pixel can thus be computed vectorially.

7.6.8 Fourier Transform

While discussing spatial filtering with low-pass and high-pass filters, we considered the image as consisting of two types of components with respect to the spatial scales. In Fourier transform, this idea is extended to many spatial scales. Earlier, image enhancement was described by manipulating the spatial features in the spatial domain (x,y) of the image. But as a mandatory requirement, Fourier transform requires that the image should be first transformed from the spatial domain to the frequency domain. Conceptually this amounts to generating a continuous function through the discrete DN values as they are plotted along each row and column of pixels of the image. The undulations of this function along any given row or column of pixels can be mathematically described as a superposition of an infinite series of sine and cosine waves (of the fundamental and higher harmonic components) with various amplitudes, frequencies and phases. Now the Fourier transform of these sinosoidal waves results in the computation of the amplitude and phase for each of the spatial frequency in the image. These Fourier components can be displayed in a two-dimensional scatter plot known as a Fourier spectrum. The lower frequencies in the scene are plotted at the center of the spectrum and progressively the higher frequencies are plotted outwards. Features aligned horizontally in the original image result in vertical components, and those aligned vertically in the original image result in horizontal components in the Fourier spectrum.

Once the Fourier spectrum of an image is known, it is possible to regenerate the original image through the application of an inverse Fourier transform. Thus Fourier spectrum of an image can be used to enhance the quality of an image with the help of the low and high frequency block filters. For this, the filter is directly applied on the Fourier spectrum and then inverse Fourier transform is performed to get the required result. This process is presented in the flow diagram Fig.7.23.

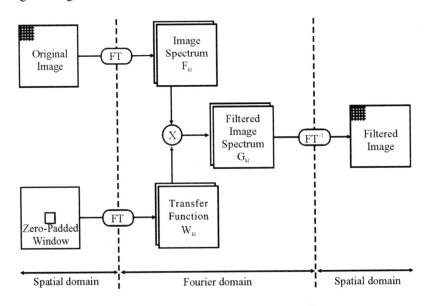

Fig. 7.23 : Flow diagram of Fourier Transform filter operation.

Since Fourier transform technique is applicable to global situation for which processing is time consuming, a fast Fourier transform (FFT) may be applied which is now available to the users as a result of the modern technological advancements.

7.6.9 Power Spectrum

Power spectrum is the square of the Fourier amplitude function. This is a useful tool for fractal analysis of satellite images. Since the spatial frequency power spectrum localizes information about global patterns in the spatial domain, it is useful as a diagnostic tool for global periodic noise, or as a pattern recognition tool for the global spatial patterns.

7.7 GROUND TRUTH COLLECTION TO SUPPORT IMAGE CLASSIFICATION

Collateral data already available for the scene under investigation plus the information collected from some meticulously designed ground investigations over several training sites of the scene form the ground truth.

Ground truths are used for remote sensing data calibration/correction, interpretation of target properties, training and verification. Training sites for ground truth collection should be so chosen that they are as accessible as possible. Ground truth collection for dynamic surface properties like soil moisture should be planned in synchrony with the satellite pass.

7.7.1 Data Calibration/Correction

Signals from surface features received by the space-borne sensors are modified not only by the intervening atmosphere and the surface topography but also by the variations in the angle of illumination of the incident irradiance and the angle of viewing by the sensors. Data calibration is therefore needed to apply necessary correction to the data. As a consequence the image quality can be improved leading to better feature identification and classification.

7.7.2 Interpretation of Target Properties

Ground experiments are needed to establish the links between the radiative and surface physical properties which are useful in the correct image interpretations. For example, surface reflectance (albedo) and surface temperature which can be determined from remote sensing are used to make some interpretations about the surface properties such as snow depth, water content, plant moisture stress, soil moisture, phytoplankton concentration in water bodies etc.

7.7.3 Training

For the identification of surface features like crop type, tree species, disease in vegetation, perched aquifers, ore deposits etc., training sites are located with the help of a global positioning system (GPS) (alternately with the help of a Toposheet of the area under study and its FCC) during ground truthing, and their corresponding spectral signatures are identified on the remote sensing imagery.

7.7.4 Verification

The performance of the thematic land use / land cover classification using remote sensing technique over a masked area (a sub-division, a district, a state etc.) is judged by estimating the classification accuracy. This verification is done by ground observations. It is to be noted that the sample areas used for verification are not the ones used for the training sites. Like the training sites, the sample areas used for verification should be distributed on the entire study area following an appropriate statistical sampling procedure.

7.8 THEMATIC IMAGE CLASSIFICATION AND INFORMATION EXTRACTION

After digital image processing and image enhancement, we need to classify the image with respect to spatial distribution of identifiable earth surface features. Normally multispectral data (primary or derived) are used to perform such classification.

In order to assign a group of pixels a land-cover class, we follow the technique of their pattern recognition which can be visualized in three different ways : (i) spectral pattern recognition, (ii) spatial pattern recognition and (iii) temporal pattern recognition.

7.8.1 Spectral Pattern Recognition

This refers to the family of classification procedures that utilizes pixel-by-pixel spectral information as the basis of automated land-cover classification.

7.8.2 Spatial Pattern Recognition

This involves categorization of image pixels on the basis of their spatial relationship with the pixels surrounding them. Spatial classifiers may consider such aspects as image texture, feature size, shape, directionality, context and repetition.

7.8.3 Temporal Pattern Recognition

This uses time as an aid in feature identification. For example, in agricultural crop surveys, distinct spectral and spatial changes during a growing season can help discrimination on multi-date imageries that would be otherwise impossible from any single –date imagery.

Image classification leads to the production of thematic maps from the imagery. Thus information extraction over a given area becomes possible which can be displayed in the GIS format for generating the necessary land-use related management strategies.

For thematic classification of an image, usually three important steps are followed :

(i) **Feature Image Extraction** – depending on the earth surface features to be classified, the feature image may be the multispectral image itself, or one of its transformed image obtained using the necessary spatial or spectral transform.

(ii) **Training Sites Identification** – these are groups of pixels to be used for training classifiers to recognize certain classes of land surface features. Training set identification may be either supervised or unsupervised.

(iii) **Labeling** – for this discriminant functions are assigned to the entire feature space which are applied to label all pixels to generate the required number of classes or categories. In case of working with unsupervised training sets, the classified features must be selectively supervised.

The Sensing

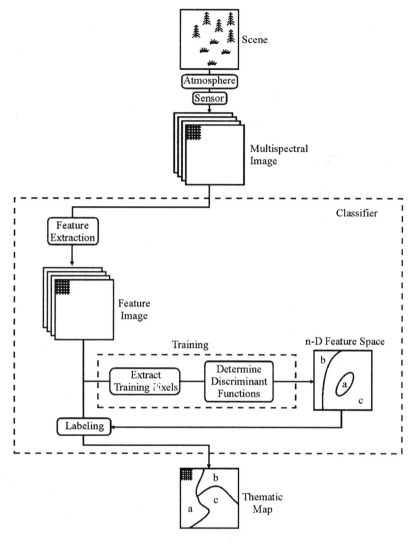

Fig. 7.24 : Flow diagram showing the Processes of Digital Image Classification (after Schowengerdt, 1997)

The above steps are presented in the flow diagram Fig-7.24. The land-use/land-cover classification hierarchy adopted by NRSA, which can be worked out at levels – I and II using the interpretation keys (tone, size, shape, texture, pattern, association, location, aspect, shadow and resolution) and the collateral data are given in Table-10.2.

7.8.4 Hard and Soft Classification

Classification algorithms are associated with a likelihood function for assignment of a class label to each pixel. In a hard classification the class label is selected with the maximum likelihood. The feature space decision boundaries for a hard classification are well defined. However, if the likelihood values are retained, allowing for multiple labels at each pixel, then a soft classification results. The feature space decision boundary in this case becomes fuzzy.

7.8.5 Supervised Classification

For any image classification, the themes of interest are dictated by the type of application. Thus themes of agricultural classes differ from those of forestry and geological classes.

There are three basic steps in supervised classification. First is the training stage in which numerical data are collected from training sites on the spectral response patterns of the land-cover categories. Second is the classification stage in which unknown pixels are compared to the spectral patterns of the training sites and they are assigned a class which looks most likely to a similar category. A number of classifiers used at the classification stage are minimum-distance –to-means classifier, parallelepiped classifier and Gaussian maximum likelihood classifier. The third is the output stage when results are presented in form of maps, tables of area data, and digital data files.

In supervised classification, the prototype pixel samples (training sites) are labeled by ground truths. Here care should be taken to see that the training site is a homogeneous area of the respective class. Of course, at the same time it should include a range of variability for the class. Thus more than one training sites per class is usually considered.

As the supervised training is specified by the desired map themes and not by the characteristics of the data itself, the technique does not guarantee that the classes will actually be distinguishable from one another. Thus a separability analysis is performed on training data to estimate the expected error in the classification for various feature combinations. This exercise may suggest for dropping some of the initially chosen features before the classification is extended to cover the full image.

For finding the inter-class separability, some simple measures of the separation of the means may be tried to start with. These constitute the city block, Euclidean distance and angular distance. However, these do not account for overlap of class distributions due to their variances.

The normalized city block appears better than the city block measure in the sense that it is not only proportional to the separation of class means but also inversely proportional to their standard deviations. However, if the class means are equal, it will be zero regardless of the class variances.

Mohalanobis separability measure is a multivariate generalization of Euclidean measure for normal distributions. However, it becomes always zero when the class means are equal.

The divergence and Bhattacharyya distance measures avoid this problem. Divergence becomes zero only if the class means and covariance matrices are equal. However, the problem still remains in both these measures as they increase without bound for large class separations, and do not asymptotically converge to one, as does the probability of correct classification.

The transformed divergence, based on the ratio of probabilities for classes a and b, does exhibit this behavior. The Jeffries – Matusita distance depending on the difference between the probability functions for a and b, is similar to the probability of correct classification for large class separations, but requires more computation than the transformed divergence. Table-7.8 presents the formulae of the different feature space separability measures described above.

Table 7.8 : Separability distance measures between the distributions of two different (a, b) feature space (Duda and Hart, 1973; Swain and Davis, 1978; Richards, 1993)

Separability measure	Formula
City Block	$L_1 = |\mu_a - \mu_b| = \sum_{k=1}^{n} |m_{ak} - m_{bk}|$
Euclidean	$L_2 = \|\mu_a - \mu_b\| = [(\mu_a - \mu_b)^T (\mu_a - \mu_b)]^{1/2}$
	$= [\sum_{k=1}^{n} (m_{ak} - m_{bk})^2]^{1/2}$
Angular	$\theta = \arccos[(\mu_a^T \mu_b)/(\|\mu_a\| \|\mu_b\|)]$
Normalized City Block	$NL_1 = \sum_{k=1}^{n} [(|m_{ak} - m_{bk}|)/(C_{ak}^{1/2} - C_{bk}^{1/2})/2]$
Mahalanobis	$MH = [(\mu_a - \mu_b)^T \{(C_a + C_b)/2\}^{-1} (\mu_a - \mu_b)]^{1/2}$
Divergence	$D = (1/2) \operatorname{tr}[(C_a - C_b)(C_b^{-1} - C_a^{-1})] +$
	$(1/2) \operatorname{tr}[(C_a^{-1} - C_b^{-1})(\mu_a - \mu_b)(\mu_a - \mu_b)^T]$
Transformed Divergence	$D^t = 2[1 - e^{-D/8}]$
Bhattacharyya	$B = (1/8) MH + (1/2) \ln[\{|(C_a + C_b)/2|\}/(|C_a\| \|C_b|)^{1/2}]$
Jeffries-Matusita	$JM = [2(1 - e^{-B})]^{1/2}$

7.8.6 Unsupervised Classification

In unsupervised classification, to create a comprehensive set of classes that span the full data space, large training sample is used. Competitive algorithms, such as K-means clustering algorithm, find an optimal partitioning of the data distribution into the required number of subdivisions. The final mean vectors resulting from the clustering will be at the centroids of each subdivisions.

In unsupervised classification the classifier identifies the distinct spectral classes present in the image data, many of which might not be initially apparent to the analyst applying a supervised classification. Likewise, the spectral classes in a scene may be so numerous that it would be difficult to train on all of them. In the unsupervised approach, they are found automatically. However, it is advisable that these (unsupervised) classified data must be compared with some form of reference data to determine the identity and informational value of the spectral classes generated.

The unsupervised training is not concerned with the homogeneity of the image sites. Heterogeneous sites can be purposely chosen to ensure that all classes of interest and their respective within-class variabilities are included. The assignment of identifying labels to each cluster may be done after training or after classification of the full image. Because unsupervised training does not require any information about the area, beyond that in the image itself, it can be useful to delineate relatively homogeneous areas as potential supervised training sites. In fact, in supervised classification we impose the ground truth knowledge on the analysis to constrain the classes and their characteristics, whereas in unsupervised classification an algorithm determines the inherent structure of the data, unconstrained by external knowledge.

In the supervised approach we define useful information categories and then examin their spectral separability, whereas in the unsupervised approach we determine spectrally separable classes and then determine their informational utility. This shows that the supervised and unsupervised classifications are complementary to each other.

7.8.7 Minimum-distance-to-means Classifier

To classify earth's surface features in the imagery of the study area, we have to choose a number of uniform training sites of each of the known feature classes. Then the pixel DN values of the training sites belonging to each of the feature classes are presented in a scatterogram. For ease of illustrating the concept, a two band scatterogram (e.g., for MSS band-3 and band-4) is presented in Fig. 7.25a. In the next step the mean of the DN values in each spectral band for various feature classes are calculated and plotted in the scatterogram shown in Fig. 7.25b. Here the sign $\otimes$ indicates the mean spectral vectors (coordinates) of different feature classes, namely water, forest, hay, corn, sand and urban. In the final step, the unknown pixels, say for example P and Q are chosen and their spectral coordinates are plotted. Now find out the distance of these unknown pixel coordinates from the various class mean coordinates. The unknown pixels are then classified depending on their minimum distance to the mean coordinates of a certain class. Thus this algorithm for classification is called the minimum-distance-to-means classifier. It is a simple classifier, no doubt, but it is not always found successful which is apparent from Fig. 7.25. Here the unknown pixel P is classified as corn, but although the unknown pixel Q has the minimum distance to the mean of class – sand, it is found to be inside the scatter points of the urban class creating confusion. In order to complete the classification of the imagery of the entire study area, the above procedure is extended to all the unknown pixels. For the classification of multispectral imagery (Fig.7.26), the steps described above can also be extended to multidimensional coordinate space to compute the means of different classes present, and also to find out the minimum distance of the unknown pixels (to be classified) to the means of these classes.

7.8.8 Level-slice Classifier

It is one of the simplest classifiers, sometimes also called the box classifier. Just like in the minimum-distance-to-means classification scheme, here also we first obtain the scatterogram of different surface feature classes and compute their class means. Then the decision boundary of the box classifier is taken to be (mean ± δ) in each spectral dimension (here δ is the standard deviation). Thus in two dimensional space the box classifier is a rectangle as shown in Fig. 7.27a. Now if an unlabeled pixel coordinates fall within any one of these classified boxes, then the pixel is assigned to that class. The box classifier algorithm also isolates an unlabeled class consisting of all pixels that do not fall within any of the designated class boxes, thereby preventing overestimation of class coverage by the outliers. However, this classifier also gives rise to conflicts if a pixel vector falls within two or more boxes as shown in Fig. 7.27a. In this figure it is seen that the unlabeled pixels P and Q which were assigned to the

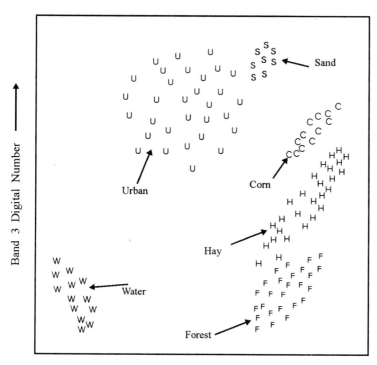

Fig. 7.25(a) : Minimum-distance-to-means classifier: Scatterogram in two-dimensional spectral space showing pixel data from selected training sites. (after Lillesand and Kiefer, 1987)

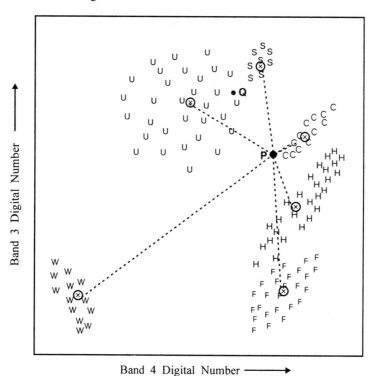

Fig. 7.25(b) : Minimum-distance-to-means classification strategy. (after Lilles and Kiefer, 1987)

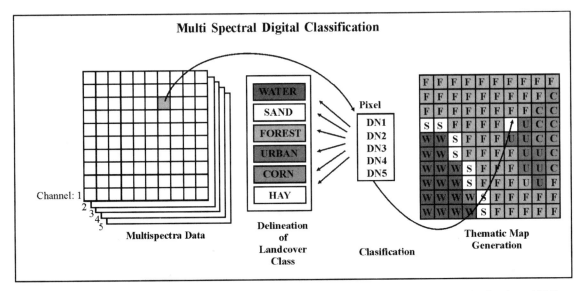

Fig. 7.26 : Multispectral digital classification strategy (Radhakrishnan, Varadan and Diwakar, 1992)

classes corn and sand respectively according to the minimum-distance-to-means classifier, have now been assigned the classes hay and urban respectively. Overlapping of decision boundary boxes belonging to classes corn and hay, and hay and forest, resulting in confusion in the box classifier algorithm, is due to the fact that there is a high covariance between the spectral bands (MSS-3 and MSS-4 in the present case) for the features corn, hay and forest, as a result of which the rectangular decision regions fit the category training data very poorly. The problem can be partially tackled by employing stepped decision region boxes as shown in Fig. 7.27b.

7.8.9 Parallelepiped Classifier

The level-slice classifier or the box classifier whose decision boundary is a rectangular box in two dimensions is, in fact, a special case of the parallelepiped classifier. Since high covariance between spectral bands for most earth's surface features is often the rule rather than exception, the multispectral DN values are not typically aligned with the multidimensional data axes. Thus the level-slice classifier does not fit multispectral remote sensing data well. However, the true parallelepiped classifier permits multidimensional boxes that can not only align with the data clusters but also offers faces that are parallelograms rather than rectangles. The improvement in parallelepiped classification algorithm over the box classifier algorithm is schematically shown in Fig. 7.28 in two dimensions.

7.8.10 Maximum Likelihood Classifier

Maximum likelihood classifier considers a probability model to determine the decision boundaries for different classes present in a remote sensing imagery. To do this it demands that the training data statistics for each class in each band are normally distributed (i.e., Gaussian by nature) so that the distribution of class response pattern can be completely described by the mean vector and the covariance matrix. In fact, when a large number of representative training pixels are taken from each class, and the class histogram is drawn for individual bands, it is found

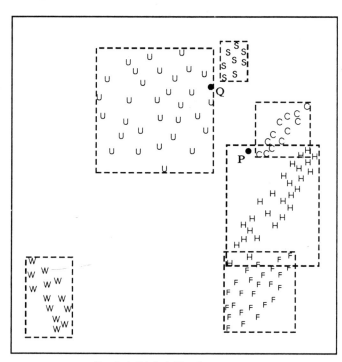

Fig. 7.27(a) : Level-slice/Box/Parallelepiped classifier :

Level-slice/Parallelepiped classification strategy. (after Lollesand and Kiefer, 1979)

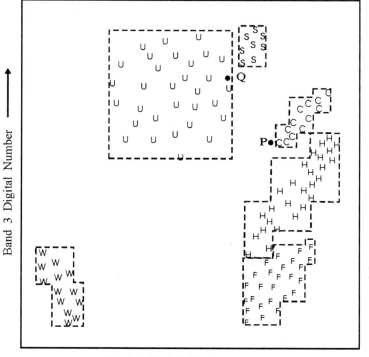

Fig. 7.27(b) : Parallelpiped classification strategy with stepped decision region boundaries. (after Lillesand and Kiefer, 1979)

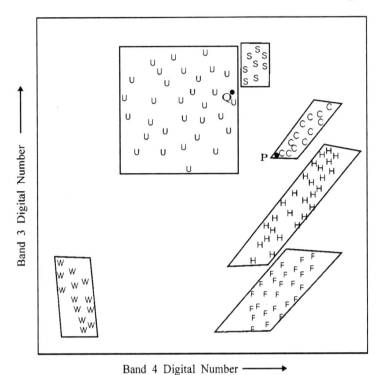

Fig. 7.28 : True parallelepiped classification strategy in two-dimensions

to be reasonably Gaussian. Thus these histograms are used as approximations to compute the continuous probability density function of an infinite sample of data. In case the training data shows bimodal or trimodal histograms in a single band, the data are not found suitable for maximum likelihood classifier. However, in such cases the individual modes are found to behave like Gaussian distribution and hence the individual mode histograms should be considered for separate classes. Thus having computed the probability density function (histogram) for each class in each band, as shown in Fig. 7.29a, and the class mean vectors (M_i) and class covariance matrices (Cov_i), we can run a program to compute the statistical probability of a given pixel characteristics (DN values) for taking decision to assign the pixel an appropriate land cover class. The formula used for calculating the statistical probability of the unlabeled pixel is the following :

$$p_i = [-0.5 \ \log_e \det(Cov_i)] - [0.5(X - M_i)^T \ (Cov_i^{-1}) \ (X - M_i)] \quad (7.55)$$

where X is the DN values of the unlabeled pixel given by

$$X = \begin{bmatrix} DN_1 \\ DN_2 \\ DN_3 \\ \dots \\ \dots \end{bmatrix}$$

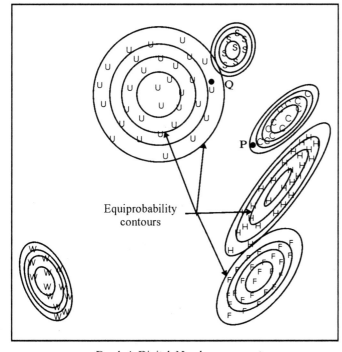

Fig. 7.29(a) : Gaussian maximum likelihood classifier :
Probability density functions defined by maximum likelihood classification strategy. (after Lillesand and Kiefer, 1987)

Fig. 7.29(b) : Equiprobability contours defined by maximum likelihood classification strategy. (after Lillesand and Kiefer, 1987)

For this, first the statistical probability p_1 is computed for an unlabeled pixel X taking its DN values, and the mean and covariance matrix of class-1 (say for water) using the above mentioned formula. Similarly the statistical probability p_2 is then computed for the same unlabeled pixel X taking again its DN values, and the mean and covariance matrix of class-2 (say for forest this time). This process should continue for all the classes identified in the imagery. After this process is finished, the unlabeled pixel X is assigned the appropriate class following the following decision rule :

If $p_i \geq p_j$, for all possible classes j ≠ i then assign the pixel to class i. (7.56)

This process should now continue for all other unlabeled pixels of the imagery to assign their appropriate classes.

In fact, maximum likelihood classifier produces ellipsoidal equiprobability contours in the scatterogram which are nothing but the decision boundaries as shown in Fig. 7.29b. The ellipticity of these decision boundaries for different classes also shows the sensitivity of the maximum likelihood classifier to the band covariance. As a result, we can see in Fig. 7.29b that the pixel P which was included under corn by minimum-distance-to-means classifier and was accommodated in the class hay by the block classifier is finally assigned to the class corn again by the maximum likelihood classifier, supporting the parallelepiped classification algorithm.

7.8.11 K-means Clustering Algorithm

This is one of the commonly used clustering algorithm to determine the natural spectral groupings present in a data set. This approach accepts from the analyst the number of clusters to be located in the scene and accordingly arbitrarily seeds (locates) that number of cluster centers in the multidimensional measurement space. Each pixel in the image is then assigned to the cluster whose arbitrary mean vector is the closest. After all pixels have been classified in this manner, revised mean vectors of each of the clusters are computed. These revised means are then used as the basis to reclassify the image data. The procedure continues until there is no significant change in the location of the class mean vectors between successive iterations of the algorithm. Once this point is reached the analyst determines the land-cover identity of each spectral class.

Because the K-means clustering approach is iterative, it is computationally intensive. Therefore, it is often applied only to image sub-areas rather than to the full scenes. Such sub-areas are often referred to as unsupervised training areas and should not be confused with the training sites used in the supervised classification. Whereas supervised training areas are located in regions of homogeneous cover type, the unsupervised training areas are chosen to contain numerous cover types at various locations throughout the scene. This ensures that all spectral classes in the scene are represented somewhere in the various sub-areas. These areas are then clustered independently and the spectral classes from the various areas are analyzed to determine their identity. They are subjected to a pooled statistical analysis to determine their spectral separability and normality.

7.8.12 Hybrid Supervised-Unsupervised Classification

The supervised training does not always guarantee to produce class signatures that are numerically separable in the feature space. Similarly the unsupervised training does not necessarily produce classes that are meaningful to the analyst. Thus it is supposed that a hybrid (combined) approach can have the potential to satisfy both the requirements. In this hybrid approach, unsupervised training is first made on the data to produce an unlabeled cluster map of the training area. Here a large number of clusters (≥ 50) may be used for adequate data representation. Ground truth data obtained from field survey or aerial photographs are then used to evaluate the map and assign appropriate class labels to these clusters. Often it may be necessary to subdivide, combine or modify some of the clusters so as to make them correspond with the ground truth data. Now the labeled cluster map can be taken as the final classified map, or else the labeled cluster data be used for the supervised image classification.

7.8.13 Artificial Neural Network (ANN) Classification

Artificial neural network classification is a nonparametric technique for image classification. In this technique, the rule of the game is found out in an iterative way by minimizing the error criterion on labeling of the training data.

There are many variants of the artificial neural network. Only the flow diagrms of a basic network is shown in Fig.7.30. The simple network shown here has three layers : the input layer, the middle hidden layer and the output layer. The input layer nodes (i) simply work as an interface for the input data (p_i) and they do not possess any data processing function. The input patterns represent the features used for classification. For example, they may be the multispectral vectors of the training pixels, one band per node. The middle hidden layer and the output layer have processing elements at each node. Each of these processing nodes does some summation and transformation operations on the data received at its site. In the illustrated network, each processing node (j) of the hidden layer receives data (p_i) from all the input nodes and performs a summation operation incorporating the required weightage parameter (w_{ji}) to give rise to a value S_j. In the next step, when the transformation operation is carried on this S_j it produces an output h_j.

Operations in hidden layer :
$$S_j = \sum_i w_{ji} \, p_i \tag{7.55}$$
$$h_j = f(S_j) \tag{7.56}$$

Similarly each processing node (k) of the output layer on receiving data from each hidden layer (h_j) undertakes the summation operation incorporating the weightage parameter (w_{kj}) giving rise to a value S_k. When subjected to transformation operation this S_k produces the final output o_k from this node.

Operations in output layer :
$$S_k = \sum_j w_{jk} \, h_j \tag{7.57}$$
$$o_k = f(S_k) \tag{7.58}$$

The often used transformation function is a sigmoidal function given by

$$f(S) = \frac{1}{1+e^{-S}} \tag{7.59}$$

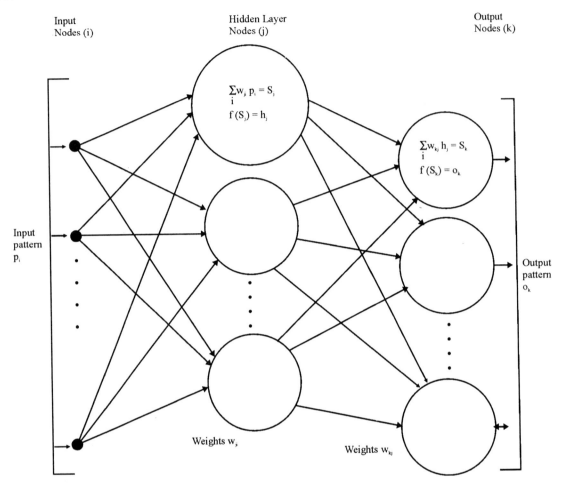

Fig. 7.30 : Schematic diagram of artificial neural network structure and function

The capability of the artificial neural network to discriminate classes of land surface features depends on the weightage parameters (w_{ji}, w_{kj}) used by the network. During training these weightage parameters are iteratively adjusted to produce a configuration from which the class patterns of interest can be distinguished. This is achieved by the back-propagation algorithm (Rumelhart et al, 1986; Lippmann, 1987) that minimizes errors over the classes and optimizes the weightage parameters. For TM data (without TIR band), 20 hidden layer nodes are sufficient for discriminating 20 classes. However, for hyper spectral data of 200 bands around 1000 hidden layer nodes are needed for discriminating 10 classes by the artificial neural network technique.

7.9 SUB-PIXEL CLASSIFICATION

In a hard classification we assume that every pixel belongs to only one labeled category of the earth surface features. This is really not true. In fact, each pixel contains a mixture of several categories of surface features. Thus the net spectral signature of a pixel is the sum of the contributions from each of these categories present in the pixel.

Since the mixing proportions (class fractions) change from pixel to pixel, the resultant spectral signature also changes. By increasing the spectral resolution, some of the class fractions of the pixel may reduce, but the mixed pixels continue to exist regardless of their size or sensor resolution. This is always found to be true, because real earth surface features have spatial details with much smaller dimensions than any small ground resolution cell (GIFOV).

Sub-pixel classification which belongs to the soft classification category, will gain greater importance in the time to come with the introduction of hyperspectral sensors. In this regard a linear mixing spectral model can be worked out to find out the class fractions of the unmixing categories of surface features on pixel basis.

7.9.1 Fuzzy Set Classification

This classification technique also belongs to the soft classification category. Fuzzy set concept recognizes the fact that an entity may have partial membership in more than one class. Naturally this requires models to solve the unmixing problem in classification. Some existing fuzzy logic for remote sensing image classification algorithms are Fuzzy C-means (FCM) clustering (Bezdek et al, 1984; Cannon et al, 1986) and Fuzzy supervised Classification (Wang, 1990a, b).

7.10 HYPER-SPECTRAL IMAGE ANALYSIS

Any direct extension of the multispectral image classification technique to the hyper-spectral imagery, though theoretically permissible, yet practically it is found undesirable because of the following difficulties :

(i) traditional classifiers can not extract the higher information content of the hyperspectral data.

(ii) many more training data are required, and

(iii) computational costs for k-dimensional data are very high

However, some alternate techniques developed for hyperspectral image analysis will be outlined here below.

7.10.1 Visualization of Hyperspectral Image

This is done in three ways, namely

(i) **Displaying the spatial spectrogram :** In this we extract a line of pixels from the hyperspectral image data, and display the pixels in the x-axis and the k-band spectral values of each pixel as gray levels in the y-axis.

(ii) **Displaying the statistics image :** Here the class covariance matrix is displayed as a pseudo-color map where the colors indicate the degree of covariance between bands (from –ve to +ve values). The covariance matrix distinguishes the classes with similar mean signatures. Maximum likelihood classifier may be fruitfully employed in this case for classification of the hyperspectral image.

(iii) **Dynamic visualization through spectral movies :** Here the image bands (say, there are 240 hyperspectral bands each of 10nm width) are displayed sequentially in rapid succession @ 30 bands/sec (video frame rate). The main advantages of such a video display are in the visual detection of the characteristic signatures and also in the quick previewing of the data for bad bands.

7.10.2 Training for Classification

As a general rule, for any classifier sufficient number of training pixels are required to identify the class accurately. If a maximum likelihood classifier is used with Gaussian class distribution, then the class mean vectors and the covariance matrices are also required to be calculated. For hyperspectral (k-bands) data, we know that the training set for each class should contain at least $(k + 1)$ pixels in order to compute the sample covariance matrix. Moreover, as the dimensionality of the feature space increases, as in the case of hyperspectral images, an accuracy degradation in the classification known as the Hughes phenomenon sets in, and to compensate this more training samples are even found necessary for producing the desirable results.

7.10.3 Feature Extraction from Hyperspectral Data

To illustrate the usefulness of hyperspectral data in earth surface feature extraction, let us consider an example for mineral detection and identification. For this the spectrum of each pixel is divided by a reference band which is relatively free from surface absorption. This ensures a normalized spectrum for every pixel. Then the average normalized spectrum for the entire scene is calculated and it is subtracted from the normalized spectrum already produced. Surface absorption features that are below the average spectrum will now appear as negative (residuals). Thus the color composite of residual images (at specific mineral absorption bands) will be interpreted for the area of occurrence and concentration of the mineral species.

7.10.4 Spectral Fingerprints

In this feature extraction technique, one has to find out the local points of inflection (maximum slope/absorption bands) in the spectral curve using scale-space filtering. This, in fact, produces the characteristic patterns in the scale-space, known as 'fingerprints'. If the spectral data are corrected for atmospheric absorption, then the spectral absorption features will only represent the characteristics of the materials present in the pixels.

7.10.5 Absorption-band Parameters

Imaging spectrometers of high spectral resolution have the ability to identify some materials by their absorption-band characteristics. Three absorption band parameters can be derived from the spectrum. The depth of the absorption band is measured from continuum-corrected internal average relative reflectance (IARR) data. The width of the absorption band is measured at half the band depth, and the position of the absorption band is taken as the wavelength at the band absorption minimum.

7.10.6 Spectral Derivative Ratio

It is found that the ratio of the nth order derivatives of the at-remote sensor radiance data with respect to the wavelength (λ) at two wavelengths approximately equals to the ratio of the same n^{th} order derivatives of the target's spectral reflectance.

$$\frac{d^n L/d\lambda^n \big|_{\lambda_1}}{d^n L/d\lambda^n \big|_{\lambda_2}} = \frac{d^n \rho/d\lambda^n \big|_{\lambda_1}}{d^n \rho/d\lambda^n \big|_{\lambda_2}} \qquad (7.55)$$

The order of derivative n may take integral values, n = 1, 2, ...

For hyperspectral data registered in discrete narrow wavelength bands, the continuous derivatives shown in the above equation are to be relpaced by the discrete derivatives. The usefulness of this approach is that even the uncalibrated (for atmospheric absorption) radiance data can be used to compute spectral derivative signatures that will match those from the reflectance data of the target materials, leading to their identification (Philpot, 1991).

7.10.7 Classification Algorithms for Hyperspectral Data

For this the atmospheric water absorption bands should be first excluded from the hyperspectral data. This will reduce computation time and also avoid quality degradation from low signal-to-noise ratio bands in classification. Any conventional algorithm applicable for multispectral image classification can, in principle, be applied to the hyperspectral data. However, in terms of economy and quality considerations, the technique comes out to be inefficient when a 200 band hyperspectral data is subjected to processing and classification. Therefore, three restricted classification modes are outlined below for hyperspectral data collected by the imaging spectrometer. These classification modes are :

(1) Comparison of the spectrum of a single pixel or the average of a group of pixels, to all pixels in the image. This produces classes of surface features and areas associated with these classes.

(2) Matching the spectra of all pixels to the reference (calibrated/normalized) spectral reflectance library. This is helpful to identify and locate the presence of specific materials in the scene.

(3) Matching a single spectrum from the multispectral remote sensor to the reflectance library. This leads to labeling a pixel or a group of pixels. Calibration of multispectral remote sensing data is also needed in this case. It is seen that mode-1 and mode-3 described here supplement each other.

7.10.8 Binary Encoding

For practically handling large volumes of hyperspectral data, simultaneous data reduction and pattern matching are needed. In this regard, the technique of binary encoding can be adopted profitably. In this procedure a single DN threshold is chosen and values above and below the threshold are coded as 1 and 0 respectively. Thus a single bit can code the spectrum in each band. Similarly, the derivative of the spectrum can also be encoded to 0 or 1, depending on whether its value is negative or positive. The coded features are found to be less sensitive to solar irradiance and the atmospheric effects if coding is done with respect to the local spectral mean. The encoded spectra can be compared bit-by-bit using the hamming distance, defined by the number of bits that are different in the two binary numbers. A minimum distance threshold is easily used to reduce the effect of noise in the classification.

7.10.9 Spectral-angle Mapping

The hyperspectral vectors of pixels are calculated and plotted. The spectral-angle distance being independent of the magnitude of the spectral vectors, it is insensitive to the topographic variations. The classifier can therefore be applied to remote sensing data not corrected for topography, and it facilitates comparison of these data to the laboratory reflectance spectra for feature identification. The technique is equally applicable to the multispectral data of Landsat-TM, IRS-LISS and so on.

Suggestions for Supplementary Reading

These suggestions are not exhaustive but are limited to the works that will be found especially useful for developing interest and creative imagination in the field of Remote Sensing.

Books

Aggarwal, J. K. 1977. Computer Methods in Image Analysis. IEEE Press, New York.

American Society of Photogrammetry, 1960. Manual of Photographic Interpretation. Falls Church, Va.

American Society of Photogrammetry, 1980. Manual of Photogrammetry, 4th ed. Falls Church, Va.

American Society of Photogrammetry, 1983. Manual of Remote Sensing, Vol. I, R. N. Colwell (Ed.), Falls Church, Va.

Andrews, H. A. and B. R. Hunt, 1977. Digital Image restoration. Prentice-Hall, Englewood Cliffs, NJ.

Avery, T. E. and G. L. Berlin, 1985. Interpretation of Aerial Photographs, 4th ed. Burgess, Minneapolis, Minn.

Baxes, G. A. 1984. Digital Image Processing. Prentice-Hall, Englewood Cliffs, NJ.

Bernstein, R. (Ed.), 1978. Digital Image Processing for Remote Sensing. IEEE Press, New York.

Bracewell, R. 1965. The Fourier Transform and Its Applications. McGraw-Hill, New York.

Brigham, E. O. 1974. The Fast Fourier Transform. Prentice-Hall, Englewood Cliffs, NJ.

Campbell, J. B. 1987. Introduction to Remote Sensing. The Guildford Press, New York.

Castleman, K. R. 1996. Digital Image Processing. Prentice-Hall, Englewood Cliffs, NJ.

Curran, P. J. 1985. Principles of Remote Sensing, Longman, London.

Duda, R. D. and P. E. Hart, 1973. Pattern Classification and Scene Analysis. John Wiley & Sons, Inc., New York.

Ekstrom, M. P. 1984. Digital Image Processing Techniques. Academic Press, New York.

Fukunaga, K. 1990. Introduction to Statistical Pattern Recognition, 2nd ed. Academic Press, San Diego.

Ghosh, S. K. 1979. Analytical Photogrammetry. Pergamon Press, New York.

Gonzalez, R. C. and R. E. Woods, 1992. Digital Image Processing. Addition-Wesley Publishing Co., Reading, MA.

Hobbs, W. H. 1912. Earth Features and Their Meaning – An Introduction to Geology for the Student and General Reader. Macmillan Publishing Co., Ny.

Hord, Michael P. 1982. Digital Image Processing of Remotely Sensed Data. Academic Press, New York.

Jain, A. K. 1989. Fundamentals of Digital Image Processing. Prentice-Hall, Englewood Cliffs, N.J.

Jain, A. K. and R. C. Dube, 1988. Algorithms for Clustering Data. Prentice-Hall, Englewood Cliffs, NJ.

James Mike, 1985. Classification Algorithms. Cooling.

Jensen, John R., 1996. Introductory Digital Image Processing, 2nd ed. Prentice-Hall, Englewood Cliffs, New Jersey.

Lillesand, Thomas M. and Ralph W. Kiefer, 1987. Remote Sensing and Image Interpretation, 2nd ed., John Wiley & Sons, New York.

Moik, J. G. 1980. Digital Processing of Remotely Sensed Images. NASA, US Govt. Printing Office, Washington, DC.

Newton, A. R., R. P. Viljoen and T. G. Longshaw, 1981. Evaluation of Remote Sensing Methods in a Test Area Near Krugerdorp, Transvaal, II – Landsat Imagery. Trans. Geol. Soc. South Africa.

Niblack, W. 1986. An Introduction to Digital Image Processing, 3rd ed. Prentice-Hall International. Oppenheim, A. V. and R. W. Schafer, 1975. Digital Signal Processing. Prentice-Hall, Englewood Cliffs, NJ.

Pratt, W. K. 1991. Digital Image Processing, 2nd ed., John Wiley & Sons, New York.

Richards, John A., 1993. Remote Sensing Digital Image Analysis, 2nd ed. Springer-Verlag, Berlin.

Rosenfeld, A. and A. C. Kak, 1982. Digital Picture Processing, 2nd ed., Academic Press, New York.

Sabins, Floyd, F., Jr., 1987. Remote Sensing – Principles and Interpretations, 2nd ed., W. H. Freeman and Company, New York.

Schalkoff, R. J. 1989. Digital Image Processing and Computer Vision, John Wiley & Sons, New York.

Schowengerdt, R. A. 1983. Techniques for Image Processing and Classification in Remote Sensing. Academic Press, New York.

Schowengerdt, Robert A. 1997. Remote Sensing : Models and Methods for Image Processing , 2nd ed., Academic Press, San Diego.

Schurmann, J. 1996. Pattern Classification – A Unified View of Statistical and Neural Approaches. John Wiley & Sons, New York.

Swain, P. H. and S. M. Davis (Eds.), 1978. Remote Sensing : The Quantitative Approach. McGraw-Hill, New York.

Toselli, F. (Ed.), 1989. Applications of Remote Sensing to Agrometeorology, Kluwer Academic Publishers, Dordrecht, The Netherlands.

Wolf, P. R. 1983. Elements of Photogrammetry, 2nd ed. McGraw-Hill, New York / International Students Edition, Singapore.

Research Papers

American Society of Photogrammetry and Remote Sensing, 1985. SPOT Simulation Results. Photogramm. Eng. Remote Sensing (Special Issue), 51 (No.8).

American Society of Photogrammetry and Remote Sensing, 1985. Landsat Image Data Quality Analysis. Photogramm. Eng. Remote Sensing (Special Issue), 51 (No.9).

Anuta, P. 1970. Spatial Registration of Multispectral and Mulititemporal Digital Imagery using Fast Fourier Transform Techniques. IEEE Trans. Geosci. Electronics GE-8 : 353 – 368.

Anuta, P. E. 1977. Computer-assisted Analysis Techniques for Remote Sensing Data Interpretation. Geophys. 42 : 468 – 481.

Arai, K. 1992. A Supervised Thematic Mapper Classification with a Purification of Training Samples. Int. J. Remote Sensing 13 : 2039 – 2049.

Aronoff, S. 1985. The Minimum Accuracy Value as an Index of Classification Accuracy. Photogramm. Eng. Remote Sensing, 51 : 99 – 111.

Aronoff, S. 1982. Classification Accuracy : A User Approach. Photogramm. Eng. Remote Sensing, 48 : 1299 – 1307.

Augusteijn, M. F., L. E. Clemens and K. A. Shaw, 1995. Performance Evaluation of Texture Measures for Ground Cover Identification in Satellite Image by means of a Neural Network Classifier. IEEE Trans. Geosci. Remote Sensing 33 : 616 – 626.

Baraldi, A. and F. Parmiggiani, 1995. A Neural Network for Unsupervised Categorization of Multivalued Input Patterns : An Application to Satellite Image Clustering. IEEE Trans. Geosci. Remote Sensing 33 : 305 – 316.

Batson, R. M., K. Edwards and E. M. Eliason, 1976. Synthetic Stereo and Landsat Pictures. Photogramm. Eng. 42 : 1279 – 1284.

Benediktsson, J. A., P. H. Swain and O. K. Ersoy, 1990. Neural Network Approaches versus Statistical Methods in Classification of Multi-source Remote Sensing Data. IEEE Trans. Geosci. Remote Sensing 28 : 540 – 551.

Benjamin, S. and L. Gaydos, 1990. Spatial Resolution Requirements for Automated Cartographic Road Extraction. Photogramm. Eng. Remote Sensing 56 : 93 – 100.

Bernstein Ralph, 1983. Image Geometry and Rectification. In : Manual of Remote Sensing, Vol. I, 2nd ed., R. N. Colwell (Ed.). American Society of Photogrammetry, Falls Church, Va. p. 873 – 922.

Bezdek, J.C., R. Ehrlich and W. Full, 1984. FCM : The Fuzzy C – Means Clustering Algorithm. Computers and Geosciences 10 : 191 – 203.

Billingsley, Fred C. 1983. Data Processing and Reprocessing. In : Manual of Remote Sensing, Vol. I, 2nd ed., R. N. Colwell (Ed.). American Society of Photogrammetry, Falls Church, Va. p. 719 – 792.

Bischof, H., W. Schneider and A. J. Pinz, 1992. Multispectral Classification of Landsat Images using Neural Networks. IEEE Trans. Geosci. Remote Sensing 30 : 482 – 490.

Bolstad, P. V. and T. M. Lillesand, 1991. Rapid Maximum Likelihood Classification. Photogramm. Eng. Remote Sensing 57 :67 – 74.

Borel, C. C. and S. A. W. Gerstl, 1994. Nonlinear Spectral Mixing Models for Vegetative and Soil Surfaces. Remote Sensing Environ. 47 : 403 – 416.

Buchheim, M. P. and T. M. Lillesand, 1989. Semi-automated Training Field Extraction and Analysis for Efficient Digital Image Classification. Photogramm. Eng. Remote Sensing 55 : 1347 – 1355.

Byrne, G. F., P. F. Crapper and K. K. Mayo, 1980. Monitoring Land-Cover Change by Principal Component Analysis of Multi-temporal Landsat Data. Remote Sensing Environ. 10 : 175–184.

Cannon, R. L., J. V. Dave, J. C. Bezdek and M. M. Trivedi, 1986. Segmentation of a Thematic Mapper Image using the Fuzzy C – Means Clustering Algorithm. IEEE Trans. Geosci. Remote Sensing GE – 24 : 400 – 408.

Carnahan, W. H. and G. Zhou, 1986. Fourier Transform Techniques for the Evaluation of the Thematic Mapper Line Spread Function. Photogramm. Eng. Remote Sensing, 52 : 639 – 648.

Centeno, J. A. S. and V. Haertel, 1995. Adaptive Low-Pass Fuzzy Filter for Noise Removal. Photogramm. Eng. Remote Sensing 61 : 1267 – 1272.

Chavez, P. S.,Jr. and Brian Bauer, 1982. An Automatic Optimum Kernel –Size Selection Technique for Edge Enhancement. J. Remote Sensing Environ. 12 : 23 – 38.

Chavez, P. S., Jr., 1996. Image-based Atmospheric Corrections – Revisited and Improved. Photogramm. Eng. Remote Sensing 62 : 1025 – 1036.

Chen, K. S. , Y. C. Tzeng, C. F. Chen and W. L. Kao, 1995. Land-cover Classification of Multispectral Imagery using a Dynamic Learning Neural Network. Photogramm. Eng. Remote Sensing 61 : 403 – 408.

Chittineni, C. B. 1983. Edge and Line Detection in Multi-Dimensional Noisy Imagery Data. IEEE Trans. Geosci. Remote Sensing GE-21 : 163 – 174.

Civco, D. L. 1993. Artificial Neural Networks for Land-cover Classification and Mapping. Int. J. Geographical Inf. Systems 7 : 173 – 186.

Cliché, G., F. Bonn and P. Teillet, 1985. Integration of the SPOT Panchromatic Channel into its Multi-spectral Mode for Image Sharpness Enhancement. Photogramm. Eng. Remote Sensing, 51 : 311 – 316.

Colby, J. D. 1991. Topographic Normalization in Rugged Terrain. Photogramm. Eng. Remote Sensing 57 : 531 – 537.

Conese, C., M. A. Gilabert, F. Maselli and L. Bottai, 1993. Topographic Normalization of TM Scenes through the Use of an Atmospheric Correction Method and Digital Terrain Models. Photogramm. Eng. Remote Sensing 59 : 1745 - 1753.

Congalton, R. G., R. G. Oderwald and R. A. Mead, 1983. Assessing Landsat Classification Accuracy using Discrete Multivariate Statistical Techniques. Photogramm. Eng. Remote Sensing, 49 1671 – 1678.

Crist, E. P. and R. C. Cicone, 1984. Application of the Tasseled Cap Concept to Simulated Thematic Mapper Data. Photogramm. Eng. Remote Sensing 50 : 343 – 352.

Crist, E. P. and R. J. Kauth, 1986. The Tasseled Cap De-mystified. Photogramm. Eng. Remote Sensing 52 : 81 – 86.

Cross, A. M., J. J. Settle, N. A. Drake and R. T. M. Paivinen, 1991. Subpixel Measurement of Tropical Forest Cover using AVHRR Data. Int. J. Remote Sensing 12 : 1119 – 1129.

Curran, P. J. 1988. The Semivariogram in Remote Sensing : An Introduction. Remote Sensing Environ. 24 : 493 – 507.

Dejace, J., 1989. Image Processing Techniques : Filtering, Exogenous Data, Geometrical Processing. In : Applications of Remote Sensing to Agrometeorology, F. Toselli (Ed.). Kluwer Academic Publishers, Norwell, Ma. P. 57 – 128.

Dozier, J. 1981. A Method for Satellite Identification of Surface Temperature Fields of Subpixel Resolution. Remote Sensing Environ. 11 : 221 – 229.

Dozier Jeff and Alan H. Strahler, 1983. Ground Investigations in Support of Remote Sensing. In : Manual of Remote Sensing, Vol. I, 2nd ed., R. N. Colwell (Ed.). American Society of Photogrammetry, Falls Church, Va. p. 959 – 986.

Dreyer, P. 1993. Classification of Land-cover using Optimized Neural Nets on SPOT Data. Photogramm. Eng. Remote Sensing 59 : 617 – 621.

Durand, J. M. and Y. H. Kerr, 1989. An Improved Decorrelation Method for the Efficient Display of Multispectral Data. IEEE Trans. Geosci. Remote Sensing 27 : 611 – 619.

Eberlein, R. B. and J. S. Weszka, 1975. Mixures of Derivative Operators as Edge Detectors. Computer Graphics and Image Processing, 4 : 180 – 183.

Elachi, C. 1982. Radar Images of the Earth from Space. Scientific American 247 : 54.

Estes, John E., Earl J. Hajic and Larry R. Tinney, 1983. Fundamentals of Image Analysis : Analysis of Visible and Thermal Infrared Data. In : Manual of Remote Sensing, Vol. I, 2nd ed., R. N. Colwell (Ed.). American Society of Photogrammetry, Falls Church, Va. p. 987 – 1124.

Eyton, J. R. 1983. A Hybrid Image Classification Instruction Package. IEEE Trans. Geosci. Remote Sensing 49 : 1175 – 1181.

Foody, G. M. 1995. Using Prior Knowledge in Artificial Neural Network Classification with a Minimum Training Set. Int. J. Remote Sensing 16 : 301 – 312.

Foody, G. M. 1996. Relating the Land-cover Composition of Mixed Pixels to Artificial Neural Network Classification Output. Photogramm. Eng. Remote Sensing 62 : 491 – 499.

Foody, G. M. and D. P. Cox, 1994. Subpixel Land-cover Composition Estimation using a Linear Mixure Model and Fuzzy Membership Functions. Int. J. Remote Sensing 15 : 619 – 631.

Foody, G. M., M. B. McCulloch and W. B. Yates, 1995. Classification of Remotely Sensed Data by an Artificial Neural Network : Issues Related to Training Data Characteristics. Photogramm. Eng. Remote Sensing 61 : 391 – 401.

Ford, G. E., V. R. Algazi and D. I. Meyer, 1983. A Noninteractive Procedure for Land-Use Determination. Remote Sensing Environ. 13 : 1 – 16.

Frel, W. 1977. Image Enhancement by Histogram Hyperbolization. Computer Graphics and Image Processing. 6 : 286 – 294.

Friedmann, D. E., J. P. Friedel, K. L. Magnussen, R. Kwok and S. Richardson, 1983. Multiple Scene Precision Rectification of Spaceborne Imagery with Very Few Ground Control Points. Photogramm. Eng. Remote Sensing 49 : 1657 – 1667.

Full, W. E., R. Ehrlich and J. C. Bezdek, 1982. Fuzzy Q Model – A New Model Approach for Linear Unmixing. Mathematical Geology 14 : 259 – 270.

Fung, T. and E. LeDrew, 1987. Application of Principal Components Analysis to Change Detection. Photogramm. Eng. Remote Sensing 53 : 1649 – 1658.

Gao, B. C. and A. F. H. Goetz, 1990. Column Atmospheric Water Vapor and Vegetation Liquid Water Retrievals from Airborne Imaging Spectrometer Data. J. Geophys. Res. 95 : 3549 – 3564.

Garguet-Duport, B., J. Girel, J. M. Chassery and G. Pautou, 1996. The Use of Multi-Resolution Analysis and Wavelets Transform for Merging SPOT Panchromatic and Multispectral Image Data. Photogramm. Eng. Remote Sensing 62 : 1057 – 1066.

Giles, P. T., M. A. Chapman and S. E. Franklin, 1994. Incorporation of a Digital Elevation Model Derived from Stereoscopic Satellite Imagery in Automated Terrain Analysis. Computers & Geosciences 20 : 441 – 460.

Gillespie, A. R. and A. B. Kahle, 1977. The Construction and Interpretation of a Digital Thermal Inertia Image. Photogramm. Eng. Remote Sensing, 43 : 983 – 1000.

Gillespie, A. R., A. B. Kahle and R. E. Walker, 1986. Color Enhancement of Highly Correlated Images – I : Decorrelation and HIS Contrast Stretches. Remote Sensing Environ. 20 : 209 – 235.

Ginevan, M. E. 1979. Testing Land-Use Map Accuracy : Another Look. Photogramm. Eng. Remote Sensing, 45 : 1371 – 1377.

Goetz, A., G. Vane, J. E. Solomon and B. N. Rock, 1985. Imaging Spectrometry for Earth Remote Sensing. Science 228 : 1147 – 1153.

Goldberg, M. and S. Shlien, 1976. A Four – Dimentional Histogram Approach to the Clustering of Landsat Data. Canadian J. Remote Sensing, 2 : 1 – 11.

Green, William B. and Robert M. Haralick, 1983. Remote Sensing Software Systems. In : Manual of Remote Sensing, Vol. I, 2nd ed., R. N. Colwell (Ed.). American Society of Photogrammetry, Falls Church, Va. p. 807 – 839.

Gurney, C. M. and J. R. G. Townshend, 1983. The Use of Contextual Information in the Classification of Remotely Sensed Data. Photogramm. Eng. Remote Sensing, 49 : 55 – 64.

Gyer, M. S. 1992. Adjuncts and Alternatives to Neural Networks for Supervised Classification. IEEE Trans. Geosci. Remote Sensing 22 : 35 – 46.

Hara, Y., R. G. Atkins, S. H. Yueh, R. T. Shin and J. A. Kong, 1994. Application of Neural Networks to Radar Image Classification. IEEE Trans. Geosci. Remote Sensing 32 : 100 – 109.

Haralick, R. M. 1979. Statistical and Structural Approaches to Texture. Proc. IEEE 67 : 786 – 804.

Haralick, R. M., K. Shanmugan and I. Dinstein, 1973. Textural Features for Image Classification. IEEE Trans. System, Man and Cybernetics SMC-3 : 610 – 621.

Haralick, R. M., C. A. Hlavka, R. Yokoyama and S. M. Carlyle, 1980. Spectral – Temporal Classification Using Vegetation Phenology. IEEE Trans. Geosci. Remote Sensing GE-18 : 167 – 174.

Haralick, Robert M. and King-Sun Fu, 1983. Pattern Recognition and Classification. In : Manual of Remote Sensing, Vol. I, 2nd ed., R. N. Colwell (Ed.). American Society of Photogrammetry, Falls Church, Va. p. 793 – 805.

Hardin, P. J. 1994. Parametric and Nearest-neighbor Methods for Hybrid Classification : A Comparison of Pixel Assignment Accuracy. Photogramm. Eng. Remote Sensing 60 : 1439 – 1448.

Hardin, P. J. and C. N. Thomson, 1992. Fast Nearest Neighbor Classification Methods for Multi-spectral Imagery. The Professional Geographer 44 : 191 – 201.

Harris, R. 1985. Contextual Classification Post-Processing of Landsat Data using a Probabilistic Relaxation Model. Int. J. Remote Sensing, 6 : 847 – 866.

Harsanyi, J. C. and C. I. Chang, 1994. Hyperspectral Image Classification and Dimensionality Reduction : An Orthogonal Subspace Projection Approach. IEEE Trans. Geosci. Remote Sensing 32 : 779 – 785.

Heermann, P. D. and N. Khazenie, 1992. Classification of Multispectral Remote Sensing Data using a Back-propagation Neural Network. IEEE Trans. Geosci. Remote Sensing 30 : 81 – 88.

Hoffbeck, J. P. and D. A. Landgrebe, 1996. Covariance Matrix Estimation and Classification with Limited Training Data. IEEE Trans. Pattern Analysis and Machine Intelligence 18 : 763 – 767.

Holben, B. N. and C. O. Justice, 1981. An Examination of Spectral Band Ratioing to reduce the Topographic Effect on Remotely Sensed Data. Int. J. Remote Sensing 2 : 115 – 133.

Hord, R. M. and W. Brooner, 1976. Land-Use Map Accuracy Criteria. Photogramm. Eng. Remote Sensing, 42 : 671 – 677.

Howarth, P. J. and G. M. Wickware, 1981. Procedures for Change Detection Using Landsat Digital Data. Int. J. Remote Sensing, 2 : 277 – 291.

Howarth, P. J. and E. Boasson, 1983. Landsat Digital Enhancement for Change Detection in Urban Environments. Remote Sensing Environ. 13 : 149 – 160.

Hubaux, A., 1989. The Luxuriant Image – An Introduction to Remote Sensing. In : Applications of Remote Sensing to Agrometeorology, F. Toselli (Ed.). Kluwer Academic Publishers, Dordrecht, The Netherlands.. P. 57 – 128.

Hsu, S. 1978. Texture – Tone Analysis for Automated Land-Use Mapping. Photogramm. Eng. Remote Sensing. 44 : 1193 – 1404.

Huete, A. R. 1988. A Soil Adjusted Vegetation Index (SAVI). Remote Sensing Environ. 25 : 295 – 309.

Huete, A. R., D. F. Post and R. D. Jackson, 1984. Soil Spectral Effects on 4-Space Vegetation Discrimination. Remote Sensing Environ. 14 : 155 – 165.

Huete, A. R. and R. Escadafal, 1991. Assessment of Biophysical Soil Properties through Spectral Decomposition Techniques. Remote Sensing Environ. 35 : 149 – 159.

Hutchinson, C. F. 1982. Techniques for Combining Landsat and Ancillary Data for Digital Classification Improvement. Photogramm. Eng. Remote Sensing, 48 : 123 – 130.

Irons, J. R. and G. W. Petersen, 1981. Texture Transforms of Remote Sensing Data. Remote Sensing Environ. 11 : 359 – 370.

Irons, J. R., B. L. Markham, R. F. Nelson, D. L. Toll, D. L. Williams, R. S. Latty and M. L. Stauffer, 1985. The Effect of Spatial Resolution on the Classification of Thematic Mapper Data. Int. J. Remote Sensing 6 : 1385 – 1403.

Jackson, R. D. 1983. Spectral Indices in N-Space. Remote Sensing Environ. 13 : 409 – 421.

Jackson, R. D., P. N. Slater and P. J. Pinter, 1983. Adjusting the Tasseled-Cap Brightness and Greenness Factors for Atmospheric Path Radiance and Absorption on a Pixel-by-Pixel Basis. Int. J. Remote Sensing 4 : 313 – 323.

Jasinski, M. F. 1996. Estimation of Subpixel Vegetation Density of Natural Regions using Satellite Multispectral Imagery. IEEE Trans. Geosci. Remote Sensing 34 : 804 – 813.

Jasinski, M. F. and P. S. Eagleson, 1990. Estimation of Subpixel Vegetation Cover using Red – Infrared Scatterograms. IEEE Trans. Geosci. Remote Sensing 28 : 253 – 267

Jensen, J. R. 1983. Educational Image Processing : An Overview. Photogramm. Eng. Remote Sensing 49 : 1151 – 1157.

Jia, X. and J. A. Richards, 1993. Binary Coding of Imaging Spectrometry Data for Fast Spectral Matching and Classification. Remote Sensing Environ. 32 : 274 – 281.

Justice, C. O., B. L. Markham, J. R. G. Townshend and R. L. Kennard, 1989. Spatial Degradation of Satellite Data. Int. J. Remote Sensing 10 : 1539 – 1561.

Justice, C. O. and J. R. G. Townshend, 1982. A Comparison of Unsupervised Classification Procedures on Landsat MSS Data for an Area of Complex Surface Conditions in Basilicata, Southern Italy. Remote Sensing Environ. 12 : 407 – 420.

Kahle, A. B., A. R. Gillespie and A. F. H. Goetz, 1976. Thermal Inertia Imaging : A New Geologic Mapping Tool. Geophys. Res. Lett. 3 : 26

Kalayeh, H. M., M. J. Muasher and D. A. Landgrebe, 1983. Feature Selection with Limited Training Samples. IEEE Trans. Geosci. Remote Sensing, GE-21 : 434 – 437.

Kanellopoulos, I., A. Varfis, G. G. Wilkinson and J. Megier, 1992. Land-cover Discrimination in SPOT Imagery by Artificial Neural Network – A Twenty Class Experiment. Int. J. Remote sensing 13 : 917 – 924.

Kaur Ravinder, B. C. Panda and S. K. Srivastava, 1997. One-Dimensional Characterization of Natural Surfaces using N-Dimensional Spectral Information. Int. J. Remote Sensing, 18 : 406 – 416.

Kettig, R. L. and D. A. Landgrede, 1976. Classification of Multispectral Image Data by Extraction and Classification of Homogeneous Objects. IEEE Trans. Geosci. Electronics. GE-14 : 19 – 26.

Kruse, F. A. 1988. Use of Airborne Imaging Spectrometer Data to map Minerals associated with Hydrothermally Altered Rocks in the Northern Grapevine Mountains, Nevada and California. Remote

Sensing Environ. 24 : 31 – 51.

Kruse, F. A., K. S. Kierein-Young and J. W. Boardman, 1990. Mineral Mapping at Cuprite, Nevada with a 63-Channel Imaging Spectrometer. Photogramm. Eng. Remote Sensing 56 : 83 – 92.

Kruse, F. A., A. B. Lefkoff, J. W. Boardman, K. B. Heidebrecht, A. T. Shapiro, P. J. Barloon and A. F. H. Goetz, 1993. The Spectral Image Processing System (SIPS) – Interactive Visualization and Analysis of Imaging Spectrometer Data. Remote Sensing Environ. 44 : 145 – 163.

Landgrebe, D. A. 1980. The Development of a Spectral – Spatial Classifier for each Observational Data. Pattern Recognition 12 : 165 – 175.

Lavreau, J. 1991. De-hazing Landsat Thematic Mapper Images. Photogramm. Eng. Remote Sensing 57 : 1297 – 1302.

Lee, C. and D. A. Landgrebe, 1991. Fast Likelihood Classification. IEEE Trans. Geosci. Remote Sensing 29 : 509 – 517.

Lee, C. and D. A. Landgrebe, 1993. Analyzing High-Dimensional Multispectral Data. IEEE Trans. Geosci. Remote Sensing 31 : 792 – 800.

Lee, J. S. 1980. Digital Image Enhancement and Noise Filtering by use of Local Statistics. IEEE Trans. Pattern Analysis and Machine Intelligence PAMI – 2 :165 – 168.

Lee, J. 1983. Digital Image Smoothing and the Sigma Filter. Computer Vision, Graphics and Image Processing 24 : 255 – 269.

Lee, J. B., A. S. Woodyatt and M. Berman, 1990. Enhancement of High Spectral Resolution Remote Sensing Data by a Noise-Adjusted Principal Components Transform. IEEE Trans. Geosci. Remote Sensing 28 : 295 – 304.

Lee, T., J. A. Richards and P. H. Swain, 1987. Probabilistic and Evidential Approaches for Multi-source Data Analysis. IEEE Trans. Geosci. Remote Sensing GE-25 : 283 – 293.

Lippmann, R. P. 1987. An Introduction to Computing with Neural Nets. IEEE ASSP Magazine (April Issue) : 4 – 22.

Liu, Z. K. and J. Y. Xiao, 1991. Classification of Remotely Sensed Image Data using Artificial Neural Networks. Int. J. Remote Sensing 12 : 2433 – 2438.

Lo, C. P. and R. L. Shipman, 1990. A GIS Approach to Land-Use Change Dynamics Detection. Photogramm. Eng. Remote Sensing, 56 : 1483 – 1491.

McDonnell, M. J. 1981. Box Filtering Techniques. Computer Graphics and Image Processing 17 : 65 – 70.

McKeown, D. M., Jr., J. Wilson, A. Harvey and J. McDermott, 1985. Rule-based Interpretation of Aerial Imagery. IEEE Trans. Pattern Analysis and Machine Intelligence PAMI-7 : 570 – 585.

Marr, D. and E. Hildreth, 1980. Theory of Edge Detection. Proc. Roy. Soc. London 207 : 187 – 217.

Marsh, S. E., P. Switzer and R. J. P. Lyon, 1980. Resolving the Percentage Component Terrains within Single Resolution Elements. Photogramm. Eng. Remote Sensing 46 : 1079 – 1086.

Martin, L. R. G., P. J. Howarth and G. H. Holder, 1988. Multispectral Classification of Land-Use at the Rural-Urban Fringe using SPOT Data. Canadian J. Remote Sensing, 14 : 72 – 79.

Maxwell, E. L. 1976. Multivariate System Analysis of Multispectral Imagery. Photogramm. Eng. Remote Sensing, 42 : 1173 – 1186.

Mead, R. A. and J. Szajgin, 1982. Landsat Classification Accuracy Assessment Procedures. Photogramm.

Eng. Remote Sensing, 48 : 139 – 141.

Mehldau, G. and R. A. Schowengerdt, 1990. A C – Extension for Rule-based Image Classification Systems. Photogramm. Eng. Remote Sensing 56 : 887 – 892.

Mitchell, O. R., C. R. Myers and W. Boyne, 1977. A Min – Max Measure for Image Texture Analysis. IEEE Trans. Computers C – 26 : 408 - 414.

Moller-Jensen, 1990. Knowledge-Based Classification of an Urban Area using Texture and Context Information in Landsat TM Imagery. Photogramm. Eng. Remote Sensing, 56 : 899 – 904.

Munechika, C.K., J. S. Warnick, C. Salvaggio and J. R. Schott, 1993. Resolution Enhancement of Multispectral Image Data to improve Classification Accuracy. Photogramm. Eng. Remote Sensing 59 : 67 – 72.

Nichols David, 1983. Digital Hardware. In : Manual of Remote Sensing, Vol. I, 2nd ed. R. N. Colwell (Ed.). American Society of Photogrammetry, Falls Church, Va. p. 841 – 871.

Panda, B. C. and S. K. Srivastava, 1988. Multi-dimensional Discrimination of Some Rabi Crops. Proc. Natl. Symp. : Remote Sensing in Rural Development, HAU, Hisar. p. 209 – 217.

Paola, J. D. and R. A. Schowengerdt, 1995. A Review and Analysis of Back-Propagation Neural Networks for Classification of Remotely Sensed Multispectral Imagery. Int. J. Remote Sensing 16 : 3033 – 3058.

Paola, J. D. and R. A. Schowengerdt, 1995a. A Detailed Comparison of Back-Propagation Neural Network and Maximum Likelihood Classifiers for Urban Land-Use Classification. IEEE Trans. Geosci. Remote Sensing 33 : 981 – 996.

Parvis, M. 1950. Drainage Pattern Significance in Airphoto Identification of Soils and Bedrocks. Photogramm. Eng. 16 : 387 – 409.

Peli, T. and D. Malah, 1982. A Study of Edge Detection Algorithms. Computer Graphics and Image Processing. 20 : 1 – 21.

Philpot, W. D. 1991. The Derivative Ratio Algorithm : Avoiding Atmospheric Effects in Remote Sensing. IEEE Trans. Geosci. Remote sensing 29 : 350 – 357.

Piech, M. A. and K. R. Piech, 1987. Symbolic Representation of Hyperspectral Data. Appl. Optics 26 : 4018 – 4026.

Piech, M. A. and K. R. Piech, 1989. Hyperspectral Interactions : Invariance and Scaling. Appl. Optics 28 : 481 – 489.

Pratt, W. K. 1974. Correlation Techniques of Image Registration. IEEE Trans. Aerospace and Electronic Systems AES – 10 : 353 – 358.

Proy, C., D. Tanve and P. Y. Deschamps, 1989. Evaluation of Topographic Effects in Remotely Sensed Data. Remote Sensing Environ. 30 : 21 – 32.

Quarmby, N. A. and J. L. Cushnie, 1989. Monitoring Urban Land-Cover Changes at the Urban Fringe from SPOT HRV Imagery in South-East England. Int. J. Remote Sensing 10 : 953 – 963.

Radhakrishnan, K., V. Jayaraman, Geeta Varadan and B. Manikiam, 1991. Infrastructure in India for Analysis of Remotely Sensed Data. Current Science (Special Issue), 61 (Nos.3 & 4) : 266 – 271.

Radhakrishnan, K., Geeta Varadan and P. G. Diwakar, 1992. Digital Image Processing Techniques – An Overview. In : Natural Resources Management – A New Perspective. R. L. Karale (Ed.), NNRMS, Publications and Public Relations Unit, ISRO HQ, Bangalore. p. 66 – 85.

Rao, D. P., R. R. Navalgund and Y. V. N. Krishna Murthy, 1996. Cadastral Applications using IRS-1C Data : Some Case Studies. Current Science (Special Issue), 70 (7) : 624 – 628.

Reddy, B. S. and B. N. Chatterji, 1996. An FFT-based Technique for Translation, Rotation and Scale-Invariant Image Registration. IEEE Trans. Image Processing 5 : 1266 – 1271.

Richards, J. A., D. A. Landgrebe and P. H. Swain, 1982. A Means for Utilizing Ancillary Information in Multispectral Classification. Remote Sensing Environ. 12 : 463 – 477.

Richardson, A. J. and C. L. Wiegand, 1977. Distinguishing Vegetation from Soil Background Information. Photogramm. Eng. Remote Sensing 43 : 1541 – 1552.

Ritter, N. D. and G. F. Hepner, 1990. Application of an Artificial Neural Network to Land-Cover Classification of Thematic Mapper Imagery. Computers & Geosciences 16 : 873 – 880.

Rosenfeld, G. H. 1986. Analysis of Thematic Map Classification Error Matrices. Photogramm. Eng. Remote Sensing, 52 : 681 – 688.

Rosenfeld, G. H. and K. Fitzpatrick-Lins, 1986. A Coefficient of Agreement as a Measure of Thematic Classification Accuracy. Photogramm. Eng. Remote Sensing, 52 : 223 – 227.

Ru-Ye, W. 1986. An Approach to Tree-Classifier Design based on Hierachical Clustering. Int. J. Remote Sensing, 7 : 75 – 88.

Salu, Y. and J. Tilton, 1993. Classification of Multispectral Image Data by the Binary Neural Network and by Non-parametric Pixel-by-Pixel Methods. IEEE Trans. Geosci. Remote Sensing 31 : 606 – 617.

Schowengerdt, R. A. 1980. Reconstruction of Multispatial, Multispectral Image Data using Spatial frequency Content. Photogramm. Eng. Remote Sensing, 46 : 1325 – 1334.

Schreier, H., L. C. Goodfellow and L. M. Lavkulich, 1982. The Use of Digital Multidate Landsat Imagery in Terrain Classification. Photogramm. Eng. Remote Sensing, 48 : 111 – 119.

Serpico, S. B. and F. Roli, 1995. Classification of Multi-sensor Remote Sensing Images by Structured Neural Networks. IEEE Trans. Geosci. Remote Sensing 33 : 562 – 577.

Settle, J. J. and N. A. Drake, 1993. Linear Mixing and the Estimation of Ground Cover Proportions. Int. J. Remote Sensing 14 : 1159 – 1177.

Shaw, G. B. 1979. Local and Regional Edge Detectors : Some Comparisons. Computer Graphics and Image Processing. 9 : 135 – 149.

Shis, E. H.. and R. A. Schowengerdt, 1983. Classification of Arid Geomorphic Surfaces using Landsat Spectral and Textural Features. Photogramm. Eng. Remote Sensing, 49 : 337 – 347.

Singh, A. and A. Harrison, 1985. Standardized Principal Components. Int. J. Remote Sensing 6 : 883 – 896.

Skidmore, A. K. 1989. An Expert System Classifies Eucalyptus Forest Types using Thematic Mapper Data and a Digital Terrain Model. Photogramm. Eng. Remote Sensing, 55 : 1449 – 1464.

Srivastava, S. K., J. P. Sinha and B. C. Panda, 1992. A Study on Dependence of Some Biophysical Processes on Simple Ratio of Near-Infrared and Visible Hemispherical Canopy Reflectance. Ann. agric. Res. 13 : 218 – 220.

Strahler, A. H. 1980. The Use of Prior Probabilities in Maximum Likelihood Classification of Remotely Sensed Data. Remote Sensing Environ. 10 : 135 – 163.

Swain, P. H., H. J. Siegel and B. W. Smith, 1980. Contextual Classification of Multispectral Remote Sensing Data using a Multiprocessor System. IEEE Trans. Geosci. Remote Sensing, GE – 18 : 197 – 204.

Swain, P. H., S. B. Vardeman and J. C. Tilton, 1981. Contextual Classification of Multispectral Image Data. Pattern Recognition, 13 : 429 - 441.

Swann, R., D. Hawkins, A. Westwell-Roper and W. Johnstone, 1988. The Potential for Automated Mapping from Geocoded Digital Image Data. Photogramm. Eng. Remote Sensing 54 : 187 – 193.

Tateishi, R. and A. Akutsu, 1992. Relative DEM Production from SPOT Data without GCP. Int. J. Remote Sensing 13 : 2517 – 2530.

Teillet, P. M. and G. Fedosejevs, 1995. On the Dark Target Approach to Atmospheric Correction of Remotely Sensed Data. Canadian J. Remote Sensing 21 : 374 – 387.

Tom, C. H. and L. D. Miller, 1984. An Automated Land-Use Mapping Comparison of the Bayesian Maximum Likelihood and Linear Discriminant Analysis Algorithms. Photogramm. Eng. Remote Sensing, 50 : 193 – 207.

Townsend, F. E. 1986. The Enhancement of Computer Classification by Logical Smoothing. Photogramm. Eng. Remote Sensing, 52 : 213 – 221.

Townshend, J. R. G. and C. O. Justice, 1986. Analysis of the Dynamics of African Vegetation using the Normalized Difference Vegetation Index. Int. J. Remote Sensing 7 : 1435 – 1445.

Townshend, J., C. Justice, W. Li, C. Gurney and J. McManus, 1991. Global Land-Cover Classification by Remote Sensing : Present Capability and Future Possibilities. Remote Sensing Environ. 35 : 243 – 255.

Tucker, C. J. and P. C. Sellers, 1986. Satellite Remote Sensinmg of Primary Production. Int. J. Remote Sensing 7 : 1395 – 1416.

Vizy, K. N. 1974. Detecting and Monitoring Oil Slicks with Aerial Photos. Photogramm. Eng. 40 : 697 – 708.

Wang, F. 1990. Improving Remote Sensing Image Analysis through Fuzzy Information Representation. Photogramm. Eng. Remote Sensing 56 : 1163 – 1169.

Wang, F. 1990a. Fuzzy Supervised Classification of Remote Sensing Images. IEEE Trans. Geosci. Remote sensing 28 : 194 – 201.

Weismiller, R. A., S. J. Kristof, D. K. Scholz, P. E. Anuta and S. A. Momin, 1977. Change Detection in Coastal Zone Environments. Photogramm. Eng. Remote Sensing, 43 : 1533 – 1539.

Welch, R. 1982. Spatial Resolution Requirements for Urban Studies. Int. J. Remote Sensing, 3 : 139 – 146.

Welch, R. and M. Ehlers, 1988. Cartographic Feature Extraction with Integrated SIR-B and Landsat TM Images. Int. J. Remote Sensing 9 : 873 – 889.

Wessman, C. A. 1994. Estimating Canopy Biochemistry through Imaging Spectrometry. In : Imaging Spectrometry – A Tool for Environmental Observations, J. Hill and J. Megier (Eds.). Kluwer Academic Publishers, Dordrecht, The Netherlands.

Wharton, S. W. 1982. A Contextural Classification Method for Recognizing Land-Use Patterns in High Resolution Remotely Sensed Data. Pattern Recognition 15 : 317 – 324.

Wharton, S. W. 1987. A Spectral-Knowledge- Based Approach for Urban Land-Cover Discrimination. IEEE Trans. Geosci. Remote Sensing, GE-25 : 272 – 282.

Yoshida, T. and S. Omatu, 1994. Neural Network Approach to Land-Cover Mapping. IEEE Trans. Geosci. Remote Sensing 32 1103 – 1109.

Zheng, Q. and R. Chellappa, 1993. A Computational Vision Approach to Image Registration. IEEE Trans. Image Processing 2 : 311 – 326.

Radar Remote Sensing

8

8.1 INTRODUCTION

Radar stands for radio detection and ranging. It operates in the microwave and radio frequency bands of the electromagnetic spectrum. Radar is basically an active remote sensing system as it carries its own source of energy (klystron, magnetron, radio antenna) to illuminate the targets required to be monitored and characterized through processing and analysis of the radar return signals. Like the visible and infrared remote sensing, radar remote sensing can be of non-imaging and imaging types. We use scatterometers and altimeters in the non-imaging mode of radar information collection. In imaging radar, pulses of fan shaped radar beam from the antenna scans the terrain surface both in the range and azimuth directions (Fig.8.1). In order to enhance the signal strength from the terrain features of interest and to increase the ground resolution, the look angle of the antenna is kept oriented in the optimum direction which happens

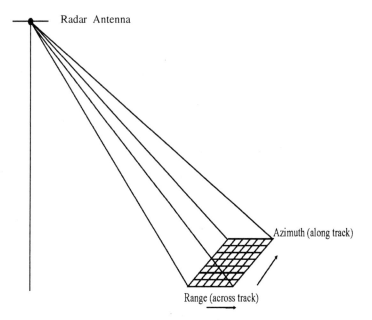

Fig. 8.1 : Terrain illumination by the radar beam. Two distinct spatial resolutions : The range and azimuth resolutions in radar remote sensing.

to be the side looking direction (and not in the nadir direction as in case of visible and infrared remote sensing). It is because of this reason that the imaging radar is called 'side looking airborne radar (SLAR). Depending on the sensor platforms the SLAR may be a side looking aircraft borne radar system or a side looking satellite borne radar system. A block diagram of the SLAR system indicating the signal transmission, reception and processing is given in Fig.4.4. The important advantages of the radar remote sensing system are that the radar signal is unaffected by the weather conditions, especially clouds, and it has day and night operation capability because it does not depend on the sunlight.

In the radar system, the antenna not only transmits the radar pulse to illuminate the terrain, but also receives the radar return from the terrain. A duplexer, which is an electronic switch, prevents the transmitted and received pulses to interfere with each other. This is achieved by blocking the receiver circuit during transmission, and the transmitter circuit during reception of the signals. A receiver amplifies the weak signal energy collected by the antenna and maintains the variations in the intensity of the pulses of radar return. The receiver also records the time delay of the return pulse indicating the position of the terrain feature on the image. The radar returns thus received by the receiver may be displayed by a CRT screen, recorded on a film for subsequent optical processing, or recorded on a magnetic tape for digital image processing. In coherent mode, the radar also measures the Doppler frequency shifts. Since radar essentially measures the time intervals or frequency shifts which can be measured much more precisely than the distance, we achieve much higher resolution of the targets in radar remote sensing in addition to the already mentioned advantages of its all weather capability independent of the daylight conditions.

While studying the terrain features from the radar images (optically or digitally processed) sufficient care should be taken to judiciously consider not only the interpretation key elements such as tone, size, shape, texure, pattern, aspect (height, shadow), site and association, but also the ground truths and collateral/ancillary information, so that a remarkably accurate interpretation of the image becomes possible.

8.2 PARAMETERS AFFECTING THE RADAR RETURN SIGNALS

The character of the radar return depends on the modification of the radar signal effected by its interaction with the target/terrain materials. Thus the parameters affecting the radar return can be grouped into two classes : the system parameters and the terrain parameters which are given below :

8.2.1 Radar System Parameters

These parameters include :
1. Frequency/wavelength,   2. Polarization,   3. Depression angle/incidence angle,
4. Look direction, and   5. Speckle (noise)

8.8.2 Terrain/Target Parameters

These parameters include :
1. Surface geometry, 2. Surface roughness, and 3. Dielectric properties

In case the terrain is an exposed soil surface, then under the terrain parameters we may further consider the following soil target parameters :

1. Soil surface roughness,
2. Moisture content/dielectric constant, and
3. Tillage practices

For terrains covered with agricultural crops, we should similarly consider the following crop target parameters :

1. Crop type,
2. Crop structure,
3. Crop height,
4. Ground cover,
5. Crop density,
6. Leaf area index,
7. Crop growth stage,
8. Leaf moisture/crop biomass,
9. Row direction,
10. Weed infestation, and
11. Salinization

8.3 RADAR WAVELENGTH

The wavelength of radar is usually expressed in centimeter and its frequency in giga hertz. The standard relationship between wavelength (λ) and frequency (ν) of electromagnetic radiation is given by

$$\lambda = c/\nu \tag{8.1}$$

where c is the velocity of light. Thus for changing the wavelength (in cm) to frequency (in GHz), we use the following conversion formula :

$$\nu \text{ (GHz)} = 30/\lambda \text{ (cm)} \tag{8.2}$$

The microwave band designations, with their associated wavelength and frequency ranges, used in remote sensing are presented in Table-8.1.

Table 8.1 : Radar band designations with associated wavelength and frequency ranges

Band designation*	Wavelength λ (cm)	Frequency ν (GHz)
K_a (0.86 cm)	0.8 – 1.1	26.5 – 40.0
K	1.1 – 1.7	18.0 – 26.5
K_u	1.7 – 2.4	12.5 – 18.0
X (3.0, 3.2 cm)	2.4 – 3.8	8.0 – 12.5
C	3.8 – 7.5	4.0 – 8.0
S	7.5 – 15.0	2.0 – 4.0
L (23.5, 25.0 cm)	15.0 – 30.0	1.0 – 2.0
P	30.0 – 100.0	0.3 – 1.0

* Wavelengths commonly used in imaging radar are shown in parentheses

8.4 SMOOTH SURFACE CRITERIA

This is an important terrain information easily obtainable from radar remote sensing. The smoothness criteria following Rayleigh's criterion is given by the relationship

$$h < \lambda/(8 \cos \theta) \rightarrow \text{Smooth surface} \quad (8.3)$$

where θ is the angle of incidence of the radar beam on the terrain.

Following this formula the terrain smoothness condition for a few radar wavelengths with two different angles of incidence is shown in Table-8.2.

Table 8.2 : Terrain smoothness criteria for some radar wavelengths with varying incidence angle

Radar Band	Wavelength (cm)	Angle of Incidence θ (degree)	Smoothness Condition (<)
X	2.3	45	0.4 cm
	2.3	20	0.3 cm
C	5.3	45	0.9 cm
	5.3	20	0.7 cm
L	23.0	45	4.1 cm
	23.0	20	3.0 cm

8.5 TERRAIN MOISTURE CONTENT AND DEPTH OF PENETRATION

Often we need information on soil moisture and leaf moisture for operational purposes. In this connection, the back scattering coefficient from the terrain surface is found to be proportional to the dielectric constant of the material of the terrain which is a function of its moisture content. The relative complex dielectric constant is given by the expression

$$\varepsilon = \varepsilon' - j \varepsilon'' \quad (8.4)$$

where ε' is the real part of the complex dielectric constant or the permittivity of the medium, and ε'' is the imaginary part of the complex dielectric constant or the conductivity of the medium. As the loss factor or loss tangent is given by $\tan\delta = \varepsilon''/\varepsilon'$, ε'' can be taken proportional to the loss factor itself. It should be made clear that although the back scattering coefficient is directly related to the dielectric constant under controlled conditions, the other factors that contributes to it include the sensor parameters such as – wavelength, incidence angle and polarization, and the terrain/target parameters like bulk density, texure, roughness and vegetation which are to be given due consideration.

The penetration of microwave/radar is found to be significant when the loss tangent ($\tan \delta$) of the medium is small. The depth of penetration of microwave radar into the terrain is expressed as the skin depth ($1/\alpha$), where

$$\alpha = (2\pi/\lambda) \; (\varepsilon'/2) \; [(1 + \tan^2\delta)^{1/2} - 1]^{1/2} \quad (8.5)$$

8.6 DEPRESSION ANGLE

Earlier we have listed some radar system parameters which affect the radar return signal from the target. The depression angle (γ) is one of such important parameters. It is defined as the

Radar Remote Sensing

angle subtended by the radar beam proceeding from the antenna to the target on the terrain to the horizontal plane passing through the antenna (Fig.4.1 and 4.2). If the terrain surface is horizontal it is found that the depression angle(γ) and the incidence angle (θ) of the radar beam on the ground are complementary, i.e. $\gamma + \theta = 90^0$. But if the terrain surface is inclined at an angle φ to the horizontal, then one can find from Fig.8.2 that the relationship between the angles of depression, incidence and inclination comes out to be $\gamma + \theta + \varphi = 90^0$ for positive slope, and $\gamma + \theta - \varphi = 90^0$ for negative slope of the terrain surface.

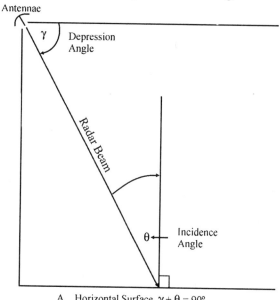

A. Horizontal Surface, $\gamma + \theta = 90°$

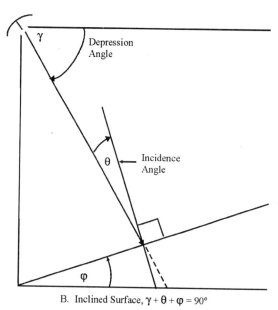

B. Inclined Surface, $\gamma + \theta + \varphi = 90°$

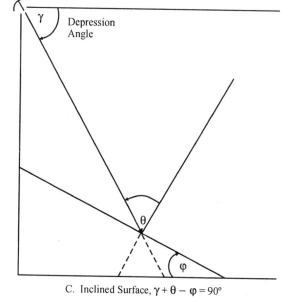

C. Inclined Surface, $\gamma + \theta - \varphi = 90°$

Fig. 8.2 : Relationship between the depression angle and the angle of incidence of the radar beam :
(A) On a horizontal surface,
(B) On an inclined surface with a positive slope
(C) On an inclined surface with a negative slope

8.7 SPATIAL RESOLUTION

Side looking airborne radar (SLAR) can be of two types : the real aperture radar (RAR) and the synthetic aperture radar (SAR). In real aperture radar the effective dimension of the antenna is taken as the actual length of the antenna, while in the synthetic aperture radar, the effective dimension of the antenna is mathematically synthesized to be many times longer than the actual size of its antenna. We have already seen in Fig.8.1 that the radar beam illuminated ground resolution cells (pixels) have two dimensions – the range dimension and the azimuth dimension. Thus while computing the spatial resolution of the real aperture radar, we have to work out the range resolution (R_r) and the azimuth resolution (R_a) separately and then multiply them to get the minimum area of the ground resolution cell (equivalent pixel). It is found that the depression angle is always higher at the near range side of an image strip than at the far range side of it. The average depression angle of an image is computed for a radar beam to the midpoint of the image strip in the range direction.

8.7.1 Range Resolution (R_r)

Range resolution is the resolution in the range direction. Consider in Fig.8.3 the points A and B in the range direction which are taken to be just resolved. By this we presume that a radar pulse of pulse length τ whose wave front strikes A first and B after a bit later, return back to the receiver antenna in a manner such that the front of the return signal from B will remain just behind the tail of the return signal from A so that these two return signals can be registered by the receiver as two distinct signals coming from the terrain feature, and under this condition, we recognize that the separation between A and B is the shortest distance between these two points which can be just seen distinct from each other. This condition is satisfied if and only if the wave fronts striking the points A and B are at a distance $\tau/2$ apart in the slant range as shown in the figure. Since the pulse length τ is normally measured in unit of time, to express it in distance measure we have to multiply it by the velocity of light c. Thus the resolution in the slant range is given by

$$\text{Slant range resolution} = BB' = \frac{\tau c}{2} \quad (8.6)$$

However, as the range resolution R_r is conventionally measured on the ground range and not in the slant range, we work out the formula for the range resolution (that is AB in the present case) using the simple geometrical relationship : $BB' = AB \cos\gamma$ (Fig.8.3), namely

$$R_r = AB = BB'/\cos\gamma$$

Now replacing the value of BB' from Eqn.8.6 we get the final expression for the range resolution as

$$R_r = (\tau c)/(2 \cos\gamma) \quad (8.7)$$

From Eqn.8.7, it is clear that the range resolution becomes better in the far range compared to the near range as the depression angle of the far range becomes less than that in the near range, and correspondingly the value of $\cos\gamma$ becomes more in the far range than in the near range. The range resolution also improves by using shorter pulse length rather than the longer ones. However, reduction in pulse length correspondingly reduces the energy of the pulse. Thus in real aperture radar the pulse length is retained at a compromised level so as to get a sufficiently strong radar return from the terrain.

Radar Remote Sensing   165

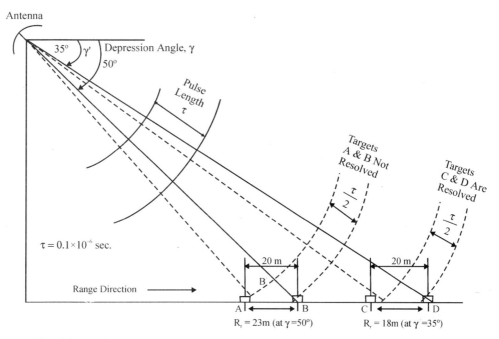

Fig. 8.3 : Radar resolution in the range direction (Rr) (after Sabins, Jr. 1987)

8.7.2 Azimuth Resolution (R_a)

Azimuth resolution is the resolution in the azimuth direction which is along the direction of flight of the sensor or along the track direction. The width (in the track direction) of the terrain strip illuminated by the radar beam actually determines the azimuth resolution. We say that two points A and B on the ground are just resolved in the azimuth direction (Fig. 8.4), if and

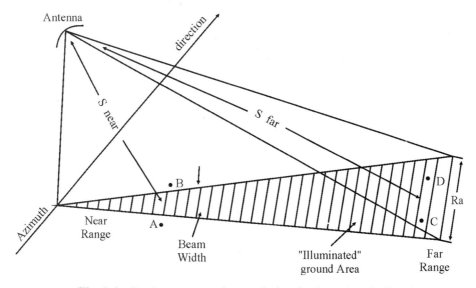

Fig. 8.4 : Real aperture radar resolution in the azimuth direction.

only if they are separated by a distance which is just greater than the beam width. Thus in the figure we see that the points A and B in the near range are clearly resolved, whereas two other points C and D in the far range whose distance between them is also the same as that between A and B, are not resolved, being both of them well within the beam width. This is because the fan shaped radar beam is narrower in the near range and wider in the far range. The angular beam width is directly proportional to the wavelength of the radar beam. The azimuth resolution is also found to be a function of the length of the antenna and the length of the slant range. Similar to the Rayleigh's resolution criteria, the smallest distance that can be just resolved in the azimuth direction is given by

$$R_a = (0.7\, S\, \lambda)/D \tag{8.8}$$

where S is the slant range length, λ is the wavelength of the radar beam and D is the length of the radar antenna. Eqn.8.8 shows that the azimuth resolution improves by using smaller wavelength of electromagnetic radiation. But it is not desirable to go towards shorter wavelengths without sacrificing the all weather capability of the radar system. Azimuth resolution also improves by increasing the antenna length. However, a long real aperture antenna offers practical engineering problems to keep it stable in space-borne observation platforms. Moreover, the azimuth resolution is found to worsen with the increase of the slant range length, i.e. in the far range side. This restricts us to confine to a narrow swath which is an uneconomic proposition. Ideally what we want to achieve is an azimuth resolution which is almost constant from the near range to the far range. In fact, this has been made possible in the development of the synthetic aperture radar (SAR) described below.

8.7.3 Synthetic Aperture Radar (SAR)

The computation of range resolution is same for the real aperture radar and the synthetic aperture radar. However, these two basic radar systems differ primarily in the methods of computation of the azimuth resolution. While the real aperture radar (RAR), known as the brute force system, uses an antenna of maximum possible length to produce a narrow angular beam width in the azimuth direction (Fig.8.4), the synthetic aperture radar (SAR) uses rather a small antenna to transmit a broad beam (Fig.4.3), and mathematically synthesize a very long antenna that would produce a very narrow beam resulting in a much higher constant azimuth resolution from the near range to the far range.

Fig.4.3a shows that as the radar antenna moves in the azimuth (flight) direction, it observes the relative motion of a target through the antenna beam (position A,B,C) in the opposite direction. Thus the radar return signals from the target gets modified due to Doppler effect. The Doppler frequency shifts of the radar return as the target object moves through the radar beam positions A, B, C is also shown in Fig.4.3a. A Laser system records the Doppler frequency shifts along with the phases of the radar return from each target and produces a holographic film. This holographic film when played back it produces an image film in which targets can be resolved as if it had been observed by a very long antenna mathematically synthesized from the Doppler shift data. Fig.4.3b shows how the synthetically lengthened antenna produces the effect of a narrow beam of constant width in the azimuth direction from the near range to the far range positions. Summarizing, we find that the synthetic aperture radar is a signal processing radar. It is a coherent radar. By processing the Doppler frequencies generated by the motion of the antenna relative to the targets, it can achieve a resolution far better than that can be achieved by a conventional real aperture radar. As mentioned earlier, the range resolution (R_r) of the synthetic aperture radar remains unaffected, being dependent on the pulse length and depression angle as per Eqn.8.7. Thus finally we can compute the area of the ground resolution cell by taking the product of the range resolution and the azimuth resolution.

$$\text{Ground resolution cell} = R_r \times R_a \tag{8.9}$$

8.8 SATELLITE RADAR SYSTEMS

The successful development of the synthetic aperture radar has made it possible to acquire radar images from space-borne observation platforms such as Seasat and Space Shuttle of NASA, Earth Resource Satellite (ERS) of the European Space Agency, Radarsat of Canada and Earth Resource Satellite (ERS) of Japan and so on. The descriptions of the side looking airborne radar presented earlier applies to the satellite systems equally well. We present below only the salient features and characteristics of Seasat, SIR-A and SIR-B radar systems.

8.8.1 Seasat

This unmanned satellite (Fig.8.5) was launched by NASA in June 1978 for the study of oceanic phenomena. Unfortunately it ceased operation due to a major electrical failure in October 1978. However, during its short life, Seasat collected a vast wealth of information both on global ocean and land surface. The global ocean surface average topography (digital elevation/terrain model) derived from the Seasat altimeter, and the average global winds derived from the Seasat scatterometer are worth mentioning. Table-8.3 presents the characteristics of Seasat SAR system.

Table 8.3 : Characteristics of Seasat SAR system (1978)

Characteristics	Values
Orbit inclination	108^0
Wavelength	L-band (23.5 cm)
Spatial resolution	25 m
Latitude coverage	72^0 N – 72^0 S
Altitude	790 Km
Image swath	100 Km
Depression angle γ	$67^0 - 73^0$
Average depression angle γ	70^0
Smoothness criterion	h < 1.0 cm
Roughness criterion	h > 5.7 cm
Polarization	HH

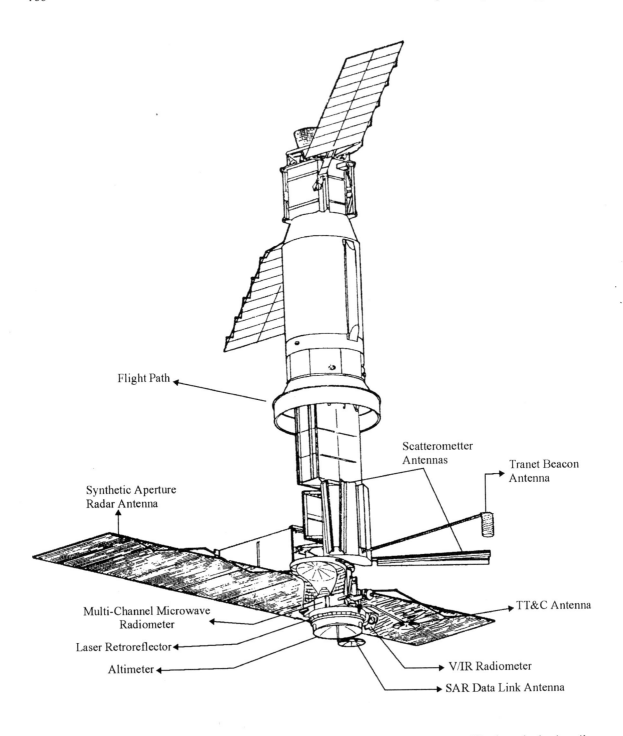

Fig. 8.5 : Seasat (June, 1978–October, 1978), The unmanned NASA microwave satellite launched primarily to monitor oceanic phenomena

8.8.2 Shuttle Imaging Radar (SIR-A)

Shuttle imaging radar was launched by NASA in November 1981. In this manned Space Shuttle mission (Fig.8.6 and 8.7) the image data were recorded as holographic film. When these data are processed in the lab it revealed the terrain features better than the Seasat image. However, SIR-A image lacks the expression of the Pacific Ocean currents and the ocean roughness, and also the clarity of Santa Barbara coast oil slicks which were prominently displayed in the Seasat images. Table 8.4 presents the characteristics of SIR-A system.

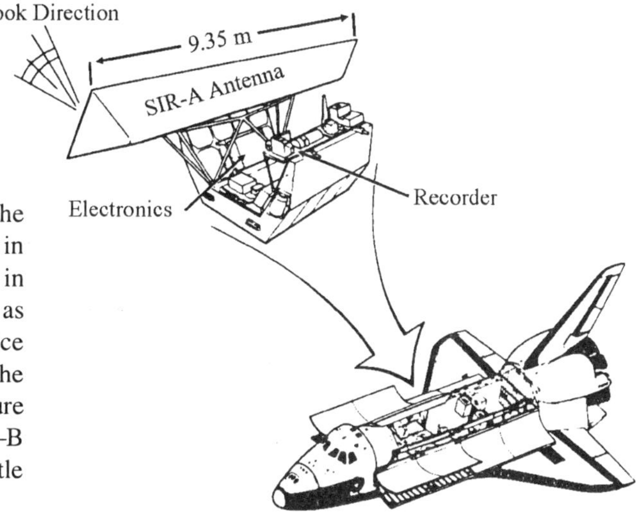

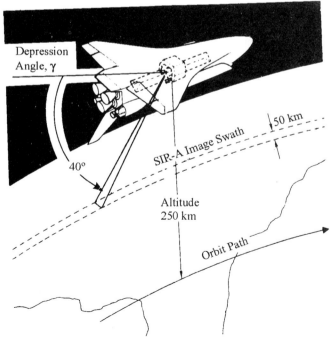

Fig. 8.6 : SIR–A was launched by NASA as the first shuttle imaging radar Expriment in November, 1981. SIR–B launched in October, 1984 was essentially same as SIR–A only with an additional device to change the depression angle of the radar antenna. The synthetic aperture radar (SAR) antenna of SIR–A/SIR–B was compactly placed inside the shuttle cargo bay. (after Sabins, Jr. 1987)

Fig. 8.7 : The SAR antenna operates when the Shuttle is in an inverted attitude as shown in this figure. (after Sabins, Jr. 1987)

Table 8.4 : Characteristics of Shuttle Imaging Radar (SIR-A) System (1981)

Characteristics	Values
Orbit inclination	38°
Wavelength	L-band (23.5 cm)
Spatial resolution	38 m
Latitude coverage	50° N – 35° S
Altitude	250 Km
Image swath	50 Km
Depression angle γ	37° – 43°
Average depression angle γ	40°
Smoothness criterion	h < 1.5 cm
Roughness criterion	h > 8.3 cm
Polarization	HH

8.8.3 Shuttle Imaging Radar (SIR-B)

SIR-B was launched by NASA in October 1984. In this manned Space Shuttle mission, the radar antenna was modified to change the depression angle (γ) during the mission within 30° – 75°. Most of the SIR-B data were recorded digitally during its three orbit passes at depression angles 32°, 45° and 62°. This was done to evaluate the changes in radar return as a function of depression angle for different terrain types and to produce stereo images based on the parallax provided by the difference in the viewing geometry. Though terrain images obtained from data of each orbit pass are in black-and-white ones, when each of the three orbit passes were projected together, then a color image was obtained which provided better discrimination of land cover types. Table-8.5 presents the characteristics of SIR-B system.

Table 8.5 : Characteristics of Shuttle Imaging Radar (SIR-B) System (1984)

Characteristics	Values
Orbit inclination	57°
Wavelength	L-band (23.5 cm)
Spatial resolution	25 m
Latitude coverage	58° N – 58° S
Altitude	225 Km
Image swath	40 Km
Depression angle γ	30° – 75° (variable)
Average depression angle γ	52°
Smoothness criterion	h < 1.2 cm
Roughness criterion	h > 6.8 cm
Polarization	HH

Suggestions for Supplementary Reading

These suggestions are not exhaustive but are limited to the works that will be found especially useful for developing interest and creative imagination in the field of Remote Sensing.

Books

Allen, T. D., 1983. Satellite Microwave remote Sensing. Halsted Press, New York.

American Society of Photogrammetry, 1983. Manual of Remote Sensing, Vol. I, 2nd ed., R. N. Colwell (Ed.), Falls Church, Va.

Berkowitz, R. S., 1965. Modern Radar. John Wiley & Sons, New York.

Cook, C. E. and M. Bernfeld, 1967. Radar Signals : An Introduction to Theory and Application. Academic Press, New York.

Curlander, J.C. and R. N. McDonough, 1991. Synthetic Aperture radar : Systems and Signal Processing. John Wiley & Sons, New York.

Deepak, A. (Ed.), 1980. Remote Sensing of Atmospheres and Oceans. Academic Press, New York.

Elachi Charles, 1988. Spaceborne Radar Remote Sensing : Applications and Techniques. IEEE Press, New York.

Harger, R. O., 1970. Synthetic Aperture Radar Systems : Theory and Design. Academic Press, New York.

Levine, S. H., 1960. Radargrammetry. McGraw-Hill, New York.

Lillesand, Thomas M. and Ralph W. Kiefer, 1987. Remote Sensing and Image Interpretation, 2nd ed., John Wiley & Sons, New York.

Rihaczek, A. W., 1969. Principles of High-Resolution Radar. McGraw-Hill, New York.

Sabins, Floyd, F., Jr., 1987. Remote Sensing – Principles and Interpretations, 2nd ed., W. H. Freeman and Company, New York.

Skolnik, M. I., 1980. Introduction to Radar Systems, 2nd ed., McGraw-Hill, New York.

Toselli, F. (Ed.), 1989. Applications of Remote Sensing to Agrometeorology. Kluwer Academic Publishers, Dordrecht, The Netherlands.

Trevett, William J., 1986. Imaging Radar for Resource Surveys. Chapman and Hall.

Ulaby, F. T., R. K. Moore and A. K. Fung, 1981. Microwave Remote Sensing : Active and Passive, Vol. I : Microwave Remote Sensing Fundamentals and Radiometry. Artech House, Dedham, Mass.

Ulaby, F. T., R. K. Moore and A. K. Fung, 1982. Microwave Remote Sensing : Active and Passive, Vol. II : Radar Remote Sensing and Surface Scattering and Emission Theory. Artech House, Dedham, Mass.

Ulaby, F. T., R. K. Moore and A. K. Fung, 1986. Microwave Remote Sensing : Active and Passive, Vol. III : From Theory to Applications. Artech House, Dedham, Mass.

Research Papers

Alpers, W. R., D. B. Ross and C. L. Rufenach, 1981. On the Detectability of Ocean Surface Waves by Real and Synthetic Aperture Radar. J. Geophys. Res. 86 : 6481 – 6498.

Atlas, D., C. Elachi and W. E. Brown, 1977. Precipitation Mapping with an Airborne SAR. J. Geophys. Res. 82 : 3445 – 3451.

Barrick, D. E., M. W. Evans and B. L. Weber, 1977. Ocean Surface Currents Mapped by Radar. Science 198 : 138 – 144.

Barrett, A. and V. Chung, 1962. A Method for Determination of High Altitude Water Vapor Abundance from Ground-based Microwave Observations. J. Geophys. Res. 67 : 4259 – 4266.

Bernstein, R. L. 1982. Sea Surface Temperature Estimation Using the NOAA-6 Satellite Advanced Very High Resolution Radiometer. J. Geophys. Res. 87 : 9455 – 9465.

Bernstein, R. L., G. H. Born and R. H. Whritner, 1982. Seasat Altimeter Determination of Ocean Current Variability. J. Geophys. Res. 87 : 3261 – 3268.

Blom, R. and C. Elachi, 1981. Spaceborne and Airborne Imaging Radar Observation of Sand Dunes. J. Geophys. Res., 86 : 3061 – 3070.

Blom, R. and M. Daily, 1982. Radar Image Processing for Rock Type Discrimination. IEEE Trans. Geosci. Remote Sensing, GE-20 : 343 – 351.

Blom, R., R. G. Crippen and C. Elachi, 1983. Detection of Subsurface Features in Seasat Radar Images of Means Valley, Mojave Desert, California. Geology 12 : 346 – 349.

Brown, W. E., Jr., C. Elachi and T. W. Thompson, 1976. Radar Imaging of Ocean Surface Patterns. J. Geophys. Res. 31 : 2657 – 2667.

Brown, W. M. and L. J. Porcello, 1969. An Introduction to Synthetic Aperture Radar. IEEE Spectrum, 6 : 52 – 62.

Bryan, M. L. 1979. The Effect of Radar Azimuth Angle on Cultural Data. Photogramm. Eng. Remote Sensing, 45 : 1097 – 1107.

Bush, T. F. and F. T. Ulaby, 1978. An Evaluation of Radar as a Crop Classifier. Remote Sensing Environ. 8 : 15 – 36.

Cheney, R. E. and J. G. Marsh, 1981. Seasat Altimeter Observations of Dynamic Topography in the Gulf Stream. J. Geophys. Res. 86 : 473 – 483.

Elachi, C., 1980. Spaceborne Imaging Radar : Geologic and Oceanographic Applications. Science, 209 : 1073 – 1082.

Elachi, C. 1982. Radar Images of the Earth from Space. Scientific American, December Issue.

Elachi Charles, 1983. Microwave and Infrared Satellite remote Sensing. In : Manual of Remote Sensing, Vol. I, 2nd ed., R. N. Colwell (Ed.). American Society of Photogrammetry, Falls Church, Va. p. 571 – 650.

Elachi, C. and J. Apel, 1976. Internal Wave Observations made with an Airborne Synthetic Aperture Imaging Radar. Geophys. Res. Lett. 3 : 647 – 650.

Elachi, C., T. Bicknell, R. L. Jordan and C. Wu, 1982. Spaceborne Imaging Synthetic Aperture Radar : Technique, Technology and Applications. Proc. IEEE, 70 : 1174 – 1209.

Fedor, L. S. and G. S. Brown, 1982. Wave Height and Wind Speed Measurement from the Seasat Altimeter. J. Geophys. Res. 87 : 3254 – 3260.

Grody, N. C. 1976. Remote Sensing of Atmospheric Water Content from Satellites using Microwave Radiometry. IEEE Trans. Antennas Propag. AP-24 : 155 – 162.

Hofer, R., E. G. Njoku and J. W. Waters, 1981. Microwave Radiometric Measurements of Sea Surface Temperature from the Seasat Satellite : First Results. Science 212 : 1385.

Jain, A. 1977. Determination of Ocean Wave Heights from Synthetic Aperture Radar Imagery. Appl. Phys. 13 : 371 – 382.

Jensen, H., L. C. Graham, L. J. Porcello and E. N. Leith, 1977. Side Looking Airborne Radar. Scientific American, 237 : 84 – 95.

Jones, W. L., L. C. Schroeder, D. H. Boggs, E. M. Bracalante, R. A. Brown, G. J. Dome, W. J. Pierson and F. J. Wentz, 1982. The Seasat – A Satellite Scatterometer : The Geophysical Evaluation of Remotely Sensed Wind Vectors Over the Ocean. J. Geophys. Res. 87 : 3297 – 3317.

Lee, J. S. 1981. Speckle Analysis and Smoothing of Synthetic Aperture Radar Images. Computer Graphics and Image Processing, 17 : 24 – 32.

Liu, C. T. 1983. Tropical Pacific Sea Surface Temperatures Measured by Seasat Microwave Radiometer and by Ships. J. Geophys. Res. 88 : 1909 – 1916.

Moore, Richard K., 1983. Imaging Radar Systems. In : Manual of Remote Sensing, Vol. I, 2nd ed., R. N. Colwell (Ed.). American Society of Photogrammetry, Falls Church, Va. p. 429 – 474.

Naraghi, M., W. Stromberg and M. Daily, 1983. Geometric Rectification of Radar Imagery using Digital Elevation Models. Photogramm. Eng. Remote sensing, 49 : 195 – 199.

Njoku, E. G. 1982. Passive Microwave Remote Sensing of the Earth from Space – A Review. Proc. IEEE, 70 : 728 – 749.

Penfield, H., M. M. Litvak, C. A. Gottlieb and A. E. Lilley, 1976. Mesospheric Ozone Measured from Ground-based Millimeter-wave Observations. J. Geophys. Res. 81 : 6115 – 6120.

Rosenkranz, P. W., M. J. Komichak and D. H. Staelin, 1982. A Method for Estimation of Atmospheric Water Vapor Profiles by Microwave Radiometry, J. Appl. Meteorol. 21 : 1364 – 1370.

Schaber, G. G., G. L. Berlin and W. E. Brown, 1976. Variations in Surface Roughness Within Death Valley, California – Geologic Evaluation of 25 cm Wavelength Radar Images. Geol. Soc. Am. Bull. 87 : 29 – 41.

Shanmugan, K. S., V. Narayanan, V. S. Frost, J. A. Stiles and J. C. Holtzman, 1981. Textural Features for Radar Image Analysis. IEEE Trans. Geosci. Remote Sensing GE-19 : 153 – 156.

Simonett, Davis S. and Robert E. Davis, 1983. Image Analysis – Active Microwave. In : Manual of Remote Sensing, Vol. I, 2nd ed., R. N. Colwell (Ed.). American Society of Photogrammetry, Falls Church, Va. p. 1125 – 1181.

Staelin, D. H. 1969. Passive Remote Sensing at Microwave Wavelengths. Proc. IEEE, 57 : 427 – 439.

Staelin, D. H. 1981. Passive Microwave Techniques for Geophysical Sensing of the Earth from Satellites. IEEE Trans. Antennas Propag. AP- 29 : 683 – 687.

Towmend, W. F., 1980. The Initial Assessment of the Performance Achieved by the Seasat Radar Altimeter. IEEE J. Oceanic Eng. OE-5 : 80 – 92.

Ulaby, F. T. 1976. Passive Microwave Remote Sensing of the Earth's Surface. IEEE Trans. Antennas Propag. AP – 24 : 112 – 115.

Ulaby, F. T., A. Aslam and M. C. Dobson, 1982. Effects of Vegetation Cover on Radar Sensitivity to Soil Moisture. IEEE Trans. Geosci. Remote Sensing, GE – 20 : 476 – 481.

Ulaby, Fawwaz T. and Keith R. Carver, 1983. Passive Microwave Radiometry, In : Manual of Remote Sensing, Vol. I, 2nd ed., R. N. Colwell (Ed.). American Society of Photogrammetry, Falls Church, Va. p. 475 – 516.

Ulaby, F. T. 1982. Radar Signatures of Terrain : Useful Monitors of Renewable Resources. Proc. IEEE. 70 : 1410 – 1428.

Waters, J. W., W. J. Wilson and F. I. Shimabukuro, 1976. Microwave Measurement of Mesospheric Carbon Monoxide. March Issue : 1174 – 1175.

Westwater, E. R., J. B. Snider and A. V. Carlson, 1975. Experimental Determination of Temperature Profiles by Ground-based Microwave Radiometry. J. Appl. Meteorol. 14 : 524 – 539.

Westwater, E. R. and N. C. Grody, 1980. Combined Surface- and Satellite-based Microwave Temperature Profile Retrieval. J. Appl. Meteorol. 19 : 1438 – 1444.

Wing, R.S. 1971. Structural Analysis from Radar Imagery. Modern Geology, 2 : 1–21.

Zwally, H. J. and P. Gloersen, 1977. Passive Microwave Images of the Polar Regions and Research Applications. Polar Record VII – 44, 18 : 431 – 450.

Global Positioning System (GPS) — 9

9.1 WHAT IS GPS ?

Global Positioning System (GPS) is a satellite based navigational aid. Essentially it is a radio-positioning navigation and time transfer system. It provides accurate information on position, velocity and time of an object or a platform at any moment, anywhere on the globe. The system's service is available worldwide with all weather capability. However, this positioning service can be obtained by any user, only if he has a suitable GPS receiver.

The NAVSTAR Global Positioning System which was established in 1973, is maintained by the USA Department of Defence. In the present state of development, it is taken as one of the most accurate and highly dependable space-borne navigational aid . The orbital configuration of 21 active satellites comprising the GPS satellite constellation is shown in Fig.9.1.

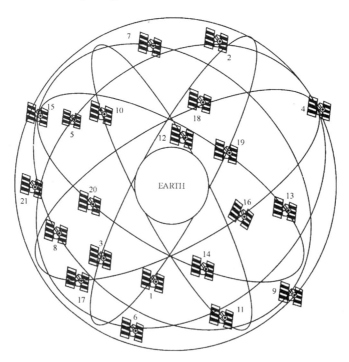

Fig. 9.1 : The constellation of active GPS satellites.

9.2 FUNCTIONAL SEGMENTS OF THE GLOBAL POSITIONING SYSTEM

Global Positioning System has three major functional segments as presented in Fig.9.2. They are : 1. Space Segmant 2. Ground Control Segment, and 3. User Segment

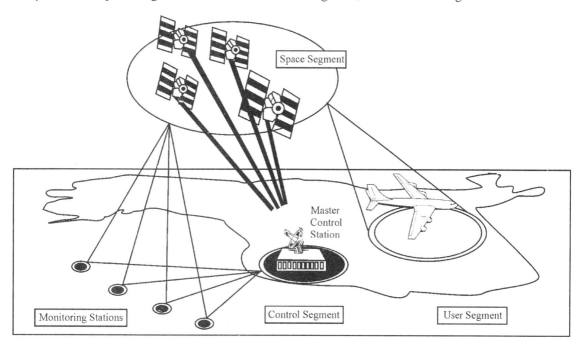

Fig. 9.2 : Basic segments of GPS : Space segment, Control segment (Master control station, monitoring stations) and User segment

9.2.1 The Space Segment

The space segment of the Global Positioning System consists of a constellation of 24 satellites, each weighing 850 Kg. These satellites are placed in six orbits, so that four satellites are accommodated in each orbit at an altitude of 20,185 Km from the surface of earth. Out of these 24 GPS satellites 21 are operational ones and 3 are kept as active spares so that they can be inducted into the operational fleet immediately in case any one of the operational satellites becomes nonfunctional. Each orbital plane of the satellites has been kept inclined at 55° to the equatorial plane of the earth and they are displaced by 120° in longitude. The orbital period of each GPS satellite has been fixed at 11: 56: 09 hrs (~ half-a-day). This disposition of the satellite constellation guarantees that at any given time anywhere on the globe at least 4 or more than 4 satellites will be in view of the user's GPS receiver (which is also called GPS Sensor Unit, GPSSU). We assume that the satellites, to be in view, must be more than 5° above the horizon of the user.

The ephemeris of the GPS satellites is known very precisely. Each GPS satellite carries four highly stable and precision atomic clocks. The GPS satellites periodically go on sending signals on their positional coordinates, time of transmission of the signal and a separate signal for establishing the range between the satellite and the user receiver.

9.2.2 The Ground Control Segment

The ground control segment of the NAVSTAR GPS consists of the Master Control Station and the Monitoring Stations, the Receiver Antennas being distributed at different locations all over the globe. Periodically the processors of these Ground Stations of known latitude, longitude and altitude, distributed worldwide, update the clock information and accurately compute the satellite ephemeris using a dedicated software. This has helped the user receiver to compute its position and velocity by measuring the clock-timing differentials and the Doppler shifts. These Ground Monitoring Stations also track all the visible satellites and their information are processed at the Master Control Station to determine the satellite orbits and restore the satellites to their specified orbits, if necessary, through telecommands. In general, the Control Segment manages the entire GPS satellite constellation.

9.2.3 The User Segment

The user segment may be any navigating system like an aeroplane or a ship carrying the GPS receiver to compute its position and velocity at any given time from the GPS satellite information with respect to their position, time of message transmission and the distance between the satellite and the user receiver. The user segment may also include a surveyor or a remote sensing specialist armed with the GPS receiver to compute the latitude, longitude, altitude and the survey time of a location of interest or a training site for the identification and classification of land-surface features from remote sensing imagery of that area around the time of the survey.

9.3 HOW DOES GPS WORK ? : THE BASICS OF GPS FUNCTIONING

By now we are aware of the fact that GPS helps us to find out our position in the 3-D space anywhere at any time with the help of a GPS receiver. This is technically achieved in several steps as follows.

9.3.1 Satellite Ranging

It has already been mentioned that each GPS satellite is provided with four highly accurate atomic clocks (each costing about hundred thousand US dollars) which are also used on ground to keep the time standard all over the world. While the International Standard Time is maintained at Paris, in India the National Standard Time is maintained at the National Physical Laboratory, New Delhi. All the GPS satellites go on generating and sending coded signals periodically with very small time period. So to get the distance (range) of a GPS satellite from the GPS receiver (user), one has to multiply the time taken for the signal to travel from the satellite to the receiver by the velocity of light. However, it is difficult to find out exactly when the radio signal left the satellite and proceeded outwards. This problem has been solved by the technique of Code Synchronization.

9.3.2 Synchronization of Codes

The design parameters of the GPS satellites and the GPS receiver are such that both of them generate the same signal code exactly at the same time. This helps us to set the receiver to perfectly synchronize itself with the GPS satellites to start with. Now that the satellite and

Global Positioning System (GPS)

the receiver are generating the signal codes in perfect synchronization, what we want to do is to receive the codes from the satellite and look back to find out how long ago the receiver generated the same code. Thus we get the differential time needed for the radio signal code to travel from the satellite to the receiver (user) and hence the range (distance) of the receiver from the satellite is calculated.

9.3.3 Computation of the 3-D Location of the User GPS Receiver : The Method of Triangulation (Geometrical Method)

So far we know how to calculate the distance (range) of the GPS receiver from several GPS satellites. To ellucidate the geometrical concept of triangulation, let us assume that our GPS receiver has, at a particular time t, ranged four satellites whose distances from it are R_1, R_2, R_3 and R_4 respectively (Fig.9.3a,b,c,d). Now take the 1st satellite as the center and draw a hollow sphere with radius R_1. This shows that the receiver is restricted to be somewhere on the surface of this sphere (Fig.9.3b). Next draw a hollow sphere of radius R_2 taking the 2nd satellite as the center. This sphere will intersect the 1st hollow sphere in a circle. Thus the GPS receiver's position is further narrowed down to a circle (Fig.9.3c). Then take the 3rd satellite as the center and draw a hollow sphere of radius R_3. This sphere will intersect the previous two hollow spheres (i.e. the circle) at two points only. Thus the GPS receiver position

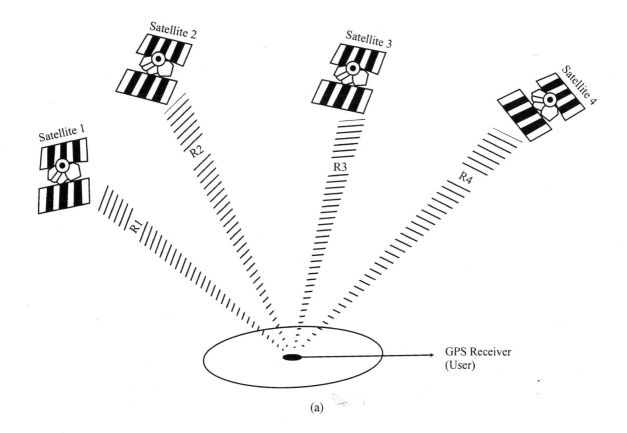

(a)

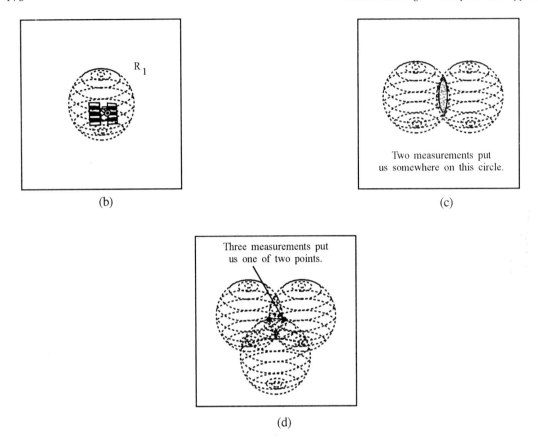

Fig. 9.3 : GPS satellite ranging : Locating the positional coordinates of the user (GPS receiver) with 4 pseudo-range values by the method of triangulation (Geometrical approach).
(a) Pseudo-ranges of 4 GPS satellites from the GPS receiver, (b) Positioning the GPS receiver in space with the help of one pseudo-range, (c) Positioning the GPS receiver with the help of two pseudo-range measurements, and (d) Positioning the GPS receiver with the help of three pseudo-range measurements.

is still narrowed down to one of these two locations in the universe (Fig.9.3d). Finally, taking the 4th satellite as the center, draw a hollow sphere of radius R_4. The surface of this hollow sphere will now pass only through one of the above mentioned two locations. Thus following triangulation technique geometrically, we can unerringly pin down the coordinates of the GPS receiver position in the universe.

9.3.4 Algebraic Computation of the 3-D Location of the GPS Receiver

Just as the astronomers have computed the ephemeris of the members of the solar system, similarly the exact ephemeris of the artificial satellites forming the constellation of GPS are also computed and known. This amounts to say that the 3-D location of the GPS satellites with respect to the geocentric coordinate system is precisely known at any given time. Thus when our GPS receiver receives a radio signal code from the GPS satellite, it not only knows

Global Positioning System (GPS)

at what time the signal was sent from the satellite, but also knows what was the latitude, longitude and altitude of that satellite at that instant of time with respect to the geocentric coordinate frame.

In case the GPS receiver is precisely synchronized with respect to the aforementioned atomic clocks on board the satellites, the timing signals from only three satellites would be sufficient to pin down the receiver's three position coordinates (x,y,z). However, synchronization of the receiver clock is seldom possible. Under the circumstance, timing signals from four satellites are actually required for a three dimensional position determinationm of the GPS receiver. The extra timing signal is used to solve the receiver clock's bias error (c_B), which in turn corrects the inaccurate range estimates. Mathematically, this amounts to solving the following four simultaneous equations for the four unknowns : x, y, z and c_B :

$$(x - x_1)^2 + (y - y_1)^2 + (z - z_1)^2 + (R_1 - c_B)^2 = 0$$
$$(x - x_2)^2 + (y - y_2)^2 + (z - z_2)^2 + (R_2 - c_B)^2 = 0$$
$$(x - x_3)^2 + (y - y_3)^2 + (z - z_3)^2 + (R_3 - c_B)^2 = 0$$
$$(x - x_4)^2 + (y - y_4)^2 + (z - z_4)^2 + (R_4 - c_B)^2 = 0 \qquad (9.1)$$

Here (x_1, y_1, z_1), (x_2, y_2, z_2), (x_3, y_3, z_3) and (x_4, y_4, z_4) are the coordinates of four GPS satellites tracked by the receiver, and R_1, R_2, R_3 and R_4 are their respective distances from the GPS receiver. Thus to get the accurate position of the GPS receiver by solving the above mentioned simultaneous algebraic equations, at least four satellites must be tracked by the receiver. The GPS receiver routinely solves the simultaneous algebric equations (9.1) using a dedicated software available with it.

9.3.5 Accuracy of Positional Measurement by GPS Receiver

Although a regular GPS receiver offers accuracy in positional measurement of the order of tens of meters, one should not infer that the system is not capable of delivering anything better. This is due to the fact that the accuracy of the GPS is more controlled by the US - DoD rather than the GPS itself. In fact, with the sophisticated high precision atomic clocks on board the GPS satellites, the minimum measurable time differential is of the order of centimeter which speaks volumes of the potential positional accuracy attainable by the GPS receiver.

Let us now consider the following technical facts to understand how one can control the accuracy of GPS measurements :

The signal transmitted from a GPS satellite to be received by the user with the help of a GPS receiver consists of two carrier frequencies called L1 frequency and L2 frequency. The L1 frequency is of 1575.42 MHz and the L2 frequency is of 1227.6 MHz. The L1 frequency is further modulated with a precision code (P-code) and a coarse acquisition code (C/A-code). Each GPS satellite is assigned with a unique C/A-code along with its P-code.

The precision code (P-code) is made available only to authorized users such as high precision military users. The precision code is protected by encryption techniques so that none other than the authorized ones can be able to use this code. On the other hand, the coarse acquisition code (C/A-code) is freely available for the civilian users and here the accuracy is meticulously brought down by introducing arbitrary errors in the measurements of position.

Both the P-code and C/A-code are pseudorandom binary codes. The randomness in these codes is only apparent because actually the binary pulses are generated following precise mathematical relationships with total predictability. The L1 and L2 frequencies are also modulated with a 50-bit per second data transmission providing satellite orbit, system time, satellite's clock behavior and the status information of all satellites to the ground control facilities.

As regards the usefulness of the pseudorandom codes, we find that these codes open up the possibility to synchronize the receiver clocks with the high precision atomic clock on board the satellite. The pseudorandom code helps in the use of cheaper, low-power satellites. It also helps in using small antennae on the GPS receivers. The pseudorandom code also simplifies the system's functioning by providing a way to have all the satellites work on the same frequency without interfering with each other. Here each satellite is allotted just a different code only.

Finally and most importantly, the pseudorandom code gives the Department of Defence, USA (DoD-USA) a method of restricting the access of the user to the system. In case the DoD changes the code, then one can not use the Global Positioning System any more.

9.3.6 Errors in GPS

Practically speaking, typical satellite clock error is about 0.5m and the ephemeris error is also about 0.5m resulting in the receiver error of about 1m. Ground monitoring stations not only regularly track the satellites but also apply corrective measures so that the satellites go on moving on dedicated orbits and the satellite ephemeris is maintained highly accurate. Any error creeping in the atomic clock is also corrected by the ground control stations.

However, for GPS there are other errors too. Ionospheric error is one of them. While the satellite signal (electromagnetic by nature) passes through the ionosphere the signal interacts with the charged particles and thereby its motion slows down. This results in error in the range calculation. Consequently there will be error in the position determination of the user GPS receiver. Fortunately, since the structure of the ionosphere is almost uniform, systematic corrections are applied for the ionospheric effect to maintain the accuracy of GPS measurements.

After ionosphere the satellite signal passes through the atmosphere of air and water vapor. Although the signal interacts with the atmospheric constituents to some extent, due to the non-uniformity of the atmosphere structure, no meaningful corrections are applied to remove the atmospheric error.

Sometimes the GPS signal gets scattered from other objects and arrive at the receiver antenna after travelling a zig-zag path. Here there is a possibility that this scattered signal will interfere with a straight line signal. This multi-path interference results in error of the receiver. To avoid this type of error, good receivers used advanced digital signal processing techniques. Also by antenna design which is a very sophisticated art, this multi-path interference effect can be minimized.

Global Positioning System (GPS)

There is an international error in GPS measurement called the Selective Availability (S/A) which the DoD (USA) can put into the system to make it useless for any hostile party. This is done by the process called Scrambling during which artificial clock and status message errors are introduced into the satellites. However, only permitted military users with the secret codes can unscramble the system to derive results with full accuracy.

9.3.7 Improving Efficiency in Location Finding

This can be done by selecting the satellites with an advantageous geometry. Widely separated satellites with almost perpendicular crossing angles are preferred to compute the receiver's 3-D coordinates much better. With more GPS satellites in view, only some sophisticated receivers can select out those satellites which will lead to better computational measurements.

GPS receivers are of two types : sequential and continuous with one or more number of receiving channels. The single channel receiver is slow in making a measurement. It is not efficient in measuring velocity and hence handling dynamic conditions like that of the aircraft, ship and so on. Thus single channel receiver should be motionless to get an accurate location measurement. By using a two-channel receiver, instantaneous precise speed, better tracking and better noise removal can be achieved. A three-channel receiver offers much better speed and accuracy. In case one of the channels is temporarily not working, the other two channels can navigate a dynamic system. With four simultaneously navigating channels, the GPS receiver generates continuous position solutions which can be internally cross-checked for better accuracy. This continuously measuring receiver can maintain its stability during fast movements and does not get confused by rapid accelerations as some sequential receivers do.

Another alternative to get ultra precise position accuracy and efficiency with GPS is to use a Differential Global Positioning System (DGPS). Essentially a DGPS consists of two or more receivers working together, one of which acts as the reference for others. The reference receiver is placed at a known location whose position is known precisely. As the reference receiver gets positioning information from the satellites, it computes its position coordinates based on these satellite data and then compares with its actual location coordinates. Thus any discrepancy is taken as due to the ionospheric and atmospheric errors. The other receivers which are in the general vicinity of the reference receiver are also assumed to have the same errors. Thus essentially the reference receiver calculates the corrections and transmits them to the other receivers in the area. They in turn apply those corrections to their positions. The end result is a very very precise and accurate position determination, even if those receivers are moving.

9.4 APPLICATIONS OF GPS

All conceivable applications of the GPS stem from the fact that a suitable GPS receiver on a platform, either stationary or moving, can compute its positional coordinates (latitude, longitude and altitude) and velocity at any given time with the help of signals from a number of GPS satellites as already described in the above paragraphs.

The main purpose of GPS is to help in navigation and positioning of remote sensing satellites. Under ideal conditions the position of a space-borne platform can be determined to an accuracy of a few meters. With the development and refinement of suitable technologies (hardware and software), and their incorporation in the GPS, the accurate position information required to support the satellite image correction, rectification, registration, processing and classification are now obtained in a routine manner. In fact, such pixel position information now reduces the dependency on ground survey for identification of the ground control points (GCPs) for this very purpose.

GPS has come to stay as an indipensable unit with our most important navigational systems such as aircraft and ships. With the help of GPS the ships and aeroplanes successfully navigate on their earmarked routes.

With the increased positional accuracy of GPS, this system is now being incorporated as an integral part of the air traffic alert and collision avoidance system.

Increased accuracy of GPS also helps in reducing the separation between aircraft during enroute phase without compromising safety. In fact, this possibility of reduced aircraft separation optimizes the routes (Fig.9.4). Under the circumstance, considerable fuel savings can be achieved when the airplanes follow the optimized routes with shorter flight paths.

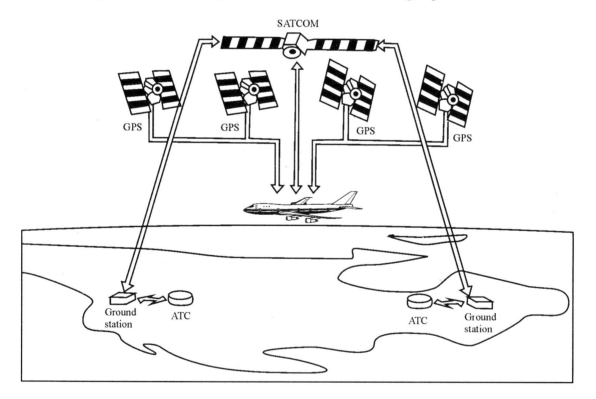

Fig. 9.4 : GPS–based Automatic Dependent Surveillance (ADS) to identify aircraft position. This technique may lead to the reduced aircraft separation and route optimization.

Reduced separation between aircraft also allows more aircraft to use the air space.

The GPS capability may also improve access of some remote airports which are not supported by modern navigational aids.

In the field of aviation, the accuracy requirements are extremely stringent for vertical position than the lateral one at any given altitude during a precision landing or takeoff. In this direction, incorporation of DGPS will hold the key for the state-of-the-art aircraft of the near future for safety landing and departure (Fig. 9.5).

GPS has also become an important search and rescue aid for aircraft and ships. In case of a disaster, they send SOS message for help at the location of the disaster, and accordingly immediate action is undertaken for the search and rescue operation.

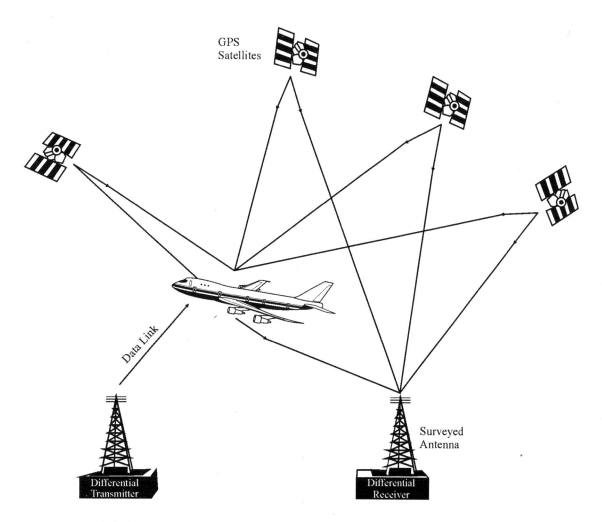

Fig. 9.5 : Differential GPS ground stations for improving aircraft's position accuracy.

High precision GPS applications dominate in the field of Defence. GPS helps the military in troop movement, navigation, siting targets (with the help of GIS), search and rescue.

Now-a-days, GPS-based guided missiles are also being used in wars, increasing the possibility of striking the target and saving time (duration of war) and money. Recently an US made prototype of Robotic Bombing X-45 Aircraft dropped a 250 pound inert smart bomb on a pre-programmed target coordinates that hit the target within inches.

GPS receivers have come to us as handy survey instruments. For example, if we want to draw the outline of a study area (say, a township), then we take the hand-held GPS receiver, move on the boundary of the study area, and at short distances monitor the positional coordinates and save them in the system. Once the polygon survey is completed, we download the data from the GPS receiver to a Computer supported by GIS software and take a print of the map to any desired scale. This eliminates the tedious conventional survey work.

GPS can work as a reliable guide for a tourist who arrives at an unknown place whose GIS is, of course, known. Aided by this global information system, the GPS receiver can now guide the tourist to a hotel, take him to places of interest for sight seeing and also guide him back to his hotel without any mistake.

Based on the basics of GPS functioning, one can think of its numerous other applications for constructive and destructive purposes.

Suggestions for Supplementary Reading

These suggestions are not exhaustive but are limited to the works that will be found especially useful for developing interest and creative imagination in the field of Remote Sensing.

American Society of Photogrammetry, 1983. Manual of Remote Sensing, Vol. I, 2nd ed., R. N. Colwell (Ed.). Falls Church, Va.

Kendal Brian, 1987. Manual of Avionics : An Introduction to the Electronics of Civil Aviation, 2nd ed., BSP Professional Books, Oxford, London, Edinburgh.

Principles of Geographic Information System (GIS)   10

10.1 INTRODUCTION

Geographic information system is a tool for handling geographic (spatial and descriptive) data. It is an organized collection of computer hardware, software, geographic data and the personnel designed to efficiently capture, store, retrieve, update, manipulate, analyze and display all forms of geographically referenced information according to the user-defined specifications (Fig.10.1 and 10.2). We can visualize the real world as consisting of many geographies such as topography, land use, land cover, utilities, soils, crops, forests, water bodies, streets, districts and so on. Data on different aspects of these geographies are stored in the GIS files or data layers (Fig.10.3). Spatial data used in GIS deal with location, shape and relationship among features, whereas the descriptive data deal with the characteristics of the features. While maps form one type of format for geographically referenced information, descriptive data may be visualized as the elements representing reality in an information system. It is important to remember that a GIS does not hold maps or pictures, it holds only a database. Thus the database concept is central to a GIS and this is the main difference between a GIS and a computer mapping system which can only produce good graphic outputs. A contemporary GIS incorporates a database management system (DBMS). So the GIS has become a tool to visualize, model, analyze and query the database. The addition of geography not only adds value to an information system, but also adds its potentiality for exponential increases in scope, scale and functionality.

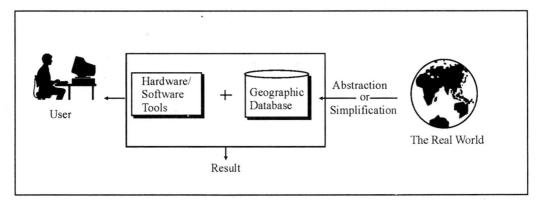

Fig. 10.1 : A simplified Diagrammatic presentation of Geographic Information System

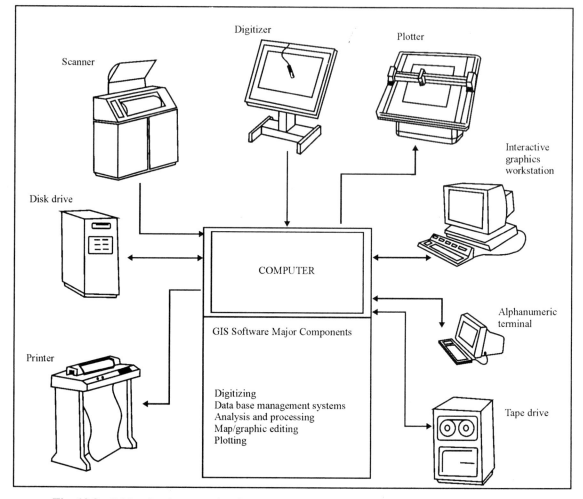

Fig. 10.2 : Major hardware and software components of Geographic Information System

A GIS links spatial data with descriptive information about a particular feature on a map. The information are stored as attributes of the geographically represented features. A GIS can also use the stored attributes to compute new information about the map features. If we want to go beyond just making pictures, we need to know three pieces of information about every feature stored in the computer, namely : (1) what it is, (2) where it is, and (3) how it relates to the other features. GIS software like ARC/INFO handles these efficiently. Thus GIS gives us the ability to associate information with a feature on a map and to create new relationships that can determine the suitability of various sites for development, evaluate environmental impacts, calculate harvest values, identify the best location for a new facility and so on. As mentioned earlier, the power of GIS lies in the link between the spatial (graphic) data and the descriptive (tabular) data. In this regard, one can find out three noteworthy characteristics of this connection :

  (i) A one-to-one relationship is maintained between features on the map and records in the feature attribute table,

Principles of Geographic Information System (GIS)

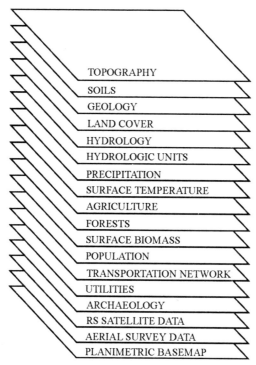

Fig. 10.3 : The real world consists of many geographies which can be represented as a number of related data layers called GIS Files. Information in GIS Database are derived from a variety of sources.

(ii) The link between the feature and the record is maintained through the unique identifier assigned to each feature, and

(iii) The unique identifier is physically stored in two places : in the files containing the x,y coordinate pairs and with the corresponding record in the feature attribute table.

The GIS software like ARC/INFO, SPANS, PAMAP, MAPS, GEOSPACE, INGIS, URAS/ MASMAP, ISROGIS and so on automatically creates and maintains this connection.

10.2 SPATIAL DATA REPRESENTATION

Spatial or geographic data can be represented by visual elements like points, lines and areas (Fig.10.4 and 10.5). The ability to visualize the data is one of the characteristics which makes geographic data processing so powerful. This is done by linking the geographic data to the visual data elements (points, lines, areas) which compose the picture.

The visual data elements can be modeled and structured in two basic ways, namely raster and vector (Fig. 10.6a, b, c, d).

10.2.1 Raster

Raster is a term which comes directly from digital display. A raster is an area on a scanning line which can be individually illuminated. Raster data represent a point, a line or an area as

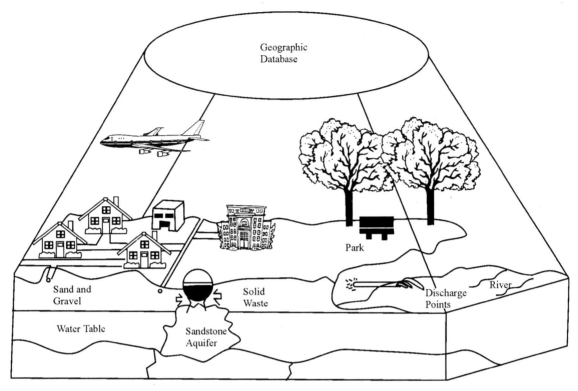

Fig. 10.4 : Geographic database includes spatial data and descriptive data

a matrix of values. The size of the cell determines the resolution of the display. A raster database requires that all values or entities be defined by a single raster or group of raster. The size of the cell defines the resolution of the database as well as the display.

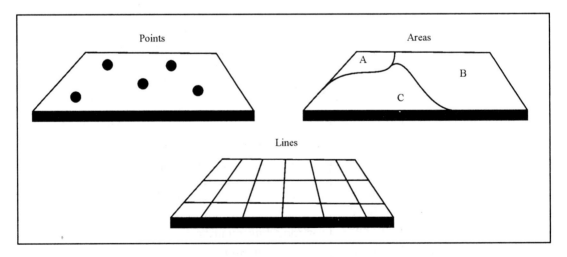

Fig. 10.5 : Geographic Data are visualized by linking them to graphic data elements : Points, Lines and Areas (Polygons)

Principles of Geographic Information System (GIS)

RASTER

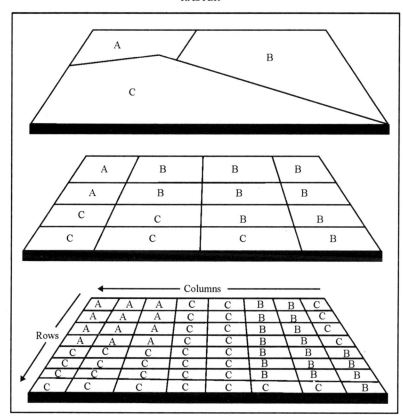

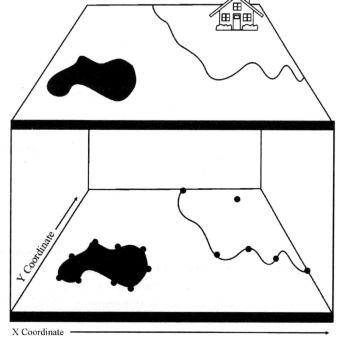

Fig. 10.6 (a) : Graphic data elements can be modeled and structured in two ways, as Rasters or Vectors.

In raster presentation, a point, line or area is represented by a matrix, the resolution of which depends on the size of the cells (grids) in the matrix.

Fig. 10.6 (b) : In vector presentation, a point is represented by cartesian coordinates (x, y), a line by a string of coordinates, and an area or polygon by a string of coordinates starting and ending at the same point

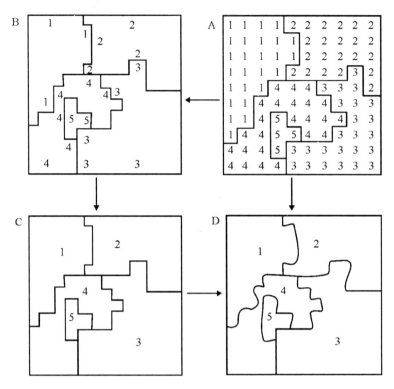

Fig. 10.6 (c): Raster to vector data conversion. Here a satellite image classified into five land use categories presented in raster format in (A) is converted to the vector format in (D) by smoothing the border lines through (B) and (C)

10.2.2 Vector

Vector is the second graphic or visual data model. A vector model defines graphic elements using basic geometry, namely a quantity which has magnitude and direction, represented by a directed line – the length representing the magnitude and whose orientation in space represent the direction.

Points are usually represented by cartesian coordinates (x,y), a line by a string of coordinates and an area or polygon by a string of coordinates starting and ending at the same point. But a vector data model for visual data allows one to define more precisely the entities in the primary data model. It also has advantages in the resolution of the display. However, this advantage is tampered by the slowness of the vector displays that require considerably more calculations. Advances in hardware and software developments have diminished this disadvantage considerably, if not eliminated it completely.

10.3 SPACE AND SPATIAL OBJECTS

Earth's surface features can be divided into a multifaceted mosaic of cultural and natural environments. The cultural environment contains objects in the landscape which are easy to recognize, namely streets, houses, utilities, business facilities and so on. Their boundaries can be determined and

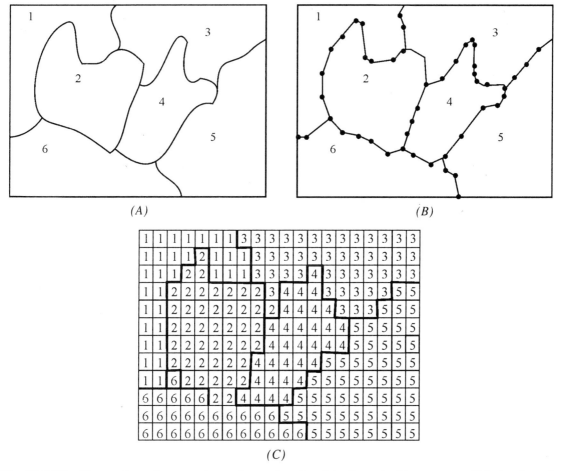

Fig. 10.6 (d) : Conversion of a map to vector and raster data formats. Here a map (A) showing six land-use categories is first digitized in vector format (B) which is subsequently converted to Raster format (C). In the raster format presentation C, the number in each cell (grid) indicates either a polygon or a thematic class.

drawn on a map with an appropriate symbol. A sub-classification of the cultural environment would be the activities that take place within the man-made environment, such as sales activity by geographic region or density of population within a specified geographic area.

Defining boundaries in the natural environment is much more difficult. Natural features tend to exist as continuous variables. It is the task of the scientists to classify continuous data into discrete manageable units. To cite a few examples where appropriate decisions are required are : how to define the edge of a forest, how to recognize that one has crossed a soil boundary, and how does the slope and aspect of a hill actually change as one walks across a ridge, and so on.

Representing observables as discrete two-dimentional data requires an understanding of both space and spatial objects. While working with GIS, one needs to be able to define space.

In this regard one can harness the idea that a space can not exist without an object and conversely there is no object without its surrounding space. A space containing an object and the boundary of an object defined by the space surrounding it are complementary to each other (Fig.10.7).

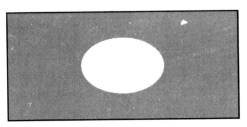 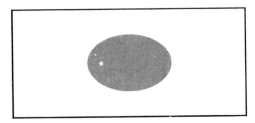

Fig. 10.7 : A space containing an object and the boundary of an object defined by the space surrounding it present the duality of space and spatial object in the sense that one does not exist without the other.

Storing information about the landscape requires that the space be classified into a series of objects or features, which can be defined discretely in the computer system.

10.4 REPRESENTING REALITY

The easiest approach to the problem of representing complex reality as discrete objects is to define a spatial boundary, the area of interest, and then divide its landscape into a set of thematic layers. Each of these layers presents a separate classification of similar data (Fig. 10.8). In a natural environment, the primary layers might be geology, soils, vegetation, topography, hydrology and so on. For the man-made environment, the main layers are buildings, streets, land use zoning and other administrative or political units. Activities that occur within this environment can include things like : how many people work or live within a certain land use or within some administrative boundary, how many cars have been sold by a dealer within the city limits, or what is the composition of the population that lives within a certain geographic area.

In GIS the spatial data capture is done by manual or automatic digitization. As mentioned earlier the geographic data model recognizes spatial objects into thematic layers. A layer provides a means of further information. It is appropriate to separate and maintain individual layers in a spatial database under the following conditions :

The components of the database may come from different sources and scales, managed and maintained by separate departments. For example, it may be logically convenient to separate a parcel layer which may be managed by the planning department, from the street center line layer which may be managed and maintained by the engineering department.

Some components of the real world change more rapidly than others. It may be appropriate to separate those layers that have different update cycles and are updated by different groups. For example, zoning may change on a monthly basis while administrative boundaries change only once in a year or more.

Different components of the real world are required for particular types of analyses. It may be appropriate to separate layers that are standard input into a particular analysis.

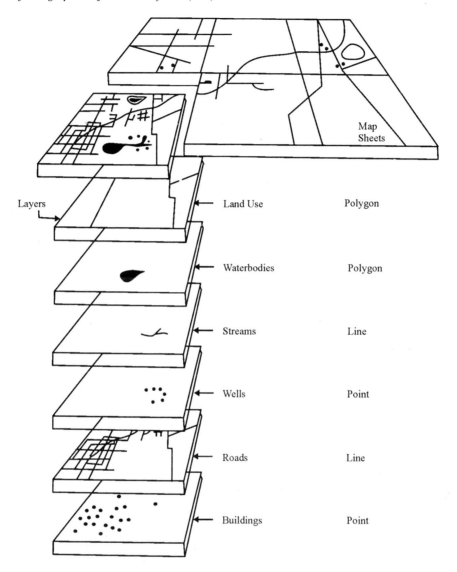

Fig. 10.8 : Presentation of saptial data. The diagram illustrates a set of geographic layers, each describing a theme of related data in terms of discrete objects

10.5 SPATIAL DATA AS COVERAGE

As mentioned earlier, there are three classes of spatial objects that can exist in two-dimensional space : points, lines and polygons. In ARC/INFO, these objects are referred to as feature classes and can be stored separately in a structure referred to as a coverage. A coverage is a collection of similar features that have been abstracted from the real world. There are three types of coverage which correspond to the fundamental spatial objects : point coverage, line coverage and polygon coverage. Each coverage is composed of a set of files – some store spatial data while others store attribute data. The table below (Table-10.1) highlights important files associated with point, line and polygon coverage. These files are used to store information about spatial objects – their locations and attributes.

Table 10.1 : Files associated with point, line and polygon coverage

Feature Class	Spatial Data	Attribute Data
Points	LAB-Point coordinates	PAT– Point Attribute Table
Lines	ARC- Arc coordinates	AAT- Arc Attribute Table
Polygon	PAL- Polygon Arc List	PAT- Polygon Attribute Table

A thematic layer may contain more than one coverage type. For example, the hydrology theme may include both lakes and streams. In terms of their storage in GIS, the lakes may be placed in a polygon coverage and the river net work in a line coverage. Alternately, both objects may be stored in the same coverage. The decision regarding layers and coverage are complex and are reflected in the spatial database design – a process dependent on factors associated with the use and maintenance of the data.

10.6 SPATIAL OBJECTS

In an automated environment, there must be a mechanism to locate and describe the geographic phenomena. In a GIS, geographic features are abstracted to become points, lines or polygons. In ARC/INFO, a point is an elemental unit, a line is made up of a string of points and a polygon is made up of a line or a series of lines.

10.6.1 Points

A point is a single coordinate (x,y) representing a geographic feature that is too small to be seen as an area. If it is an isolated object, such as a light house, it has no geometric property. It only has the property of location, which is defined by its coordinate. In an ARC/INFO database, a point is numbered and stored with a single coordinate (x,y) denoting its location and a series of attributes describing its characteristics (Fig.10.9). Each of a map feature is represented as a record in the database. A point has a coordinate location and attributes which are stored in the point attribute table.

10.6.2 Lines

Lines representing linear features are usually stored in the ARC/INFO database with descriptive attributes as well as a string of coordinates (x,y). Every line is stored as an arc and has two end points called nodes. Generally a line begins or ends where it meets another line. These kinds of connections are defined by arc-node topology. Each arc has a string of coordinates and attributes which are stored in the arc attribute table. These attributes include the geometric property of length and the topological properties of connectivity and adjacency (from-node, to-node) (Fig.10.10).

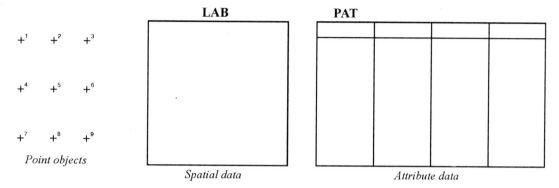

Fig. 10.9 : Representation of point object in GIS database (Point coverage)

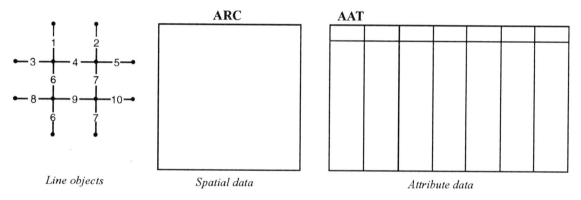

Fig. 10.10 : Representation of line objects in GIS database (Line coverage)

10.6.3 Polygons

In a polygon coverage we have all the three types of spatial objects. As before, the nodes define the ends of arcs, and the boundary of each polygon is defined by a list of connecting arcs. In order to assign descriptive attributes, each polygon must have a label point (a point feature within a polygon as opposed to a point feature by itself). Polygon attributes are stored in the polygon attribute table (Fig.10.11).

Polygons are identified by label points and their boundaries are defined by a series of connected arcs and nodes. Boundary arcs possess certain properties on either side (left polygon, right polygon) and at either end (from-node, to-node) (Fig.10.12).

All polygons have the geometric properties of area and perimeter and the topographic property of containment. These properties and relationships are stored in several related data files. There is one-to-one relationship between a polygon in a coverage and a record in the polygon attribute table. Because space is infinite, there will always be an outside or universal polygon, which defines the space extending beyond the area under study.

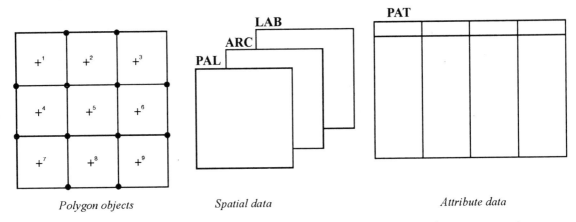

Fig. 10.11 : Representation of polygon objects in GIS database (Polygon coverage)

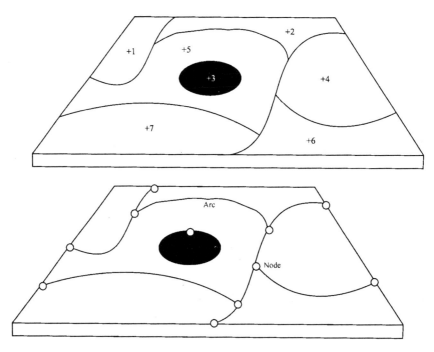

Fig. 10.12 : Identification of Polygons : Polygons are identified by label points and their boundaries are defined by a series of connected arcs and nodes

10.7 RELATIONSHIPS BETWEEN SPATIAL OBJECTS

While describing the location of geographic features, the most common means of doing so is by relating the feature to some other observable features. The type of relationships which may be used to describe locations are : to the right of, to the left of, at the end of, connected to, inside of, bounded by and next to. These are useful descriptions that are easy for us to interpret and understand, but not so to the computer.

Spatial objects are not independent entities, they exist within space and exhibit relationships to each other. The formal study of spatial relationships is the field of topology. After data have been entered as spatial objects, coverage topology is constructed. Constructing topology verifies data, identifies any errors, and establishes spatial relationships between the objects. Typically, we focus on three topologic properties : (1) adjacency, (2) connectivity and (3) containment (Fig.10.13).

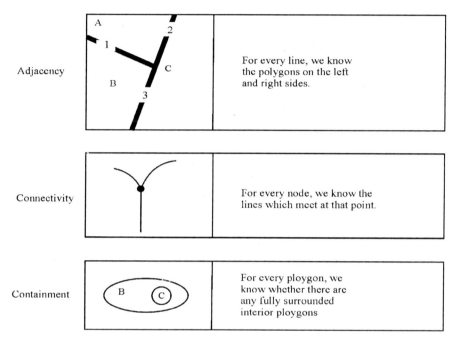

Fig. 10.13 : Basic elements of relationships between spatial objects

10.7.1 Adjacency

Adjacency means adjoining, the state of being next to. Adjacent polygons are two polygons that share a common boundary. Polygons, lines and nodes can have the property of adjacency. Consider the following examples :

Polygon-polygon

In Fig.10.14 one can find all areas adjacent to water bodies, i.e. find all polygons adjacent to a polygon type-A. This query will be translated into an operational definition- find all the polygons which share a common arc with a water body.

In Fig.10.15, if we assume a variety of land use polygons, then the question can be refined to find all forest stands adjacent to lakes, i.e. find all polygons of type-A adjacent to polygon type-B.

Polygon-polygon (Arc)

In Fig.10.16, instead of finding adjacent polygons, one can search for all arcs that separate adjacent polygons (the geographic duality of adjacency). The adjacency operation here is to

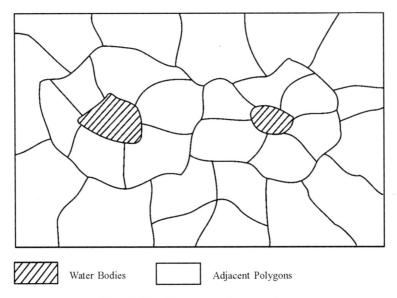

Fig. 10.14 : Polygon-polygon adjacency

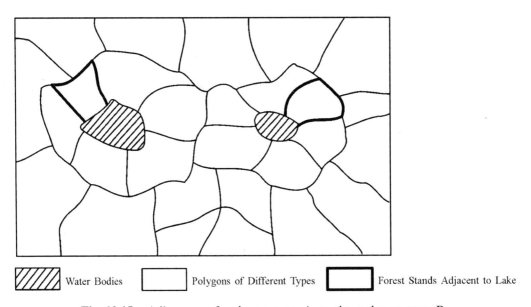

Fig. 10.15 : Adjacency of polygon type A to the polygon type B

find all lake shore adjacent to forest stands, i.e. all lines (arcs) that separate polygon type-A from polygon type-B.

Polygon-polygon (Node)

In Fig.10.17, another type of adjacency is represented by three or more polygons sharing a common point rather than a common boundary. Find points where water meets with pine stands

Principles of Geographic Information System (GIS)

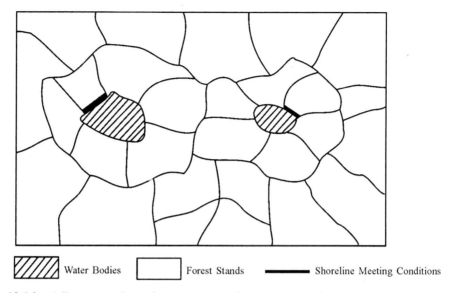

Fig. 10.16 : Adjacency : Arcs that separate polygon type A from polygon type B

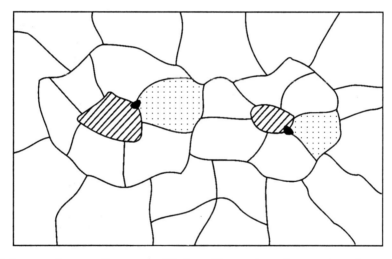

Fig. 10.17 : Polygon-polygon adjancency : Nodes adjacent to polygons type A, type B and type C

and open meadow, i.e. find all points (nodes) that are adjacent to polygon type-A, type-B and type-C. The coverage query, its geographical dual, would be to locate all polygons adjacent to a particular type of node.

Line-line

In Fig.10.18, it is also possible to investigate adjacency of lines. Identifying adjacent lines involves the examination of arcs and nodes that form a linear network. It is possible to ask a GIS to find all stream sections adjacent to sections with a flow volume greater than 100 cubic meters per second. In other words, find all lines adjacent to lines of type-A.

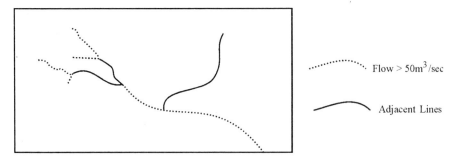

Fig. 10.18: Line-line adjancency : Lines adjacent to lines of type A.

Line-node

The geographic duality of adjacent lines is finding line segments that are adjacent to a node. For example, in Fig.10.19, one can find all stream sections adjacent to a confluence, i.e. find all lines that are adjacent to node-B.

Point-point

For points, adjacency is not applicable and is replaced by the concept of near and nearest neighbor.

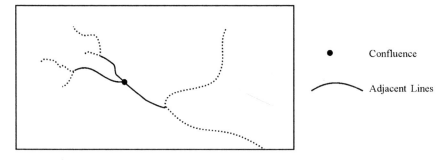

Fig. 10.19: Lines adjacent to a node (Line-Node Adjacency)

10.7.2 Connectivity

Given the layered view of reality, each property can be associated with a particular spatial object or type of space. The connectivity concept is most commonly considered a property in the line coverage environment, whereas the adjacency concept is primarily found in the polygon coverage environment. Containment refers to the "within" quality and is most identifiable with the polygon coverage.

The topological property of connectivity acts on a linear network. The task is to find out which arcs are connected to other arcs. The following examples illustrate this. Fig. 10.20 shows all power lines in the network serviced by a particular transformer. Not only are these lines connected to each other, they are also coded for the directionality. If the transformer explodes, we find that the service of all lines originating at or beyond node-A will be interrupted. In Fig.10.21 the shortest path between a child's school and his or her house is shown. It is to find the shortest route through the network from point-A to point-B.

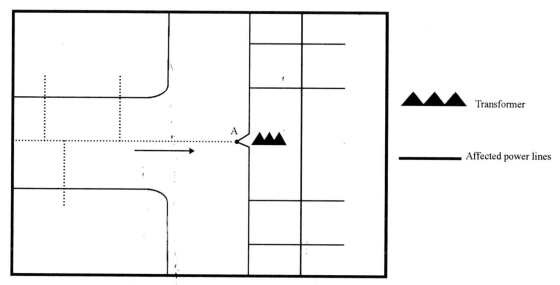

Fig. 10.20 : Power line network illustrating connectivity

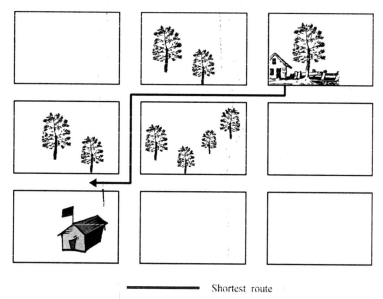

Fig. 10.21 : Connectivityt : Road Nework to find the shortest route between a child's school and his/her home

In the first example, the requirement is to trace through a directional network and find all segments that meet the particular condition (e.g. located beyond a particular node). The second example is seeking an optimum solution through the network, that is, the segments that make up the shortest route. In either case, the topological data structure enables the search from one arc to the next through an intervening node. The target for connectivity can be either a node or an arc.

The concept of connectivity between polygons can be applied to those problems where the objective is to find the shortest path through a landscape composed of different types of land cover (i.e. forest, shrubs, swamp etc.). Each kind of cover presents a different degree of resistance to travel. In such cases to move from polygon-A to polygon-B, the search would consider the connectedness of areas and the path of least resistance, plus directionality.

10.7.3 Containment

Polygon-Polygon

Containment addresses query about the "within" condition. For example, one may ask whether a polygon in Fig. 10.22 is completely within a second polygon? The classic example is to find all the islands within the lake.

By definition, islands are completely surrounded by water body. The arc that separates the land polygon from the water polygon has a common node (the from-node is the same as the to-node). There are, however, numerous more complex examples. Imagine two types of forest stands surrounded by a swamp. From a forester's perspective, the two stands are inaccessible because of the swamp land, and yet there is no longer a single arc separating the land from the surrounding water features. Additional examples could include a complex series of nested polygons.

In previous examples, we have been considering the geographic dual. Applied here, the query could have been worded in one of the two ways :

    Identify islands of trees within a swamp, or

    Identify all swamps containing islands of trees.

Both questions will result in the same answer (i.e. find swamps with treed islands).

In ARC/INFO, polygon containment relationships are stored in the polygon arc list (PAL) file. These are necessary for accurate area measurement of polygons that contain other polygons.

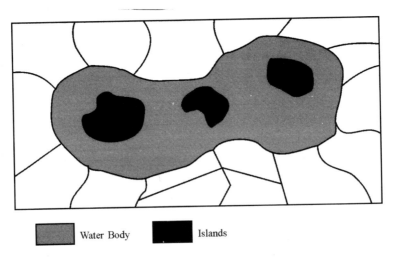

Fig. 10.22 : Containment : Polygon within a Polygon

Principles of Geographic Information System (GIS)

Line-line

It is possible to apply the concept of containment to lines. In the illustration presented in Fig. 10.23, should we regard line segment B to be within the line composed of segments A,B,C ?

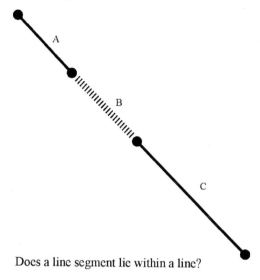

Does a line segment lie within a line?

Fig. 10.23 : Line-line containment

Point-Line, Point/Line-Polygon

In developing an understanding of the containment property, we return to the concept of layers and the objective of simplifying reality. Each layer is identified on the basis of its spatial objects and their thematic attributes. Containment addresses the question of objects lying within other objects. Since we have three object types, there are nine possible pair relationships.

To answer questions about lines within polygons, or points within polygons, we go back to the multiple layer data model. If the data in question reside in separate layers, the data layers must first be recombined before the containment property can be addressed. This will let us ask question such as "which lines lie within an area ?" or "which points lie within an area?" This process is known as overlay analysis (Fig.10.24).

The topological relationships that ARC/INFO stores have been selected for their importance in overlay analysis, spatial measurement and commonly required queries on data stored in the database. For example, point-line adjacency, point-line containment and line-line containment are not specifically stored. Instead, a variety of different operators exist in ARC/INFO which let us discover if these relationships exist.

10.8 GIS FUNCTIONS

10.8.1 Data Input Functions

These functions are required to convert the data from the existing form to the form that is suitable for use in the GIS. Georeferenced data are normally obtained from paper maps, table

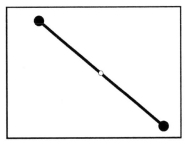
Does a point lie within a line?

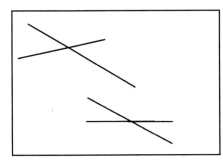
Do these lines lie within an area?

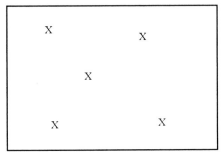
Do these points lie within an area?

Fig. 10.24 : Containment of a point within a line, and points and lines within a polygon.

of attributes and associated attribute data, air photos and satellite imageries. Fig. 10.25 highlights the process of converting the conventional data to the GIS database. Paper maps convey locations by graphic symbols (points, lines and polygons) and the attributes by color, symbol and pattern. But the GIS conveys locations by graphic symbols (points, lines and polygons), establishes and retains spatial relationships mathematically following the logic of topology.

10.8.2 Data Management Functions

These functions include those needed for the storage and retrieval of data from the GIS database. The GIS has the capability to read the data in a flexible and logical manner, to search and identify specific items or attributes, and to display these information in a spatial context. While GIS possesses many other capabilities that contribute to the functionality of the system as a whole, the query, retrieval and display functions specially give the system its practical value and superiority over other methods of handling multiple data set problems.

10.8.3 Data Manipulation and Analysis Functions

These functions determine the information that can be generated by the GIS.

Manipulation of Spatial Data

Often it becomes necessary to transform some of the original spatial data sets into a geometry which is better manageable, accurate and consistent with the other data sets already present or to be encoded in the system. Some examples of common data manipulations are presented in Fig. 10.26. These transformations are needed since the available data sets are rarely consistent in scale, projection, spatial accuracy, orientation and coverage.

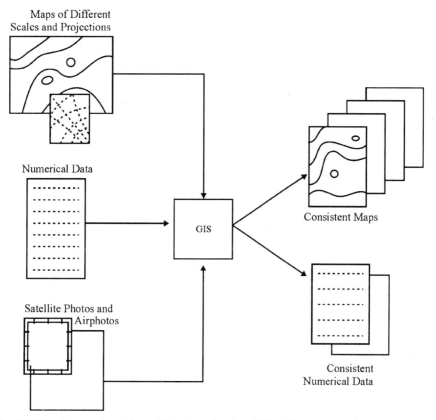

Fig. 10.25: Integration of satellite data in the GIS with maps and attribute data.

Map Overlays

Map overlay function creates new map layers with an existing one. In this process features of each coverage are intersected to create new output features. Attributes from the input layers are also combined in the overlay process to describe new output features. Fig. 10.27 and 10.28 illustrate the map overlay process in which the dominant mathematical operations are intersection and union.

Map Dissolve

Map dissolve function deletes the boundaries between adjacent polygons having the same attribute values for a specified feature. Dissolve operation also retains only the required features by clipping the unwanted polygons from the map which is also illustrated in Fig. 10.27.

Buffers

These are polygons created around points, lines and polygons (either within or without) which are illustrated in Fig. 10.29. Buffering function finds applications in zoning and impact assessment. For example, a 10 kilometer buffer around the site of a construction aggregate (representing a point buffer) may represent the economic and environmentally sound transport zone. A 200 meter line buffer along the river course may represent the protected zone where cutting trees

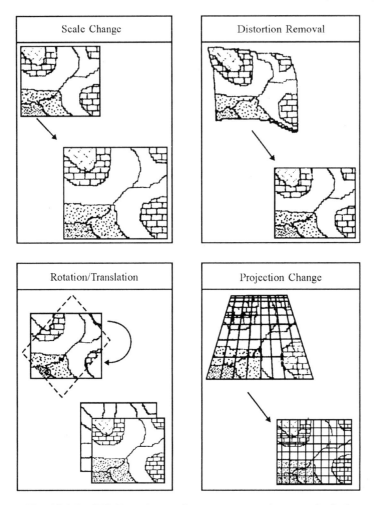

Fig. 10.26 : GIS Functions : Some common data manipulations

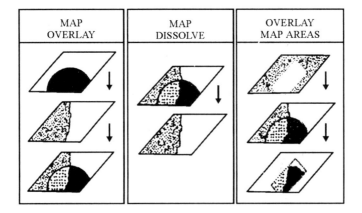

Fig. 10.27 : GIS Functions : Map overlay and dissolve

Principles of Geographic Information System (GIS)

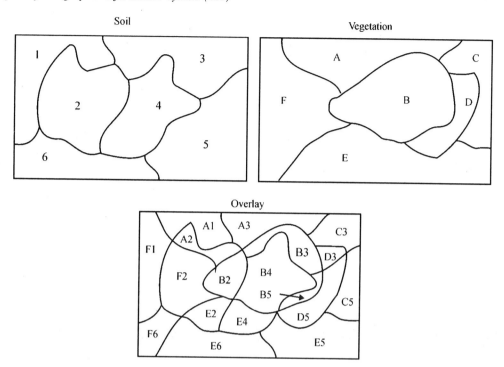

Fig. 10.28 : Overlay of two complex soil and vegetation maps

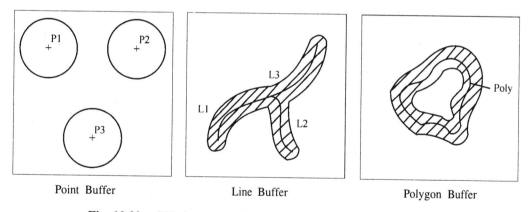

Fig. 10.29 : GIS functions : Buffers around points, lines and polygons

should be prohibited as an erosion control measure. Similarly a polygon buffer representing 5-10 minutes walking distance from a proposed city park is an useful planning application.

10.8.4 Data Output Functions

The GIS output or reporting functions are more concerned with the quality, accuracy and ease of operation rather than the capabilities available in the system. The outputs (reports) may be in the form of CRT display, maps, listings, data files or text in hard copy and so on. CRT display during interactive data processing and map development is an important operational requirement. Paper map output helps in practical field use, distribution, convention and compatibility. Listings are required for supporting lists or statistical tables for reporting purposes. Similarly digital data file storage and transfer using tapes, disks and telecommunications media has presently assumed an important archive output requirement. These various GIS outputs are presented in Fig. 10.30.

Thus far we have only touched upon the basic principles of GIS to give the reader a feeling on what it is, how it captures and manipulates the spatial data and what sorts of problems it can handle for us. Within this brief introduction, we would also like to give you an idea on how remote sensing and geographic information system can be linked together to derive optimum benefit out of both.

Fig. 10.30 : GIS outputs

10.9 LINKAGE BETWEEN REMOTE SENSING AND GIS

At present there is not an iota of doubt about the unique status maintained by both remote sensing and geographic information system in solving large number of complex problems of the society at operational levels. While geographic information system is considered as a powerful tool for the management and analysis of spatial data, the remote sensing system has demonstrated to be a powerful tool for the collection and classification of the spatial data. Thus both the systems are found to be complementary to each other. With the development in satellite sensors of very high resolution, of the order of a meter (for example, the Ikonos data), the compatibility of the geographic information system and the remote sensing system has become stronger day by day to tackle practical problems of any desired scale.

10.9.1 Remote Sensing Inputs to Geographic Information System

Geographic information system requires accurate, digital, polygonal, point or network data sets as its inputs to be processed further depending on the problems. For the remote sensing system, on the other hand, the digital remote sensing scene represents the raw data. The classified remote sensing output, expressed in polygons or other forms, is considered to be highly processed data products. For satisfying the requirements of GIS, these cellular, classified image data have to be transformed into a standardized set of polygons presented in the vector format typical of the digitizer output. Once presented in this form, the output of the remote sensing system becomes acceptable by the GIS as layers of spatial information to be processed by its input phase.

10.9.2 GIS Input to Remote Sensing System

Remote sensing data analysis requires collateral/ancillary data for development/improvement of its classification schemes. These valuable information are usually available in the GIS in usable form for the remote sensing system. Thus substantial improvements in the accuracy of classification of remote sensing imagery can be made using ancillary data on terrain topography, soils, climate and so on. It is now clear that not only the reverse flow of data (ancillary) from GIS to remote sensing system improves the classification of remote sensing imagery, it in turn makes the already improved remote sensing inputs more usable by the GIS.

10.10 IMAGE BASED INFORMATION SYSTEM

In principle, it is found that there is no difficulty to transfer pre-processed digital files from a remote sensing system to the GIS. And in practice, there is no real technical problem in the transformation of the classified pixel-based remote sensing scenes into a set of homogenous polygons defined in a vector data structure. These realizations gave rise to a hybrid image-based information system (IBIS) first conceptualized and developed by the Jet Propulsion Laboratory of California Institute of Technology. Image-based information system has the twin capabilities of an image processing system and a geographic information system. With the in-built hardware and software to efficiently handle the data on real complex systems, the image-based information system is also unique in the sense that it is the only raster-based geographic information system of any size ever produced.

Suggestions for Supplementary Reading

These suggestions are not exhaustive but are limited to the works that will be found especially useful for developing interest and creative imagination in the field of Remote Sensing.

Books

American Society of Photogrammetry, 1983. Manual of Remote Sensing, Vol. I, 2nd ed., R. N. Colwell (Ed.), Falls Church, Va.

Artsybashev, E. S. 1983. Forest Fires and Their Control. Oxonian, New Delhi (Original Russian, 1974).

Ball, Desmond and Babbage Ross, 1989. Geographic Information Systems : Defence Applications. Brassey's Australia - Pergamon Press.

Brown, A. A. and K. P. Davis, 1973. Forest Fire : Control and Use. McGraw-Hill, New York.

Burrough, P. A. 1986. Principles of Geographical Information Systems for Land Resource Assessment. Clarendon Press, Oxford.

Gautam, N. C. 1993. Fundamentals of Geographic Information System. Pink Publishing House, Subhash Nagar, Mathura.

Jenny, Hans, 1980. The Soil Resource : Origin and Behavior. Springer-Verlag, New York.

Karale, R. L. (Ed.), 1992. Natural Resources Management – A New Perspective. NNRMS, ISRO, Department of Space, Government of India, Bangalore.

McArthur, R. H. and E. O. Wilson, 1967. The Theory of Island Biogeography. Princeton University Press, Princeton, New Jersey.

McHarg, I. 1969. Design with Nature. The Natural History Press, Garden City, New York.

MacLean, Ann L. (Ed.), 1994. Remote Sensing and Geographic information Systems : An Integration of Technologies for Resource Management - A Compendium. American Society of Photogrammetry and Remote Sensing.

Raper, J. (Ed.), 1989. Three Dimensional Applications in Geographical Information Systems. Taylor & Francis, London, New York, Philadelphia.

Samet, H. 1989. The Design and Analysis of Spatial Data Structures. Addition-Wesley Publishing Company, Inc.

Star, J. and J. Estes, 1990. Geographic Information Systems : An Introduction. Prentice-Hall, Englewood Cliffs, New Jersey.

Swain, P. H. and S. M. Davis, 1978. Remote Sensing : The Quantitative Approach. McGraw-Hill, New York.

Research Papers

Archibald, P. D. 1987. GIS and Remote Sensing Data Integration. Geocarto International 3 : 67 – 73.

Burgan, R. E. and M. B. Shasby, 1984. Mapping Broad-area Fire Potential from Digital Fuel, Terrain and Weather Data. J. Forestry, 82 : 228 – 231.

Chrisman, N. R. 1987. Design of Geographic Information Systems Based on Social and Cultural Goals. Photogramm. Eng. Remote Sensing, 53 : 1367 – 1370.

Deuker, K. J. 1979. Land Resource Information Systems : A Review of Fifteen YearsExperience. Geoprocessing 1 : 105 – 128.

Ehlers, M. 1990. Remote Sensing and Geographic Information Systems : Towards Integrated Spatial Information Processing. IEEE Trans. Geosci. Remote Sensing 28 : 763 – 766.

Ehlers, M., G. Edwards and Y. Bedard, 1989. Integration of Remote Sensing with Geographic Information Systems : A Necessary Evolution. Photogramm. Eng. Remote Sensing, 55 : 1619 – 1627.

Iverson, L. R. and P. G. Risser, 1987. Analyzing Long Term Changes in Vegetation with Geographic Information System and Remotely Sensed Data. Adv. Space Res. 7 : 183 – 194.

Johnston, K. 1987. Natural Resource Modeling in the Geographic Information System Environment. Photogramm. Eng. Remote Sensing 53 : 1411 – 1415.

Junkin, B. G. 1982. Development of Three-Dimensional Spatial Displays using a Geographical based Information System. Photogramm. Eng. Remote Sensing, 48 : 577 – 586.

Marble, Duane F. and Donna J. Peuquet, 1983. Geographic Information Systems and Remote Sensing. In : Manual of remote Sensing, Vol. I, 2nd ed., R. N. Colwell (Ed.). American Society of Photogrammetry, Falls Church Va. p.923 – 958.

Meaille, R. and L. Wald, 1989. A Geographical Information System for Some Mediterranean Benthics Communities. Int. J. Geogra. Information Systems, 4 : 77 – 86.

Nellis, M. D., K. Lulla and J. Jensen, 1990. Interfacing Geographic Information Systems and Remote Sensing for Rural Land-Use Analysis. Photogramm. Eng. Remote Sensing, 56 329 – 331.

Radhakrishnan, K., S. Adiga, Geeta Varadan and P. G. Diwakar, 1996. Enhanced Geographic Information System Application using IRS-1C Data : Potential for Urban Utility Mapping and Modelling. Current Science (Special Issue), 70 (7) : 629 – 637.

Shelton, R. C. and J. E. Estes, 1981. Remote Sensing and Geographical Information Systems : An Unrealized Potential. Geoprocessing, 1 : 395 – 420.

Steinitz, C. and H. J. Brown, 1981. A Computer Modelling Approach to Managing Urban Expansion. Geo-Processing 1 : 341 – 375.

Williams, D., R. Nelson and C. Dottavio, 1985. A Georeferenced Landsat Digital Database for Forest Insect Damage Assessment. Int. J. Remote Sensing 6 : 643 – 656.

Application Areas of Remote Sensing and Geographic Information System $\boxed{11}$

There can be almost endless applications of the powerful remote sensing and geographic information techniques to tackle problems related to land surface, sea surface and atmospheric features and processes. However, within the scope of the present book, it is decided to highlight some problems on typical application areas with the necessary hints on methodology, leaving the users to work out themselves the rest to satisfy their curiosity and to feel confident on tackling such problems.

11.1 AGRICULTURAL APPLICATIONS

11.1.1 Crop Area Estimation

Assuming a favorable weather and non-limiting agricultural input conditions, crop area, at district, state and country levels, becomes the first and foremost information required to assess the crop production during the season. So to estimate the area covered by the major crops of the season, say for a district, one has to procure satellite remote sensing data for at least three phenological phases of the crops (establishment phase, maximum vegetative phase and active reproductive phase). GPS-based ground truth collection should also be done during these three satellite pass periods. At least 10% of the study area should be covered to collect the ground truths of crop fields. This helps in the accuracy of crop classification and correspondingly in the crop area estimation.

11.1.2 Crop Stress Monitoring

Adverse effects of weather, attacks by pests and diseases, lack of appropriate agricultural inputs and improper crop management practices result in crop stress. Crop stress in large areas can be detected by periodic monitoring the crop health index like NDVI from satellite remote sensing imageries. Once crop stress symptoms are found in substantial areas on the imageries, the exact type of stress can be ascertained by actual field visits and remedial measures are suggested for recovering the crops from their respective stresses. This monitoring work helps in stabilizing crop production in the concerned stress affected areas.

11.1.3 Crop Production Forecasting

This is extremely important for the national level policy planning. For this purpose, the crop production forecast of the state/nation should be made available to the policy planners at least one to one-and half months before the harvest, that is at the active reproductive phase of the crop. In order to generate this information, pixel-based crop health index (NDVI) imageries are to be transferred from remote sensing image analysis system to the geographic information system in raster or vector format. Collateral information on GIS should include land and soil characteristics, weather parameters during the season and availability of agricultural inputs like fertilizer and irrigation. With these data a suitable model, validated from similar historical data along with the corresponding crop yield figures, can be run to forecast crop production during the season.

11.2 WATER RESOURCES RELATED APPLICATIONS

11.2.1 Assessment of Land-surface Water

Land-surface water refers to water available in ponds, inland lakes, wetlands, streams, rivers and river dams including the ones used for hydroelectric power generation. These fresh water sources supply water for irrigation, industry and domestic consumption. Estimation of land-surface water in volume unit has two observable parameters, namely the areal spreads of water surface and the corresponding depths of water at incremental steps of 10cm or so. Remote sensing technique is ideal for estimating the areal spread of the water bodies using visible, near infrared and microwave bands. Use of radar (active microwave) remote sensing becomes essential during rainy season as microwave is transparent to thick clouds and has the day and night operational capability. Thus in rainy days, active flood plane mapping can be easily done using radar remote sensing data. Volume of water in the water body can be directly estimated by remote sensing technique using the reservoir's topographic data of the driest period using the scanning altimeter, as it was done by the Seasat SAR for the ocean surface. However, in absence of altimeter data from remote sensing satellites, the incremental depths of the water reservoir should be collected (ground truth) with the corresponding areal spread of water surface which will be helpful in computing the water volume. The altimeter data and/or water depth versus its areal spread data are to be transferred to GIS so that they can be used for several other applications. For example, from these data in conjunction with collateral data on rainfall in the catchment area, one can find out the amount of siltation (loss of capacity) of the dams/reservoirs in a certain time interval (years). This can be further confirmed from the measurements of turbidity of the reservoir water during different times of the year, particularly during the rainy seasons.

11.2.2 Assessment of Subsurface Water

Honestly speaking, subsurface water can not be directly assessed by remote sensing techniques with the observation of spectral signatures. However, geologists try to infer indirectly the subsurface hydrological conditions by monitoring some of the relevant surface indicators through visible, near-infrared and thermal infrared spectral bands. These surface indicators include geological/landscape features such as shallow sands/gravel and the rock types/structures, soils/soil moisture anomalies, stream flow characteristics, discontinuous ice cover on streams, differential snow melts, springs, vegetation types, vegetation distribution and so on. The interpretation keys like the characteristic shape, pattern, tone and texure of the surface features are usually considered to infer shallow aquifers under sand and gravel dominated areas with reasonable confidence. Surface soil moisture distribution may be exploited to infer the subsoil water using remote sensing technique in the thermal infrared channel. In this connection it is found that the radiometric temperature has a predictable relationship with the depth of shallow aquifers. Usually whenever a subsurface thermal anomaly exists indicating a shallow water table, then a surface temperature difference is manifested. For the detection of these shallow aquifers thermal imagery under calm (low wind speed) pre-dawn condition is found to be ideal. Monitoring soil thermal inertia/apparent soil thermal inertia is also a profitable proposition for the search of shallow aquifers.

Regarding the detection of heterogeneous aquifers, one has to search the satellite imageries for lineaments with fractures linked with topographic depressions, anomaly on soil tones and vegetation on a particular land surface.

In all subsurface water resource monitoring/mapping studies it is mandatory to take into consideration the ground truth information collected by various conventional ground water exploration techniques before bringing out the final ground water map. The validated ground water resource map obtained through remote sensing technique can be transferred and stored in the GIS for further utilization/modeling work.

At this stage it is desirable to indicate that for all aquifer mapping studies, manual interpretation of the FCCs are preferable, although high resolution remote sensing data can also be used in the digital image processing and classification mode. Time plays a decisive role in the success of subsurface water resource monitoring. For this, one has to take advantage of the best time for obtaining the satellite imagery of the survey site. The most opportune times are those that correspond to low sun-elevation angle, maximum area of bare soil, soil moisture pattern, drainage pattern, snow-melt pattern, lithologies in glaciated terrain, native vegetation density/pattern, area beneath deciduous vegetation, desert vegetation, irrigated crops versus dry-land farming, small lakes and ponds.

11.3 WEATHER AND CLIMATE RELATED APPLICATIONS

Under this category, we have handpicked only a few typical land-surface processes which can be profitably studied by remote sensing and GIS techniques. These are the processes of desertification, deforestation, urbanization and land use/land cover changes.

11.3.1 Desertification

According to the UN Conference on Environment and Development, Agenda-21 (Earth Summit,1992), desertification is defined as "land degradation in arid, semi-arid and dry sub-humid regions resulting from various factors, including climatic variations and human activities". Aridity of a region is characterized by the ratio of the mean annual precipitation (P) to the mean annual potential evapotranspiration (PET), using modified Thornthwaite formula. As per this index, the climatic zones are classified as given below (Table 11.1) :

Table 11.1 : Climatic classification on the basis of Aridity Index

Climatic Zone	P/ (PET)	% of World covered
Hyper-arid	< 0.05	7.5
Arid	0.05 – 0.20	12.5
Semi-arid	0.21 – 0.50	17.5
Dry sub-humid	0.51 – 0.65	9.9
Humid	> 0.65	39.2
Cold	> 0.65	13.4

Thus it is found that 39.9% of the world land area is subjected to varying degrees of desertification where a sustained decline in yield of useful crops takes place due to environmental changes. Among the land subjected to desertification, the arid areas receive insufficient rainfall (< 200 mm) to sustain a good agricultural crop, but it can support sufficient vegetation for pastoral activity. Semi-arid regions get 200-600 mm rainfall where drought resistant crops can be grown. Dry sub-humid areas receive an annual rainfall up to 800 mm which supports a number of agricultural crops, but they are still susceptible to desertification due to climate and bio-geophysical degradation processes.

Although at regional scale, satellite remote sensing offers the most cost effective technique for identification, monitoring and mapping of the desertification process, yet to-date no satisfactory operational remote sensing program has been developed for this purpose. As we know, desertification monitoring work aims at generating information on the sustained decline of productivity of useful crops from a dry area. This complex information can only be computed through an ingenious modeling technique. The variables on which the output of this model depends may be carefully chosen from the interdependent parameters such as insufficient rainfall, decrease in relative humidity, prevailing drought condition, decrease in vegetative cover, overgrazing areas, increase in fallow land, increase in albedo and so on. To start with, the dry region should be divided into a number of 1 x 1Km grids for the ease of surveillance and ground truth collection on the driving parameters of the desertification process. This useful databank together with the classified satellite data, all at periodic intervals, when stored in GIS format, helps one to prioritize areas on the basis of their degrees and rates of desertification. In fact, this forms the basis of regional planning to combat desertification. Suitable combative measures designed to arrest and reverse the desertification process can accordingly be demonstrated at specific sites through pilot projects which can be managed by governmental agencies or by the communities themselves.

11.3.2 Deforestation

Forest cover influences its local climate. It also helps to check soil erosion leading to sustainability in agricultural productivity. Thus deforestation monitoring and forest resources assessment at periodic intervals are very much needed at regional and hemispherical levels for planning appropriate strategies to arrest any likely future climate change.

This work can best be done using satellite remote sensing technique coupled with GIS. From high resolution satellite imageries one can not only estimate the total forested area, but can also find out the open and close forest areas. Data from periodic classified satellite imageries can show us whether the forest area is decreasing with time. These classified image data should be transferred to GIS where other collateral data of the study area are also kept in different layers. Finally the analysis of GIS data sets will indicate not only the degree of deforestation but also indicate the leading causes for such deforestation processes. Thus planning for arresting deforestation processes can be made on sound scientific footings.

11.3.3 Urbanization

Like deforestation, urbanization and industrialization also influence the local weather and climate. Urbanized areas such as metropolishes and megapolishes present themselves as heat islands.

Thermal anomalies and surface roughness anomalies between urban and surrounding areas can be easily monitored and quantified by high resolution satellite remote sensing using visible and thermal infrared (VHRR) sensors. Classified satellite image data can be transferred to GIS where ground-based data including meteorological data are stored for the purpose of modeling and policy planning. Seasonal variation of urban heat island can also be studied similarly using remote sensing and GIS techniques.

11.3.4 Land Use/Land Cover Changes

By land cover we normally mean the earth's surface features which are not affected by human activities. However, the land cover does change by natural forces like those associated with weather and climate, seismic and volcanic activities, glaciologic activities and so on. Land use, on the other hand, refers to changes on the earth's surface features brought about by the intense human activities like agriculture, urbanization, industrialization, hydroelectric and thermal power generation, transportation network activity and so on.

With the availability of satellite remote sensing data, land use/land cover classification system has been well developed by now. National Remote Sensing Agency, Department of Space, Government of India has come up with two levels of land use/land cover classification schemes. Six classes of land use/land cover are included in the Level-I classification, while in Level-II classification, each of the afore-mentioned six classes are further classified to generate altogether twenty two classes of distinguishable land use/land cover. These are given in the table (Table-11.2).

For land-use/land-cover classification, as usual, satellite remote sensing data of the study area, acquired by the multi-spectral thematic mappers (of Landsat, IRS, SPOT and so on) are procured and processed with the help of the image processor using the required softwares like ERDAS, EASI/PACE and so on. The enhanced images are then classified as per the above mentioned classification scheme using GPS-based ground truth data. The classified map is then transferred to GIS which will occupy several layers in the database, each layer being on a specific class.

Table 11.2. : Land-use/Land-cover Classification System

Level-I	Level-II
1. Built-up Land	1.1 Built-up Land
2. Agricultural Land	2.1 Crop Land
	(i) kharif
	(ii) rabi
	(iii) summer
	2.2 Plantation
	2.3 Fallow
3. Forest	3.1 Evergreen/Semi-evergreen forest
	3.2 Deciduous forest
	3.3 Degraded or scrub land
	3.4 Forest blank
	3.5 Forest plantation
	3.6 Mangrove
4. Wastelands	4.1 Salt affected land
	4.2 Waterlogged land
	4.3 Marshy/Swampy land
	4.4 Gullied/Ravinous land
	4.5 Land with or without scrub
	4.6 Sandy area (coastal & Desertic)
	4.7 Barren Rocky/Stony waste/ Sheet rock area
5. Water Bodies	5.1 River/Stream
	5.2 Lake/Reservoir/Tank/Canal
6. Others	6.1 Shifting cultivation
	6.2 Grassland/Grazing land
	6.3 Snow covered/Glacial area

To study land-use/land-cover changes, it is required to take sequential (once in a few years) remote sensing data on the study area and transferring the classified land-use/land-cover information to GIS. Now by overlapping the sequential classified maps on GIS, one can detect and estimate not only the change in the land-use/land-cover pattern with time, but can also compute the rate of land-use/land-cover change in the different classified areas.

11.4 FOREST AND RANGELAND RELATED APPLICATIONS

Here only two typical applications are briefly described. They are (a) forest fire hazard zone mapping and (b) rangeland resources monitoring.

11.4.1 Forest Fire Hazard Zone Mapping

Forest fires cause extensive damage to our natural environmental resource base. It brings enormous economic loss to a state/nation.

High resolution satellite remote sensing offers one of the best techniques for forest fire detection, mapping and damage assessment in almost near real time and cost effective manner. Remote sensing coupled with GIS can also be effectively used for forest fire hazard zone mapping. In this situation, it is possible to combine the data of a number of relevant variables to run a model to forecast fire hazard zones which can be validated by ground truth data both on the forest and on the fire history of the existing destroyed forest zones. This type of study forms a crucial step in the forest management, particularly for forest fire prevention and suppression.

Two essential components of the forest fire are the source of fire and the availability of fuel sources for burning. As regards the source of fire, it may be a natural source like lightning strike or fire by friction of dry plant parts (like dry bamboo stems) during their violent swings under windy conditions. Forest fire can also be human induced. The glaring example of this type of forest damage is for shifting cultivation by the tribal people living in the forests.

The spot of forest fire can be detected by high resolution satellite imageries using infrared scanners (AVHRR detectors). Remote sensing technique can also be profitably exploited for quick evaluation and mapping of the damaged area through classification (supervised and unsupervised) and change detection imageries.

As regards the modeling work for forest fire hazard zone mapping is concerned, the data on the variables are kept on layers in the GIS. These variables are ranked on the basis of their contribution to the forest fire damage. Accordingly the major variables are ranked in the order : vegetation, slope, aspect, proximity to road/rail and elevation. Then these variables are assigned suitable weightage factors (g) out of 256 (0 to 255) depending on their positive contribution to the forest fire. After this is done, each of these variables is further sub-classified as of high, medium and low contributions towards forest fire and assigned coefficients (k) of 0, 1 and 2 respectively. The data of each of these variables form separate layers in GIS for modeling purposes.

Fuel class (inflammability) oriented vegetation mapping is done by high resolution satellite remote sensing technique from the combined supervised and unsupervised classification of satellite imageries using visible and near-infrared sensors. These imageries also provide data on proximity to road/rail from the location of forest fire hazard. Slope and aspect of the fire hazard zone can be obtained from the digital elevation model (DEM) worked out from remote sensing/GIS software using available elevation data from existing topographic sheets. Elevation of the hazard zone can also be obtained by the microwave altimeter scanning from satellite platforms or by ground measurements using GPS. In addition to the data on the above mentioned variables, some important collateral data on weather (particularly on insolation), water availability and fire history are also to be collected and kept in separate layers of GIS for taking useful decisions.

A logical model can now be run to compute the fire hazard ranking index (H) of a location using the above mentioned parameters :

$$H = 1 + g_v k_v + g_s k_s + g_a k_a + g_r k_r + g_e k_e \qquad (10.1)$$

where the subscripts v, s, a, r and e refer to vegetation, slope, aspect, roads and elevation respectively. Water and urban land are assigned the fire hazard index zero and hence they are dropped from the model and 1 is added to the equation to avoid pixels having zero values. The final ranking index value higher than 255 is reduced to 255 to maintain the output image in a 8-bit range.

11.4.2 Rangeland Resource Monitoring

In this work we are expected to monitor the changes of the rangeland resource base for evaluating its impact on environment, livestock and wildlife. Essentially this problem has two aspects – monitoring ecological range conditions and seasonal range conditions. For this purpose the base line information about the current range condition (or of any past date from which high resolution remote sensing data are available) are to be obtained and data layers are to be created in GIS. Subsequently similar appropriate observations and measurements are to be made to find out the changes in the rangeland conditions. Both remote sensing and GIS play crucial roles in this important study.

Monitoring Ecological Range Conditions

This part of the work requires :

(i) identification of plant species which can be labeled as increaser, decreaser or invader depending on its response to changing local environmental conditions,

(ii) measurement of foliar cover (open/close canopy), plant density and plant height,

(iii) quantification of the changes in the above mentioned plant parameters at intervals of one year or more,

(iv) detection and quantification of the surface soil characteristics like soil moisture, and

(v) to discriminate between changes in soil and vegetation that are induced by management, seasonal conditions and heavy grazing.

Remote sensing and GIS techniques play the major role in this work. While high resolution satellite remote sensing data in visible and near infrared bands help in identifying and classifying plant communities from their characteristic spectral signatures and assessing crop stress through spectral indices (simple ratio/NDVI), the thermal infrared band helps to assess soil and canopy temperatures. Radar remote sensing helps in estimating soil moisture changes of the rangeland. Time sequence classified remote sensing data are to be transferred to GIS layers to detect changes in different rangeland features with time. As usual the success of remote sensing classification depends on a quantitative ground truth information on the rangeland. Meteorological data form a set of important ancillary data layers in GIS for assessing the changes in the rangeland conditions.

Monitoring Seasonal Rangeland Conditions

This part of the work demands quantitative seasonal data generation on :

(i) plant development and forage production in the range, ranch and grassland, and forage quality,

(ii) location of available forage, and

(iii) the degree of forage utilization by livestock and wildlife.

Remote sensing and GIS techniques also play a major role in this work. While satellite remote sensing technique helps in the quantitative estimation of the required data mentioned in the above list, GIS plays the role of storage of these data in layers, including the soils and meteorological data, and in data manipulation during model execution and decission making processes.

For estimating the phenological development of forage in the range, remote sensing observations are taken at the outset of plant growth, at peak green foliage development and at the termination of growth period, and images based on spectral indices such as simple ratio (IR/R), NDVI and transformed vegetation index (TVI) are developed, since these indices indicate vegetation vigour. Range biomass can then be assessed as proportional to the product of vegetation vigour and the corresponding vegetation cover. The constant of proportionality has to be evaluated from ground data on the range. The range biomass is taken to be reliable if its value ranges between 500 to 4,000 Kg/ha. Lower values are not taken as useful because of the dominance of soil reflectance under this condition. For evaluating seasonal changes, one can develop change detection images with respect to a normal season data at the above mentioned three phenological stages and then rate the change as normal, below normal or above normal. Thus operational potential of remote sensing and GIS techniques for applications in appropriate rangeland environments can be harnessed profitably.

11.5 ENGINEERING APPLICATIONS

Under this category, we have chosen only two applications for highlighting – one on dam site selection and the other on river crossing site selection for bridges leaving many others for the readers to survey in the literature.

11.5.1 Dam Site Selection

For dam site selection the chief determinants are :

(i) economic factors,

(ii) social considerations,

(iii) engineering factors (namely topographic, hydrological, geotechnical and engineering geology), and

(iv) environmental problems and opportunities.

Some considerations on sites also depend on the size and type of the dam, namely earth dam, concrete gravity dam, arch dam and so on. Examination of the site with respect to its proneness to natural hazards like earthquake and volcanic eruption etc. is considered in all dam types as these are essential in the final decision making process.

For engineering site planning, the regional terrain database maps are first prepared and then several derivative maps are made covering site and route selection, foundation quality for light construction, waste disposal, effluent irrigation, ground water supplies, sources of construction materials and natural hazard prone lands. The scale of the maps may vary from 1:50,000 to 1:8,000.

The entire database is digitized and kept in layers in GIS format. High resolution satellite data (stereoscopic viewing) in VIS/NIR and thermal IR channels and radar imageries are analyzed, augmented by extensive field investigation (ground truthing) to raise the database on engineering factors. These data are also transferred to GIS for further data manipulation.

It is always advisable to undertake environmental impact studies before the final decision on the dam site selection is approved and green signal is given for its execution.

11.5.2 River Crossing Site Selection for Bridges

The broad determinants of the river crossing site selection for bridges are the same as those for the dam site selection. In the present case, the important geotechnical consideration is the stability of slopes leading down to and up from the water crossing.

It is advisable to collect historical data on river erosion and sedimentation. Then the back slopes are mapped by stereoscopic air photos or stereoscopic high resolution satellite imageries. On the basis of information generated from these studies, necessary field investigations are undertaken to assess the amount of river channel contraction, degree of curvature of river bend, nature of bed and bank materials including the flood flows and the flow depths.

High resolution satellite remote sensing imageries often predict the behaviour of a river from its stream type (largely from its easily identifiable channel pattern). This, in turn, gives rise to a number of interrelated parameters that should be looked for while examining the air photos or high resolution satellite imageries stereoscopically. The common interrelationships found among the fluvial features and phenomena while interpreting stream type patterns and stream behaviour from operational remote sensing studies are given in Table-11.3. These information have been often used for river crossing site selection for bridges.

11.6 GEOLOGIC HAZARD ZONE MONITORING/MAPPING

Geologic hazards include – seismic hazard, geomorphic hazard, volcanic hazard, glaciologic hazard and human induced hazard.

All these hazards are amenable to study by remote sensing and GIS techniques. Since these hazards leave the signatures in the geologic records of the area, high resolution remote sensing technique finds unique applications in the delineation and mapping of potential hazard zones. But up till now these techniques have not been able to predict when such hazards are likely to occur. High resolution remote sensing data are also useful in assessing the post event impact of a particular geologic hazard in a given area. This impact evaluation study not only brings forth the economic loss and the loss of lives in densely populated areas, but also provides warning and instructions for preparedness for the future such events.

Table 11.3 : Interrelationships between the stream type pattern/stream behaviour and the fluvial features/Processes (Read Y is related to X)

Y	X
1. Increasing tendency for sediment finding upwards	i. Increasing supply (input) of coarse sediment ii. Increasing ratio of bed-load sediment to total sediment load iii. Increasing coarseness of sediment transported iv. Increasing ratio of bank-full channel width to channel depth v. Increasing stream locations in arid, glacier ice and steep mountain valley environments vi. Decreasing proportion of over-bank versus channel bed deposition
2. Increasing channel sinuosity	i. Increasing supply (input) of coarse sediment ii. Increasing ratio of bed-load sediment to total sediment load iii. Increasing coarseness of sediment transported iv. Increasing ratio of bank-full channel width to channel depth v. Increasing stream locations in arid, glacier ice and steep mountain valley environments vi. Decreasing proportion of over-bank versus channel bed deposition
3. Increasing channel gradient	i. Decreasing supply (input) of coarse sediment ii. Decreasing ratio of bed-load sediment to total sediment load iii. Decreasing coarseness of sediment transported iv. Decreasing ratio of bank-full channel width to channel depth v. Decreasing stream locations in arid, glacier ice and steep mountain valley envronments vi. Increasing proportion of over-bank versus channel bed deposition

4. Decreasing storage regulation in watershed	i. Decreasing supply (input) of coarse sediment
	ii. Decreasing ratio of bed-load sediment to total sediment load
	iii. Decreasing coarseness of sediment transported
	iv. Decreasing ratio of bank-full channel width to channel depth
	v. Decreasing stream locations in arid, glacier ice and steep mountain valley environments
	vi. Increasing proportion of over-bank versus channel bed deposition
5. Decreasing distance from headwaters	i. Decreasing supply (input) of coarse sediment
	ii. Decreasing ratio of bed-load sediment to total sediment load
	iii. Decreasing coarseness of sediment transported
	iv. Decreasing ratio of bank-full channel width to channel depth
	v. Decreasing stream locations in arid, glacier ice and steep mountain valley environments
	vi. Increasing proportion of over-bank versus channel bed deposition

11.6.1 Seismic Hazard

In order to assess the seismic hazards of a given region by pre- and post-earthquake investigations and also to map the seismic hazard zones, one has to start collecting data on the history of seismicity of the region. These data pertain to the existence of active faults, landslides and ground failures caused by vibrations, ground deformations (tectonic uplift, subsidence, warping) and earthquake generated waves along shorelines.

Post-earthquake investigations are required to find out places of likely renewed activity and to detect, record and measure surface faults, ground failures, shoreline uplift or subsidence and effects of large waves on shorelines.

High resolution satellite remote sensing and stereoscopic aerial photographs are quite helpful in monitoring the signatures left on the geologic records due to seismic hazard and assessing the degree of damage inflicted in the area. GIS on the other hand records the relevant data on layers to work on impact assessment due to the seismic hazard. The detailed analysis of active faults and intricate displaced topography requires imageries of scales 1:10,000 or higher.

Post-earthquake faults should be studied preferably at 1:1,200 scale to show details of cracks/slip displacements. Low-sun angle aerial photograph is the most important requirement for detection of fault and ground failure studies, both in pre-earthquake and post-earthquake periods. Low-angle colour photography may be used in this case for better results.

11.6.2 Geomorphic Hazard

Geomorphic hazards include all surficial hazards such as landslides, land-subsidence features and floods.

Large scale aerial photographs and high resolution imageries are extremely useful for landslide identification and classification. This is done by recognizing abnormal surface forms, namely hummocks, lobes, head scarps and streams that resurge as large springs along the toe of the slides. Radar images usually show linear features indicating geologic structures that may be linked to the development of landslides. It is desirable to collect successive coverage of aerial photographs of the study area over the period of many years which forms the required historic database.

Land-subsidence features consist of sinkholes, karst valleys, shafts and open fissures which are easily identifiable on aerial photographs and high resolution satellite imageries from their characteristic signatures. A successive series of aerial photographs or satellite imageries covering many years can provide information on the development of these land-subsidence features.

Satellite remote sensing technique has been successfully applied for the study of floods to assess the damage and to determine the flood duration and movement of different aspects of floods (peaks and fronts, maximum flood inundation). It also delineates floodplains and the structures on them. Satellite images with repetitive passes are used for studying the flood dynamics where it persists for a long duration.

GIS data layers on geomorphic hazards help in modeling and decision making activities.

11.6.3 Volcanic Hazard

Volcanic hazards to human life and environment include hot avalanches, hot particles and gas clouds, ash falls (tephra), mud flows, lava flows and floods.

Potential volcanic hazard zones can be recognized by the characteristic historical records of volcanic activities. These characteristic signatures include distinctive nature of volcanic landforms, presence of geothermal activity in the summit crater, in the flanks and in the close proximity to volcanoes.

To create an exhaustive permanent record on a specific volcanic event or a sequence of volcanic events, all related data are to be acquired before, during and following the volcanic activity over the area. Aircraft, satellite and terrestrial imaging systems help to achieve this target. Terrestrial photographs, oblique aerial photographs, vertical aerial photographs, aerial thermographs/thermograms, satellite imageries using VIS/NIR and IR channels, radar satellite imageries and maps made from the image data including sequential topographic maps, all have been effectively used for this purpose. Other sensors used to record volcanic activities are seismic sensors, infrasonic arrays, satellite and other relay systems to transmit thermal data, seismic signals, data from tilt meter arrays and signals of volcanic gases from data collection platforms to a central receiving facility for analysis. The complete set of database can be kept in layers of GIS for modeling and decision making purposes.

Remote sensing satellite images can be effectively used to get the extent of ash deposition and lava flows from the volcanic eruption. Satellite thermography are very much used for the survey of active or dormant volcanoes in different parts of the world. From these surveys quantitative measurements of volcanic thermal emission can also be made.

Sequential aerial photography always plays an important role in determining where spraying of sea water should be made either to stem or to stop the flow of lava into the fishing port.

Remote sensing satellite observations can be profitably utilized for volcanic hazard prediction. For this, using IR sensors operating in 8-14µm band, the locations of enhanced heat flows can be identified which might possibly precede a flank of eruption. Thermal anomaly in the summit of the crater, fractures and faults in the region and bulging of the area can also be observed. Increase in thermal anomaly prior to the eruption when upper part of the bulge appear to be perforated by heat leaks can be easily detected. Satellite thermal imageries of fracture patterns (radial, ring types) around the crater (dome) are related to the location of the first and subsequent lava domes. Quantitative measurements of thermal changes associated with the sequence of eruptive events can also be made by this technique.

It is well known that large explosive volcanic eruptions cause loss of lives and properties in the populated areas around the volcano. It also affects plants and animals by noxious gas emission, deposition of tephra, flooding caused by glacier outburst and inundation by lava flows in extensive areas. Decrease in the amount of solar energy reaching the earth's surface due to the tephra dust can be monitored by polar orbiting weather satellites (NOAA). The expansion of the initial eruption cloud is easily recorded by the successive images from a geo-stationary operational environmental satellite (GOES).

Thus an impact assessment study on volcanic hazards should not only deal with economic loss and loss of lives and property in densely populated areas, but also concentrate on the far reaching impact of large explosive volcanic eruptions on the earth's climate change due to the enormous ejection of the volcanic materials into the stratosphere.

11.6.4 Glaciologic Hazard

Slow movement glacier that advances or recedes a few meters to hundred meters per year can be easily studied by acquiring periodic high resolution remote sensing satellite imageries.

In the fast movement or surge glaciers the terminus speed increases to several times its normal flow, thereby advancing several kilometers per month. Such sudden advances may cover highways, railroads and other strategic man-made structures. It may also dam the adjacent valleys producing ice-dammed lakes. These features can also be monitored by satellite remote sensing techniques because of the characteristic spectral signatures of the glacial features.

Glacier - outburst floods are found to occur from glacierized areas. Serious flooding conditions in the down stream occur mostly due to failure of the above mentioned ice-dammed lakes and as a result of subglacial geothermal/volcanic activities. Ground-based and aerial photographs and remote sensing satellite imageries can be profitably used to monitor and document the phenomena related to glacier–outburst floods.

For the study of tidal glaciers like iceberg discharge and rapid ice break-off, conducting conventional ground surveys are not possible. Therefore sequential low altitude (~5,000m) aerial photographs are acquired at intervals ranging from about few weeks to few months. From these sequential photographs the surface ice velocity vectors, the speed of flow along the centre line of these calving glaciers as a function of time, surface deformation tensors and surface height changes (in meters) are photogrammetrically determined. These data are then used to develop and validate the numerical models of glacier flow leading to predictions.

In order to acquire detailed and more exact information on glaciologic hazards, multi-probe data collection has to be undertaken. These include periodic low-altitude oblique aerial photographs, high- and low- altitude vertical aerial photographs, hydrographic soundings for determination of water depth, Mini-Sparker and Lister Boomer surveys to determine submarine morphology of the terminal moraine shoal, air-borne and surface radio-echo soundings for glacier thickness, specially enhanced remote sensing satellite images for mapping iceberg plumes and radio-echo sounding techniques for studying glaciers on cascade volcanoes to predict the magnitude of flood hazards and so on.

11.7 HUMAN-INDUCED GEOLOGIC HAZARD

Human-induced geologic hazards result when a stable or quasi-stable natural environmental condition is subjected to an abrupt change. The extent of these hazards are small compared to natural hazards and hence they are best seen on large scale aerial photographs. Aerial photographs also serve to reduce the need for tedious, time-consuming and labour –intensive conventional ground surveys.

Eor a systematic study of human-induced geologic hazard, first of all one needs to establish a reference database on GIS. Large scale (1:20,000) sequential series of ground and aerial photographs, high resolution multi-spectral remote sensing satellite scanner imageries, thermal IR imageries and satellite radar imageries are to be collected over many years in order to build up this reference database. In fact, these remotely sensed photographs are of great scientific and legal significance in helping to establish an objective and unbiased cause and effect relationship. Particularly this is found to be true in cases of (i) coastal erosion after the construction of sea-walls and artificial dunes, and (ii) movement slope faces on spoilbanks, tailing pond embankments and landfills for roads and railways.

Direction and extent of slope movement can be determined photogrammetrically by detecting the motion of survey markers placed on slope faces and taking periodic low-flying aerial photographs.

11.7.1 Coal Mine Fires

Coal mine fire may be caused by man or by natural agents like lightning strikes and spontaneous ignition.

While detection of deep mine fires in coal seams have not yet been made possible by air-borne thermal imaging techniques, the shallow coal seam fires and outcrop/surface refuse fires have been successfully monitored and mapped by these techniques. Since direct sunlight distorts the thermal imagery of coal mine fire sites, the low-lying aerial flights are normally scheduled for night or twilight to obtain thermal imageries of the location using either 3-5.5μm or 8-14μm channel of the electromagnetic spectrum.

11.7.2 Dam Failure

Observable symptoms of the dam during its pre-failure conditions, history of events leading to the dam failure and assessment of damage downstream resulting from the dam failure can all be studied comprehensibly from a series of large scale repetitive aerial and ground photographs.

In the pre-failure condition, earthen dams can be studied for their stability by monitoring the movement markers placed on the face, on abutments and on the back side of the dam by repetitive aerial photographs. The changes in the configuration of the dam can be determined in this way. Leakage through and around the dam can be similarly monitored by using thermal and moisture sensors in the aerial photography. Repetitive large scale aerial photographs can also provide a reliable record of the construction history of the dam and its spillway.

In a similar manner aerial photographs can detect and monitor the development of cracks and loose masonry blocks in the concrete and masonry dams.

Repetitive aerial photographs of the pool side behind the dam provide information on its slope stability and sedimentation. These are also useful in monitoring pool levels and finding out the pool changes with respect to runoff. High resolution satellite remote sensing technique is used as an invaluable tool for monitoring the development of flood plains downstream from the dam, particularly with reference to the large volume of spillway discharge and flooding by the failure of the dam.

Suggestions for Supplementary Reading

These suggestions are not exhaustive but are limited to the works that will be found especially useful for developing interest and creative imagination in the field of Remote Sensing.

Agricultural Applications

Books

American Society of Photogrammetry, 1983. Manual of Remote Sensing, Vol. II, 2nd ed., R. N. Colwell (Ed.), Falls Church, Va.

Blazquez, C. H. and F. W. Horn, Jr., 1980. Aerial Color Infrared Photography : Applications in Citriculture. NASA Ref. Publ. 1067, US Govt. Printing Office, Washington, DC.

Fraysse, G. 1980. Remote Sensing Applications in Agriculture and Hydrology. Published for the Commission of the European Committee, Luxenbourg.

Karale, R. L. (Ed.), 1992. Natural Resources Management – A New Perspective. NNRMS, ISRO, Department of Space, Government of India, Bangalore.

Steven, M. D. and J. A. Clark, 1990. Applications of Remote sensing in Agriculture. Butterworths, U.K.

Toselli, F. (Ed.), 1989. Applications of Remote Sensing to Agrometeorology. Kluwer Academic Publishers, Dordrecht, The Netherlands.

Research Papers

Badhwar, G. D. 1985. Classification of Corn and Soybeans using Multi-temporal Thematic Mapper Data. Photogramm. Eng. Remote Sensing, 45 : 605 – 610.

Bauer, M. E. 1975. The Role of Remote Sensing in Determining the Distribution and Yield of Crops. Advances in Agronomy, 27 : 271 – 304.

Bauer, M. E., M. M. Hixson, B. J. Davis and J. B. Etheridge, 1978. Area Estimation of Crops by Digital Analysis of Landsat Data. Photogramm. Eng. Remote Sensing, 44 1033 – 1043.

Bauer, M. E. (Ed.), 1984. AgRISTARS Issues. Remote Sensing Environ. 14 (Nos. 1 to 3).

Bauer, M. E. 1985. Spectral Inputs to Crop Identification and Condition Assessment. Proc. IEEE, 73 : 1071 –1085.

Bauer, M. E. et al. 1986. Field Spectroscopy of Agricultural Crops. IEEE Trans. Geosci. Remote Sensing, GE-24 : 65 – 75.

Blazquez, C. H., R. A. Elliot and G. J. Edwards, 1981. Vegetable Crop Management with Remote Sensing. Photogramm. Eng. Remote Sensaing, 47 : 543 – 547.

Brisco, B. and R. Protz, 1982. Manual and Automatic Crop Identification with Airborne Radar Imagery. Photogramm. Eng. Remote Sensing, 48 : 101 – 109.

Brisco, B., F. T. Ulaby and R. Protz, 1984. Improving Crop Classification through Attention to the Timing of Airborne Radar Acquisitions. Photogramm. Eng. Remote Sensing, 50 : 739 – 745.

Brown, D. E. and A. M. Winer, 1986. Estimating Urban Vegetation Cover in Los Angeles. Photogramm. Eng. Remote Sensing, 52 : 117 – 123.

Bryan, M. L. and J. Clark, 1984. Potentials for Change Detection using Seasat Synthetic Aperture Radar Data. Remote Sensing Environ. 16 : 107 – 124.

Cihlar, J., T. Sommerfeldt and B. Paterson, 1979. Soil Water Content Estimation in Fallow Fields from Airborne Thermal Scanner Measurements. Canadian J. Remote Sensing, 5 : 18 – 32.

Collins William, 1978. Remote Sensing of Crop Type and Maturity. Photogramm. Eng. Remote Sensing, 44 : 43 – 55.

Curran, P. J. 1980. Multispectral Photographic Remote Sensing of Green Vegetation Biomass and Productivity. Photogramm. Eng. Remote Sensing, 48 : 243 – 250.

Deutsch, M. and F. H. Ruggles, 1978. Hydrological Applications of Landsat Imagery used in the Study of the 1973 Indus River Flood, Pakistan. Water Resources Bull. 14 : 261 – 274.

Estes, J. E., M. R. Mel and J. O. Hooper, 1977. Measuring Soil Moisture with an Airborne Imaging Passive Microwave Radiometer. Photogramm. Eng. Remote Sensing, 43 : 1273 – 1281.

Hatfield, J. L., J. P. Millard and R. C. Goettelman, 1982. Variability of Surface Temperature in Agricultural Fields of Central California. Photogramm. Eng. Remote Sensing, 48 : 1319 – 1325.

Heilman, J. L., E. T. Kanemasu, N. J. Rosenberg and B. L. Blad, 1976. Thermal Scanner Measurement of Canopy Temperatures to estimate Evapotranspiration. Remote Sensing Environ. 5 : 137 – 145.

Holben, B. C., C. J. Tucker and C. J. Fun, 1980. Assessing Leaf Area and Leaf Biomass with Spectral Data. Photogramm. Eng. Remote Sensing, 45 : 651 – 656.

Idso, S. B., R. D. Jackson and R. J. Reginato, 1975. Estimating Evaporation : A Technique Adaptable to Remote Sensing. Science, 189 : 991 – 992.

Idso, S. B., R. D. Jackson and R. J. Reginato, 1975a. Detection of Soil Moisture by Remote Surveillance. Am. Scientist, 63 (Sept- Oct Issue).

Idso, S. B., T. J. Schmugge, R. D. Jackson and R. J. Reginato, 1975b. The Utility of Surface Temperature Measurements for the Remote Sensing of Soil Water Status. J. Geophys. Res. 80 : 3044 – 3049.

Idso, S. B. and W. L. Ehrler, 1976. Estimating Soil Moisture in the Root Zone of Crops : A Technique Adaptable to Remote Sensing. Geophys. Res. Lett. 3 : 23 – 25.

Idso, S. B., R. J. Reginato and R. D. Jackson, 1977. An Equation for Potential Evaporation from Soil, Water and Crop Surfaces Adaptable to use by Remote Sensing. Geophys. Res. Lett. 4 : 187 – 188.

Idso, S. B., R. D. Jackson and R. J. Reginato, 1978. Remote Sensing for Agricultural Water Management and Crop Yield Prediction. Agric. Water Manage. 1 : 299 – 310.

Idso, S. B., J. L. Hatfield, R. D. Jackson and R. J. Reginato, 1979. Gram Yield Prediction : Extending the Stress- Degree-Day Approach to accommodate Climatic Variability. Remote Sensing Environ. 8: 267 – 272.

Jackson, R. D., R. J. Reginato and S. B. Idso, 1977. Wheat Canopy Temperature : A Practical Tool for Evaluating Water Requirements. Water Resource Res. !3 : 651 – 656.

Jackson, R. D., S. B. Idso, R. J. Reginato and P. J. Pinter, Jr., 1981. Canopy Temperature as a Crop Water Stress Indicator. Water Resource Res. 17 : 1133 – 1138.

Jensen, J. R. 1983. Urban / Suburban Land-Use Analysis. In : Manual of Remote Sensing, Vol. II, 2nd ed., R. N. Colwell (Ed.). American Society of Photogrammetry, Falls Church, Va. p. 1571 – 1666.

Jensen, J. R. 1983. Biophysical Remote Sensing. Ann. Assoc. Am. Geographers, 73 : 111 – 132.

Jensen, J. R. and D. L. Chery, 1980. Landsat Crop Identification for Watershed Water-balance Determinations. Int. J. Remote Sensing, 1 : 345 – 359.

Justice, C. O., J. R. G. Townshend, B. N. Holben and C. J. Tucker, 1985. Analysis of the Phenology of Global Vegetation using Meteorological Satellite Data. Int. J. Remote Sensing, 6 : 1271 – 1318.

Kanemasu, E. T., J. L. Heilman, J. O. Bagley and W. L. Powers, 1977. Using Landsat Data to estimate Evapotranspiration of Winter Wheat. Environ. Manage. 1 : 515 – 520.

Kaur Ravinder and B. C. Panda, 1992. Towards Operationalization of Crop Management through Remote sensing. Proc. Natl. Symp. Remote Sensing for Sustainable Development, Lucknow. p. 377 – 382.

Kaur Ravinder, S. K. Bhadra, M. Bhavanarayana and B. C. Panda, 1998. A Numerical Technique for Delineation of Soil Mapping Units using Multispectral Remote Sensing Data. Photonirvachak : J. Indian Soc. Remote sensing, 26 : 149 – 160.

Kauth, R. J. and G. S. Thomas, 1976. The Tasseled Cap – A Graphic Description of the Spectral – Temporal Development of Agricultural Crops as seen by Landsat. Proc. 2nd Int. Symp. on Machine Processing of Remotely Sensed Data. Purdue University, West Lafayette, Ind. : Laboratory for the Applications of Remote Sensing. p. 41 – 51.

Kolm, K. E. and H. L. Case, 1984. The Identification of Irrigated Crop Types and Estimation of Acreages from Landsat Imagery. Photogramm. Eng. Remote Sensing, 50 : 1479 – 1490.

Kristof, S. J. and A. L. Zachary, 1974. Mapping Soil Features from Multispectral Scanner Data. Photogramm. Eng. 40 : 1427 – 1434.

Kumar, R. and L. F. Silva, 1977. Separability of Agricultural Crop Types by Remote Sensing in the Visible and Infrared Wavelength Regions. IEEE Trans. Geosci. Electronics, GE-15 : 42 – 49.

Ladouceur, G., R. Allard and S. Ghosh, 1986. Semi-automatic Survey of Crop Damage using Color Infrared Photography. Photogramm. Eng. Remote Sensing, 52 : 111 – 115.

Lo, T. H., F. L. Scarpace and T. M. Lillesand, 1986. Use of Multi-temporal Spectral Profiles in Agricultural Land- Cover Classification. Photogramm. Eng. Remote Sensing, 52 : 535 – 544.

Meyer, Merle P. and L. Calpouzos, 1968. Detection of Crop Disease Identifications. Photogramm. Eng. 36 : 1116 – 1125.

Milfred, C. J. and R. W. Kiefer, 1976. Analysis of Soil Variability with Repetitive Aerial Photography. Soil Sci. Soc. Am. J. 40 : 553 – 557.

Millard, J. P., R. C. Goettelman, R. D. Jackson, R. J. Reginato and S. B. Idso, 1978. Crop Water Stress Assessment using an Airborne Thermal Scanner. Photogramm. Eng. Remote Sensing, 44 : 77 – 85.

Myers, Victor I. 1983. Remote Sensing Applications in Agriculture. In : Manual of Remote Sensing, Vol. II, 2nd ed., R. N. Colwell (Ed.). American Society of Photogrammetry, Falls Church, Va. p. 2111 – 2228.

Nash, D. B. 1985. Detection of Bed-rock Topography Beneath a Thin Cover of Alluvium using Thermal Remote Sensing. Photogramm. Eng. Remote Sensing, 51 : 77 – 88.

Navalgund, R. R., J. S. Parihar, Ajai and P. P. Nageswara Rao, 1991. Crop Inventory Using Remotely Sensed Data. Current Science (Special Issues), 61 (Nos. 3 & 4) : 162 – 171.

Navalgund, R. R., J. S. Parihar, L. Venkataratnam, M. V. Krishna Rao, S. Panigrahy, M. C. Chakraborty, K. R. Hebbar, M. P. Oza, S. A. Sharma, N. Bhagia and V. K. Dadhwal, 1996. Early Results from Crop Studies using IRS-1C Data. Current Science (Special Issue), 70 (No.7) : 568 – 574.

Nellis, M. D. 1985. Interpretation of Thermal Infrared Imagery for Irrigation Water Resources Management. J. Geography, 84 : 11 – 14.

Perry, C. R. and L. F. Lautenschlager, 1984. Functional Equivalence of Spectral Vegetation Indices. Remote Sensing Environ. 14 : 169 – 182.

Panda, B. C. 1990. Philosophical Foundation of Crop Stress Detection through Remote Sensing Technique. Proc. Natl. Symp. Remote Sensing for Agricultural Applications, Indian Agricultural Research Institute, New Delhi. p. 158 – 172.

Panda, B. C. 1992. Crop Management – The Remote Sensing Approach in Indian Context. Proc. Silver Jubilee Seminar on Training and Education in Remote Sensing for Resource Management, Indian Institute of Remote Sensing, Dehra Dun. P. 180 – 182.

Panda, B. C. and R. K. Sharma, 1998. Monitoring from Space – A New Tool for Sustainable Agricultural Research. In : Wheat – Research Needs beyond 2000 A.D., S. Nagarajan, Gyanendra Singh and B. S. Tyagi (Eds.). Narosa Publishing House, New Delhi. p. 361 – 367.

Philpotts, L. E. and V. R. Wallen, 1969. IR Color for Crop Disease Identifications. Photogramm. Eng. 35 : 1116 – 1125.

Pollock, R. B. and E. T. Kanemasu, 1979. Estimating Leaf Area Index of Wheat with Landsat Data. Remote Sensing Environ. 8 : 307 – 312.

Price, J. C. 1982. Satellite Orbital Dynamics and Observation Strategies in Support of Agricultural Applications. Photogramm. Eng. Remote Sensing, 48 : 1603 – 1611.

Reginato, R. J., S. B. Idso and R. D. Jackson, 1978. Estimating Forage Crop Production : A Technique Adaptable to Remote Sensing. Remote Sensing Environ. 7 : 77 – 80.

Richardson, A. J. and C. L. Wiegand, 1977. Distinguishing Vegetation from Soil Background Information. Photogramm. Eng. Remote Sensing, 43 : 1541 – 1552.

Richardson, A. J., R. M. Menges and P. R. Nixon, 1985. Distinguishing Weeds from Crop Plants using Video Remote Sensing. Photogramm. Eng. Remote Sensing, 51 : 1785 – 1790.

Rosenthal, W. D. and B. J. Blanchard, 1984. Active Microwave Responses : An Aid in Improved Crop Classification. Photogramm. Eng. Remote Sensing, 50 : 461 – 468.

Ryerson, R. A., R. N. Dobbins and C. Thibault, 1985. Timely Crop Area Estimations from Landsat. Photogramm. Eng. Remote sensing, 51 : 1735 – 1743.

Satterwhite, M., W. Rice and J. Shipman, 1984. Using Landform and Vegetative Factors to Improve the Interpretation of Landsat Imagery. Photogramm. Eng. Remote Sensing, 50 : 83 – 91.

Sauer, E. K. 1981. Hydrology of Glacial Deposits from Aerial Photographs. Photogramm. Eng. Remote Sensing, 47 : 811 – 822.

Schmugge, T. J., P. Gloersen, T. Wilheit and F. Geiger, 1974. Remote Sensing of Soil Moisture with Microwave Radiometers. J. Geophys. Res. 79 : 317 – 323.

Schmugge, T. J. 1978. Remote Sensing of Surface Soil Moisture. J. Appl. Meteor. 17 : 1549.

Soer, G. J. R. 1980. Estimation of Regional Evapotranspiration and Soil Moisture Conditions using Remotely Sensed Crop Surface Temperatures. Remote Sensing Environ. 9 : 27 – 45.

Srivastava, S. K. and B. C. Panda, 1990. Evaluation of Characteristic Profile Models for Crop Discrimination. Proc. Natl. Symp. Remote Sensing for Agricultural Applications, Indian Agricultural Research Institute, New Delhi. p. 228 – 237.

Tarpley, J. D., S. R. Schneider and R. L. Money, 1984. Global Vegetation Indices from the NOAA-7 Meteorological Satellite. J. Climate & Appl. Meteorol. 23 : 491 – 494.

Thompson, D. R. and O. A. Wehmanen, 1980. Using Landsat Digital Data to detect Moisture Stress in Corn – Soybean Growing Regions. Photogramm. Eng. Remote Sensing, 46 : 1082 – 1089.

Tucker, C. J. 1979. Red and Photographic Infrared Linear Combinations for Monitoring Vegetation. Remote Sensing Environ. 8 : 127 – 150.

Tucker, C. J. 1980. Remote Sensing of Leaf Water Content in the Near-Infrared. Remote Sensing Environ. 10 : 23 – 32.

Tucker, C. J., J. H. Elgin, Jr. and J. E. McMurtrey III, 1979. Temporal Spectral Measurements of Corn and Soybean Crops. Photogramm. Eng. Remote Sensing, 45 : 643 – 653.

Tucker, C. J., J. R. G. Townshend and T. E. Goff, 1985. African Land-Cover Classification using Satellite Data. Science 227 : 369 – 375.

Ulaby, F. T., P. T. Batlivala and J. E. Bane, 1980. Crop Identification with L-Band Radar. Photogramm. Eng. Remote Sensing, 46 : 101 – 105.

Vlcek, J. and D. King, 1983. Detection of Subsurface Soil Moisture by Thermal Sensing : Results of Laboratory, Close-Range and Aerial Studies. Photogramm. Eng. Remote Sensing, 49 : 1593 – 1597.

Westin, F. C. 1976. LandsaT Data, Its Use in a Soil Survey Program. Soil Sci. Soc. Am. J. 40 : 81 – 89.

Westin, F. C. and G. D. Lemme, 1978. Landsat Spectral Signatures : Studies with Soil Associations and Vegetation. Photogramm. Eng. Remote sensing, 44 : 315 – 325.

Wiegand, C. L., A. J. Richardson and E. T. Kanemasu, 1979. Leaf Area Index Estimates for Wheat from Landsat and their Implications on Evapotranspiration and Crop Modeling. Agronomy J. 71 : 336 – 342.

Wong, K. W., T. H. Thornburn and M. A. Khoury, 1977. Automatic Soil Identification from Remote Sensing Data. Photogramm. Eng. Remote Sensing, 43 : 73 – 80.

Water Resource Related Applications

Books

American Society of Photogrammetry, 1983. Manual of Remote Sensing, Vol. II, 2nd ed., R. N. Colwell (Ed.), Falls Church, Va.

Engman, E. T. and R. J. Gurney, 1991. Remote Sensing in Hydrology. Chapman and Hall Publishers.

Karale, R. L. (Ed.), 1992. Natural Resources Management – A New Perspective. NNRMS, ISRO, Department of Space, Government of India, Bangalore.

Toselli, F. (Ed.), 1989. Applications of Remote Sensing to Agrometeorology. Kluwer Academic Publishers, Dordrecht, The Netherlands.

Research Papers

Abdel-Hady, M. 1971. Depth to Ground-water Table by Remote Sensing. J. Irrigation Proc. Am. Soc. Civil Eng. 97 : 355 – 367.

Brooks, R. L., W. J. Campbell, R. O. Ramseier, H. R. Stanley and H. J. Zwaily, 1978. Ice Sheet Topology by Satellite Altimetry. Nature, 274 : 539 – 543.

Byrne, G. F., J. E. Begg, P. M. Flemming and F. X. Dunin, 1979. Remotely Sensed Land-Cover Temperature and Soil Water Status – A Brief Review. Remote Sensing Environ. 8 : 291 – 305.

Choudhury, B. T., T. Schmugge, R. W. Newton and A. Chang, 1979. Effects of Surface Roughness on Microwave Emission from Soils. J. Geophys. Res. 8A : 5699 – 5706.

Eagleman, J. and W. Lin, 1976. Remote Sensing of Soil Moisture by a 21-cm Passive Radiometer. J. Geophys. Res. 81 : 3660.

Estes, J. E., M. R. Mel and J. O. Hooper, 1977. Measuring Soil Moisture with an Airborne Imaging Passive Microwave Radiometer. Photogramm. Eng. Remote Sensing, 43 : 1273 – 1281.

Heilman, J. L. and D. G. Moore, 1981. HCMM Detection of High Soil Moisture Areas. Remote Sensing Environ. 11 : 73 – 78.

Heilman, J. L. and D. G. Moore, 1982. Evaluating Depth to Shallow Ground Water using Heat Capacity Mapping Mission (HCMM) Data. Photogramm. Eng. Remote Sensing, 48 : 1903 – 1906.

Huntley, D. 1978. On the Detection of Shallow Acquifers using Thermal Infrared Imagery. Water Resources Res. 14 : 1075 – 1083.

Idso, S. B., T. J. Schmugge, R. D. Jackson and R. J. Reginato, 1975. The Utility of Surface Temperature Measurements for the Remote Sensing of Soil Water Status. J. Geophys. Res. 80 : 3044 – 3049.

Idso, S. B. and W. L. Ehrler, 1976. Estimating Soil Moisture in the Root Zone of Crops : A Technique Adaptable to Remote Sensing. Geophys. Res. Lett. 3 : 23 – 25.

Jackson, T. J., A. Chang and T. J. Schmugge, 1981. Aircraft Active Microwave Measurements for Estimating Soil Moisture. Photogramm. Eng. Remote Sensing, 47 : 801 – 805.

Khorram, S. 1981. Use of Ocean Color Scanner Data in Water Quality Mapping. Photogramm. Eng. Remote Sensing, 47 : 667 – 676.

Lathrop, R. G., Jr. and T. M. Lillesand, 1986. Use of Thematic Mapper Data to Assess Water Quality in Green Bay and Central Lake Michigan. Photogramm. Eng. Remote Sensing, 52 : 671 – 680.

Lattman, L. H. and R. R. Parizek, 1964. Relationship Between Fracture Traces and the Occurrence of Ground Water in Carbonate Rocks. J. Hydrology, 2 : 73 – 91.

Newton, R. W. and J. W. Rouse, 1980. Microwave Radiometer Measurements of Soil Moisture Content. IEEE Trans. Antennas Propag. AP-28 : 680 – 686.

Njoku, E. G. and J. A. Kong, 1977. Theory of Passive Microwave Remote Sensing of Near Surface Soil Moisture. J. Geophys. Res. 82 : 3103 – 3118.

Pratt, D. A., C. D. Ellyett, E. C. Mcglaughlin and P. McNabb, 1978. Recent Advances in the Application of Thermal Infrared Scanning to Geological and Hydrological Studies. Remote Sensing Environ. 7 : 177 – 184.

Ramamoorthi, A. S., S. Thiruvengadachari and A. V. Kulkarni, 1991. IRS-1A Applications in Hydrology and Water Resources. Current Science (Special Issue), 61 (No. 3 & 4) : 180 – 188.

Rango, A., J. Foster and V. V. Solomonson, 1975. Extraction and Utilization of Space Acquired Physiographic Data for Water Resources Development. Bull. Water Resources, 11 : 1245 – 1255.

Rango, A. and K. I. Itten, 1976. Satellite Potentials in Snow Cover Monitoring and Runoff Prediction. Nordic Hydrology, 7 : 209 – 230.

Rango, A. and J. Martinec, 1979. Application of a Snowmelt-Runoff Model using Landsat Data. Nordic Hydrology, 10 : 225 – 238.

Rango, A. and V. V. Solomonson, 1974. Regional Flood Mapping from Space. Water Resources Res. 10 : 473 – 484.

Reddy, P. R., K. Vinod Kumar and K. Seshadri, 1996. Use of IRS-1C Data in Ground Water Studies. Current Science (Special Issue), 70 (7) : 600 – 605.

Sahai Baldev, A. Bhattacharya and V. S. Hegde, 1991. IRS-1A Applications for Ground Water Targetting. Current Science (Special Issue), 61 (No.3 & 4) : 172 – 179.

Sahoo, R. N., M. Bhavanarayana, Y. V. Subba Rao and B. C. Panda, 2001. Soil Moisture Quantification through Total Information Content Index. In : Physical Methods of Soil Characterization, J. Behari (Ed.). Narosa Publishing House, New Delhi. p. 147 – 154.

Schmugge, T. J. 1978. Remote Sensing of Surface Soil Moisture. J. Appl. Meteorol. 17 : 1549 – 1557.

Schmugge, T. J. 1980. Effect of Texture on Microwave Emission from Soils. IEEE Trans. Geosci. Remote Sensing, GE-18 : 353 – 361.

Schmugge, T. J. 1983. Remote Sensing of Soil Moisture : Recent advances. IEEE Trans. Geosci. Remote Sensing, GE-21 : 336 – 344.

Schmugge, T. J., P. Gloersen, T. Wilheit and F. Geiger, 1974. Remote Sensing of Soil Moisture with Microwave Radiometers. J. Geophys. Res. 79 : 317 – 323.

Schmugge, T. J., J. M. Meneely, A. Rango and R. Neff, 1977. Satellite Microwave Observations of Soil Moisture Variations. Bull. Water Resources, 13 : 265.

Siddiqui, S. H. and R. R. Parizek, 1971. Hydrogeologic Factors Influencing Well Yields in Folded and Faulted Carbonate Rocks in Central Pennsylvania. Water Resources Res. 7 : 1295 – 1312.

Tewinkel, G. C. 1963. Water Depths from Aerial Photographs. Photogramm. Eng. 29 : 1037 – 1042.

Thiruvengadachari, S., S. Jonna, K. A. Hakeem, P. V. Raju and A. T. Jeyaseelan, 1996. Improved Water Management : The IRS-1C Contribution. Current Science (Special Issue), 70 (7) : 589 – 599.

Ulaby, F. T. 1974. Radar Measurements of Soil Moisture Content. IEEE Trans. Antennas Propag. AP-22 : 257 – 265.

Ulaby, F. T. and P. P. Batlibala, 1976. Optimum Radar Parameters for Mapping Soil Moisture. IEEE Trans. Geosci. Electronics, GS – 14 : 81 – 93.

Ulaby, F. T., P. P. Batlibala and M. C. Dobson, 1978. Microwave Backscatter Dependence on Surface Roughness, Soil Moisture and Soil Texture : Part-I : Bare Soil. IEEE Trans. Geosci. Electronics, GS-16 : 286 – 295.

Ulaby, F. T., J. Cihlar and R. K. Moore, 1974. Active Microwave Measurements of Soil Water Content. Remote Sensing Environ. 3 : 185 – 203.

Wang, J. R., J.C. Shiue and J. E. McMurtrey III. 1980. Microwave Remote Sensing of Soil Moisture Content over Bare and Vegetated Fields. Geophys. Res. Lett. 7 : 801 – 804.

Wang, J. R. and B. J. Choudhury, 1981. Remote sensing of Soil Moisture Content over Bare Field at 1.46 GHz frequency. J. Geophys. Res. 86 : 5277 – 5282.

Work, E. A. and D. S. Gilmer, 1976. Utilization of Satellite Data for Inventorying Prairie Ponds and Lakes. Photogramm. Eng. Remote Sensing, 42 : 685 – 694.

Weather and Climate Related Applications

Books

American Society of Photogrammetry, 1983. Manual of Remote Sensing, Vol. II, 2nd ed., R. N. Colwell (Ed.), Falls Church, Va.

Barrett, E. C. and L. F. Curtis, 1982. Introduction to Environmental Remote Sensing, 2nd ed., Halsted Press, Wiley, New York.

Branch, M. C. 1971. City Planning and Aerial Information. Harvard University Press, Cambridge, Mass.

Cracknell, A. P. (Ed.), 1981. Remote Sensing in Meteorology, Oceanography and Hydrology. Ellis Horwood, West Sussex, England.

Cracknell, A. P. (Ed.), 1983. Remote Sensing Applications in Marine Science and Technology. D. Reidel Publishing Co., Dordrecht, Holland.

Deepak Adarsh, 1980. Remote Sensing of Atmospheres and Oceans. Academic Press.

Estes, J. E. and L. W. Senger (Eds.), 1974. Remote Sensing : Techniques for Environmental Analysis. Hamilton, Santa Barbara, California.

Gower, J. F. R. (Ed.), 1980. Oceanography from Space. Plenum Press, New York.

Haughton, J. T., A. H. Cook and H. Charnock (Eds.), 1985. The Study of the Ocean and Land Surface from Satellites. Cambridge University Press, New York.

Holtz, R. K. 1985. The Surveillant Science : Remote Sensing of the Environment, 2nd ed. Wiley, New York.

Karale, R. L. (Ed.), 1992. Natural Resources Management – A New Perspective. NNRMS, ISRO, Department of Space, Government of India, Bangalore.

Lindgren, D. T. 1985. Land-Use Planning and Remote Sensing. Nijhoff, Dordrecht, The Netherlands.

Lintz, J. and D. S. Simonett (Eds.), 1976. Remote Sensing of Environment. Addition-Wesley, Reading, Mass.

Oke, T. R. 1978. Boundary Layer Climates. Methuen and Co. Ltd., London.

Robinson, I. S. 1985. Satellite Oceanography. John Wiley & Sons, New York.

Thorbury, W. D. 1969. Principles of Geomorphology, 2nd ed., John Wiley & Sons, New York.

Toselli, F. (Ed.), 1989. Applications of Remote Sensing to Agrometeorology. Kluwer Academic Publishers, Dordrecht, The Netherlands.

Way, D. 1978. Terrain Analysis – A Guide to Site Selection Using Aerial Photographic Interpretation, 2nd ed., Dowden, Hutchinson & Ross, Stroudsburg, Pa.

Wilson, A. G., P. H. Rees and C. M. Leigh, 1977. Models of Cities and Regions. Chichester : John Wiley.

Research Papers

Adeniyi, Peter O. 1980. Land-Use Change Analysis using Sequential Aerial Photography and Computer Techniques. Photogramm. Eng. Remote sensing, 46 : 1147 – 1164.

Adeniyi, P. O. 1983. An Aerial Photographic Method for Estimating Urban Population. Photogramm. Eng. Remote Sensing, 49 : 545 – 560.

Aronoff, S. and G. A. Ross, 1982. Detection of Environmental Disturbance using Color Aerial Photography and Thermal Infrared Imagery. Photogramm. Eng. Remote Sensing, 48 : 587 – 591.

Atwater, M. A. 1972. Thermal Effects of Urbanization and Industrialization in the Boundary Layer – A Numerical Study. Boundary-Layer Meteorology, 3 : 229 – 245.

Badhwar, G. D., R. B. MacDonald and N. C. Mehta, 1986. Satellite-Derived Leaf Area Index and Vegetation Maps as Input to Global Carbon Cycle Model – A Hierarchical Approach. Int. J. Remote sensing, 7 265 – 281.

Bornstein, R. D. 1968. Observations of the Urban Heat Island Effects in New York City. J. Appl. Meteorol. 7 : 575 – 582.

Brown, R. J. 1978. Infrared Scanner Technology applied to Building Heat Loss Determination. Canadian J. Remote Sensing, 4 : 1 – 9.

Brown, R. J., J. Cihlar and P. M. Teillet, 1981. Quantitative Residential Heat Loss Study. Photogramm. Eng. Remote Sensing, 47 : 1327 – 1333.

Carlson, T. N., J. N. Augustine and F. E. Boland, 1977. Potential Application of Satellite Temperature Measurements in the Analysis of Land Use over Urban Areas. Bull. Am. Meteorol. Soc. 58 : 1301 – 1303.

Changnon, S. A. 1978. Urban Effects on Severe Local Storms at St. Louis. J. Appl. Meteorol. 17 : 578 – 586.

Charney, J. 1975. Dynamics of Desert and Drought in the Sahel. Qr. J Roy. Meteorol. Soc. 101 : 193 – 202.

Charney, J., P. H. Stone and W. J. Quirk, 1975. Drought in the Sahara : A Biogeophysical Feedback Mechanism. Science, 187 : 434 – 435.

Chelton, D. B., K. J. Hussey and M. E. Park, 1981. Global Satellite Measurements of Water Vapor, Wind Speed and Wave Height. Nature, 234 : 529 – 532.

Chrisman, N. R. 1987. The Accuracy of Map Overlays : A Reassessment. Landscape & Urban Planning, 14 : 427 – 439.

Desai, P. S., A Narain, S. R. Nayak, B. Manikiam, S. Adiga and A. N. Nath, 1991. IRS-1A Applications for Coastal and Marine Resources. Current Science (Special Issue), 61 (No.3 & 4) : 204 – 208.

Dhir, R. P. and J. R. Sharma, 1991. IRS-1A Applications in Desertification Studies. Current Science (Special Issue), 61 (No.3 & 4) : 257 – 259.

Eriedman, S. Z. and G. L. Angelici, 1979. The Detection of Urban Expansion from Landsat Imagery. The Remote Sensing Quarterly 5 : 3 – 17.

Forster, B. C. 1985. An Examination of Some Problems and Solutions in Monitoring Urban Areas from Satellite Platforms. Int. J. Remote Sensing, 6 : 139 – 151.

Garstang, M., P. D. Tyson and G. D. Emmit, 1975. The Structure of Heat Islands. Rev. Geophys. Space Phys. 13 : 139 – 165.

Haack, B. N. 1984. L- and X-Band Like- and Cross- Polarized Synthetic Aperture Radar for Investigating Urban Environments. Photogramm. Eng. Remote Sensing, 50 : 331 – 30.

Hare, K. F. 1977. Climate and Desertification. In : Desertification : Its Causes and Consequences. Pergamon Press. P. 63 – 168.

Henderson, F. M. 1979. Land-Use Analysis of Radar Imagery. Photogramm. Eng. Remote sensing, 45 : 295 – 307.

Henderson, F. M. 1983. A Comparison of SAR Brightness Levels and Urban Land-Cover Classes. Photogramm. Eng. Remote sensing, 49 : 1585 – 1591.

Huff, F. A. and J. L. Vogel, 1978. Urban Topographic and Diurnal Effects on Rainfall in the St. Louis Region. J. Appl. Meteorol. 17 : 565 – 577.

Jensen, J. R. 1981. Urban Change Detection Mapping using Landsat Data. The American Cartographer, 8 : 127 – 147.

Jensen, J. R. 1983. Urban / Sub-Urban Land-Use Analysis . In : Manual of Remote Sensing, Vol. II, 2nd ed., R. N. Colwell (Ed.). American Society of Photogrammetry, Falls Church, Va. p. 1571 – 1666.

Jensen, J. R. and D. L. Toll, 1982. Detecting Residential Land-Use Development at the Urban Fringe. Photogramm. Eng. Remote Sensing, 48 : 629 – 643.

Karale, R. L., L. Venkataratnam, J. L. Sehgal and A. K. Sinha, 1991. Soil Mapping with IRS-1A Data in Areas of Complex Soil-Scapes. Current Science (Special Issue), 61 (No.3 & 4) : 198 – 203.

Khanna, P. and V. K. Kondawar, 1991. Application of Remote Sensing Techniques for Environmental Impact Assessment. Current Science (Special Issue), 61 (No.3 & 4): 252 – 256.

Ketchum, R. D. 1984. Seasat SAR Sea Ice Imagery – Summer Melt to Autumn Freeze-up. Int. J. Remote Sensing, 5 : 533 – 544.

Landsberg, H. E. 1956. The Climate in Towns. In : Man's Role in Changing the Face of the Earth, W. L. Thomas Jr. (Ed.). University of Chicago Press. P. 584 – 606.

Landsberg, H. E. 1970. Man Made Climate Changes. Science, 170 : 1725 – 1734.

Lillesand, R. L. 1972. Techniques for Change Detection. IEEE Trans. Computers, C- 21 : 654.

Matson, M., E. P. McClain, D. F. McGinnis and J. A. Pritchard, 1978. Satellite Detection of Urban Heat Islands. Monthly Weather Rev. 106 : 1725 – 1734.

Mintzer, O. W., F. A. Kulacki and L. E. Winget, 1983. Measuring Heat Loss from Flat-Roof Buildings with Calibrated Thermography. Photogramm. Eng. Remote Sensing, 49 : 777 – 788.

Nayak Sailesh, Prakash Chauhan, H. B. Chauhan, Anjali Bahuguna and A. Narendra Nath, 1996. IRS-1C Applications for Coastal Zone Management. Current Science (Special Issue), 70 (No.7) : 614 – 618.

Osborne, A. R. and T. L. Burch, 1980. Internal Solitons in the Andaman Sea. Science, 208 : 451 – 460.

Panda, B. C. 1992. Remote Sensing Methods for Estimating Evapotranspiration. Proc. Natl. Seminar : Soil Moisture Processes and Modelling, Agricultural Engineering Department, Indian Institute of Technology, Kharagpur. P. 172 – 177.

Panda, B. C. 1993. Dynamic Evaluation of Soil Thermal Inertia. Proc. Natl. Symp. Remote Sensing, Guwahati, Assam. P. 452 – 455.

Pease, R. W. and D. A. Nichols, 1976. Energy Balance Maps from Remotely sensed Imagery. Photogramm. Eng. Remote Sensing, 42 : 1367 – 1374.

Potter, G. L., H. W. Ellsaesser, M. C. Mac Cracken and F. M. Luther, 1975. Possible Climatic Impact of Tropical Deforestation. Nature, 258 : 697 – 698.

Pilon, P. G., P. J. Howarth, R. A. Bullock and P. O. Adeniyi, 1988. An Enhanced Classification Approach to Change Detection in Semi-Arid Environments. Photogramm. Eng. Remote Sensing, 54 : 1709 – 1716.

Price, J. E. 1979. Assessment of the Urban Heat Island Effect through the use of Satellite Data. Monthly Weather Rev. 107 : 1554 – 1557.

Rao, P. K. 1972. Remote Sensing of Urban Heat Islands from an Environmental Satellite. Bull. Am. Meteorol. Soc. 53 : 647 – 648.

Rao, D. P., N. C. Gautam, R. L. Karale and Baldev Sahai, 1991. IRS-1A Application for Land-Use / Land-Cover Mapping in India. Current Science (Special Issue), 61 (No. 3 & 4) : 153 – 161.

Rao, D. P., N. C. Gautam and Baldev Sahai, 1991. IRS-1A Application for Wasteland Mapping. Current Science (Special Issue), 61 (No.3 & 4) : 193 – 197..

Robinove, C. J., P. S. Chavez, Jr., D. Gehring and R. Holmgren, 1981. Arid Land Monitoring using Landsat Albedo Difference Images. Remote Sensing Environ. 11 : 133 – 156.

Rush, M. and S. Vernon, 1975. Remote Sensing and Urban Public Health. Photogramm. Eng. Remote Sensing, 41 1149 – 1155.

Shepard, J. R. 1964. A Concept of Change Detection. Photogramm. Eng. Remote Sensing, 30 : 649.

Terjung, W. G. 1970. Urban Energy Balance Climatology – A Preliminary Investigation of the City – Man System in Downtown Los Angeles. Geophys. Rev. 60 : 31 – 50.

Tueller, P. T. 1987. Remote Sensing Science Applications in Arid Environments. Remote Sensing Environ. 23 143 – 154.

United Nations, 1977. Desertification - An Overview. In : Desertification : Its Causes and Consequences. Pergamon Press. p. 1 – 62.

Vijayaraghavan, K. M., V. K. Gupta, Ravinder Kaur and B. C. Panda, 1990. Spectral Transmission Characteristics of the Atmosphere over Delhi during 1987 – 88. Proc. Natl. Symp. : Remote Sensing for Agricultural Applications, Indian Agricultural Research Institute, New Delhi. p. 477 – 482.

Watson, K., S. Hummer-Miller and D. L. Sawatzky, 1982. Registration of Heat Capacity Mapping Mission Day and Night Images. Photogramm. Eng. Remote Sensing, 48 : 263 – 268.

Weismiller, R. A., S. J. Kristof, D. K. Scholz, P. E. Anuta and S. A. Momin, 1977. Change Detection in Coastal Environments. Photogramm. Eng. Remote Sensing, 43 : 1533 – 1539.

Welch, R. 1980. Monitoring Urban Population and Energy Utilization Patterns from Satellite Data. Remote Sensing Environ. 9 : 1 – 9.

Welch, R. and S. Zupku, 1980. Urban Area Energy Utilization Patterns from DMSP Data. Photogramm. Eng. Remote sensing, 46 : 201 – 207.

Welch, R. 1982. Spatial Resolution Requirement for Urban Studies. Int. J. Remote Sensing, 3 : 139 – 146.

Weller, B. S. 1973. Remote Sensing and Urban Information Systems. Photogramm. Eng. 34 : 1041 – 1050.

White, J. M., F. D. Eaton and A. H. Auer, 1978. The Net Radiation Budget of the St. Louis Metropolitan Area. J. Appl. Meteorol. 17 : 141 – 155.

Forest Related Applications

Books

American Society of Photogrammetry, 1983. Manual of Remote Sensing, Vol. II, 2nd ed., R. N. Colwell (Ed.), Falls Church, Va.

Artsybashev, E. S. 1983. Forest Fires and Their Control. Oxonian, New Delhi (Original Russian, 1974)

Barrett, E. C. and L. F. Curtis, 1982. Introduction to Environmental Remote Sensing, 2nd ed., Halsted Press, Wiley, New York.

Brown, A. A. and K. P. Davis, 1973. Forest Fire : Control and Use. McGraw-Hill, New York.

Estes, J. E. and L. W. Senger (Eds.), 1974. Remote Sensing : Techniques for Environmental Analysis. Hamilton, Santa Barbara, California.

Karale, R. L. (Ed.), 1992. Natural Resources Management – A New Perspective. NNRMS, ISRO, Department of Space, Government of India, Bangalore.

Kessell, S. R. 1979. Gradient Modeling : resource and Fire Management. Springer-Verlag, New York.

Lintz, J. and D. S. Simonett (Eds.), 1976. Remote Sensing of Environment. Addition-Wesley, Reading, Mass.

National Academi of Sciences, 1970. Remote Sensing with Special Reference to Agriculture and Forestry. Washington, DC.

Sharma, M. K. 1986. Remote sensing and Forest Surveys. International Book Distributors, Dehradun.

Toselli, F. (Ed.), 1989. Applications of Remote Sensing to Agrometeorology. Kluwer Academic Publishers, Dordrecht, The Netherlands.

Research Papers

Aldrich, R. C. 1975. Detecting Disturbances in a Forest Environment. Photogramm. Eng. Remote Sensing, 41 : 38 – 48.

American Society of Photogrammetry, 1982. Panoramic Photography and Forest Remote Sensing. Photogramm. Eng. Remote Sensing (Special Issue), 48 (No.5).

Arnold,K. 1951. Uses of Aerial Photographs in Control of Forest Fires. J. Forestry 49 : 631.

Befort, W. 1986. Large Scale Sampling Photography for Forest Habitat–Type Identification. Photogramm. Eng. Remote Sensing, 52 : 101 – 108.

Burgan, R. E. and M. B. Shasby, 1984. Mapping Broad - Area Fire Potential from Digital Fuel, Terrain and Weather Data. J. Forestry, 82 : 228 – 231.

Ciesla, W. M. 1977. Color versus Color IR Photos for Forest Insect Damage from High Altitude Color IR Photos. Photogramm. Eng. 40 : 683 – 689.

Ciesla, W. M., R. A. Allison and F. P. Weber, 1982. Panoramic Aerial Photography in Forest Pest Management. Photogramm. Eng. Remote Sensing, 48 : 719 – 723.

Ciesla, W. M. 1984. Mission : Track the Gypsy from 65,000 feet. American Forests, 90 : 30 – 33 and 54 – 56.

Ciesla, W. M., D. D. Bennett and J. A. Caylor, 1984. Mapping Effectiveness of Insecticide Treatments against Pandora Moth with Color IR Photos. Photogramm.Eng. Remote Sensing, 50 : 73 – 79.

Ciesla, W. M. 1989. Aerial Photos for Assessment of Forest Decline – A Multinational Overview. J. Forestry, 87 : 37 – 41.

Cochrane, G. R. 1970. Color and False Color Aerial Photography for Mapping Bush Fires and Forest Vegetation. Proc. New Zealand Ecol. Soc. 17 : 96 – 105.

Fox, L.,III., J. A. Brockhaus and N. D. Tosta, 1985. Classification of Timberland Productivity in North-Western California using Landsat Topographic and Ecological Data. Photogramm. Eng. Remote Sensing, 51 : 1745 – 1752.

Flannigan, M. D. and T. H. Vonder Haar, 1986. Forest Fire Monitoring using NOAA Satellite AVHRR. Canadian J. Forest Res. 16 : 975 – 982.

Frank, T. D. and S. A. Isard, 1986. Alpine Vegetation Classification using High Resolution Aerial Imagery and Topoclimatic Index Values. Photogramm. Eng. Remote Sensing, 52 : 381 – 388.

Franklin, J., T. L. Logan, C. E. Woodcock and A. H. Strahler, 1986. Coniferous Forest Classification and Inventory using Landsat and Digital Terrain Data. IEEE Trans. Geosci. Remote sensing, GE-24: 139 – 149.

Gammon, P. T. and V. Carter, 1979. Vegetation Mapping with Seasonal Color Infrared Photographs. Photogramm. Eng. Remote Sensing, 45 : 87 – 97.

Heller, R. C. 1978. Case Applications of Remote Sensing for Vegetation Damage Assessments. Photogramm. Eng. 44 : 1159 – 1166.

Heller, R. C., R. C. Aldrich and W. F. Bailey, 1959. An Evaluation of Aerial Photography for Detecting Southern Pine Beetle Damage. Photogramm. Eng. 15 : 595 – 606.

Heller, Robert C. and Joseph J. Ulliman, 1983. Forest Resource Assessments. In : Manual of Remote Sensing, Vol. II, 2nd ed., R. N. Colwell (ED.). American Society of Photogrammetry, Falls Church, Va. p. 2229 – 2324.

Hopkins, P. F., A. L. Maclean and T. M. Lillesand, 1988. Assessment of Thematic Mapper Imagery for Forestry Applications under Lake States Conditions. Photogramm. Eng. Remote Sensing, 54 : 61 – 70.

Horler, D. N. H. and F. J. Ahern, 1986. Forestry Information Content of Thematic Mapper Data. Int. J. Remote Sensing, 7 : 405 – 428.

Jakubauskas, M. E., K. P. Lulla and P. W. Mausel, 1990. Assessment of Vegetation Change in a Fire Altered Forest Landscape. Photogramm. Eng. Remote sensing, 56 : 371 – 377.

Lawrence, R. D. and J. H. Herzog, 1975. Geology and Forestry Classification from ERTS-1 Digital Data. Photogramm. Eng. Remote Sensing, 41 : 1241 – 1251.

Losee, S. T. B. 1953. Timber Estimates from Large Scale Photographs. Photogramm. Eng. 19 : 752 – 762.

Maclean, G. A. and W. Krabill, 1986. Gross Merchantable Timber Volume Estimation using an Airborne Lidar System. Canadian J. Remote Sensing, 12 : 7 – 18.

Marshall, J. R. and M. P. Meyer, 1978. Field Evaluation of Small Scale Forest Resource Aerial Photography. Photogramm. Eng. Remote Sensing, 44 : 37 – 42.

Martin, F. C. 1985. Using a Geographic Information System for Forest Land Mapping and Management. Photogramm. Eng. Remote Sensing, 51 : 1753 – 1759.

Matson, M. and B. Holben, 1987. Satellite Detection of Tropical Burning in Brazil. Int. J. Remote Sensing, 8 : 509 – 516.

Meyer, M. P. and D. W. French, 1966. Forest Disease Spread. Photogramm. Eng. 32 : 812 – 814.

Meyer, M. P. and D. W. French, 1967. Detection of Diseased Trees. Photogramm. Eng. 32 : 1035 – 1040.

Miline, A. K. 1986. The Use of Remote sensing in Mapping and Monitoring Vegetational Change associated with Bushfire Events in Eastern Australia. Geocarto International, 1 : 25 – 32.

Murtha, P. A. and J. A. McLean, 1981. Extravisual Damage Detection ? Defining the Standard Normal Tree. Photogramm. Eng. Remote Sensing, 47 : 515 – 522.

Myers, B. J. and M. L. Bensen, 1981. Rainforest Species on Large Scale Color Photos. Photogramm. Eng. Remote Sensing, 47 : 505 – 513.

Myers, B. J. et al. 1984. Shadowless or Sunlit Photos for Forest Disease Detection ? Photogramm. Eng. Remote Sensing, 50 : 63 – 72.

Nelson, R. and B. Holben, 1986. Identifying Deforestation in Brazil using Multi-Resolution Satellite Data. Int. J. Remote Sensing, 7 : 429 – 448.

Roth, E. R., R. C. Heller and W. A. Stegall, 1963. Color Photography for Oak Wilt Detection. J. Forestry, 61 : 774 – 778.

Roy, P. S., C. B. S. Dutt, R. N. Jadhav, B. K. Ranganath, M. S. R. Murthy, B. Gharai, V. Udaya Lakshmi, A. K. Kandya and P. S. Thakker, 1996. IRS-1C Data Utilization for Forestry Applications. Current Science (Special Issue), 70 (No.7) : 606 – 613.

Running, S. W., D. L. Peterson, M. A. Spanner and K. B. Teuber, 1986. Remote Sensing of Coniferous Forest Leaf Area. Ecology, 67 : 273 – 276.

Savastano, K. J., K. H. Faller and R. L. Iverson, 1984. Estimating Vegetation Coverage in St. Joseph Bay, Florida with an Airborne Multispectral Scanner. Photogramm. Eng. Remote Sensing, 50 : 1159 – 1170.

Sayn-Wittgenstein, L. 1961. Recognition of Tree Species on Air Photographs by Crown Characteristics. Photogramm. Eng. 27 : 792 – 809.

Smith, J. L. 1986. Evaluation of the Effects of Photo Measurement Errors on Predictions of Stand Volume from Aerial Photography. Photogramm. Eng. Remote Sensing, 52 : 401 – 410.

Turner, B. J. and D. N. Thompson, 1982. Barrier Island Vegetation Mapping using Digitized Aerial Photography. Photogramm. Eng. Remote Sensing, 48 : 1327 – 1335.

Ulliman, J. J. and D. W. French, 1977. Detection of Oak Wilt with Color IR Aerial Photography. Photogramm. Eng. Remote Sensing, 43 : 1267 – 1272.

Worley, D. P. and H. A. Meyer, 1955. Measurement of Crown Diameter and Crown Cover and their Accuracy on 1 : 12,000 Scale Photographs. Photogramm. Eng. 21 : 372 – 375.

Unni, Madhavan N. V., P. S. Roy, R. N. Jadhav, A. K. Tiwari, S. Sudhakar, B. K.Ranganath and S. L. Dobral, 1991. IRS-1A Applications in Forestry. Current Science (Special Issue), 61 (No.3 & 4) : 189 – 192.

Young, H. E., F. M. Call and T. C. Tryon, 1963. Multi-Million Acre Forest Inventories based on Air Photos. Photogramm. Eng. 29 : 641 – 644.

Rangeland Applications

Books

American Society of Photogrammetry, 1983. Manual of Remote Sensing, Vol. II, 2nd ed., R. N. Colwell (Ed.), Falls Church, Va.

Barrett, E. C. and L. F. Curtis, 1982. Introduction to Environmental Remote Sensing, 2nd ed., Halsted Press, Wiley, New York.

Estes, J. E. and L. W. Senger (Eds.), 1974. Remote Sensing : Techniques for Environmental Analysis. Hamilton, Santa Barbara, California.

Karale, R. L. (Ed.), 1992. Natural Resources Management – A New Perspective. NNRMS, ISRO, Department of Space, Government of India, Bangalore.

Lintz, J. and D. S. Simonett (Eds.), 1976. Remote Sensing of Environment. Addition-Wesley, Reading, Mass.

Toselli, F. (Ed.), 1989. Applications of Remote Sensing to Agrometeorology. Kluwer Academic Publishers, Dordrecht, The Netherlands.

Research Papers

Carneggie, David M., Barry J. Schrumpf and David A. Mouat, 1983. Rangeland Applications. In : Manual of Remote Sensing, Vol. II, 2nd ed., R. N. Colwell (Ed.). American Society of Photogrammetry, Falls Church, Va. p. 2325 – 2384.

Dudzinski, M. L. and G. L. Arnold, 1967. Aerial Photography and Statistical Analysis for Studying Behaviour Patterns of Grazing Animals. J. Range Manage. 20 : 77 – 83.

Everitt, J. H. and P. R. Nixon, 1985. False Color Video Imagery : A Potential Remote Sensing Tool for Range Management. Photogramm. Eng. Remote Sensing, 51 : 675 – 679.

Everitt, J. H., A. J. Richardson and A. H. Gerbermann, 1985. Identification of Rangeland Sites on Small Scale (1 : 120,000) Color – Infrared Aerial Photos. Photogramm. Eng. Remote Sensing, 51 : 89 – 93.

Everitt, J. H. et al., 1985. Leaf Reflectance – Nitrogen – Chlorophyll Relations in Buffelgrass. Photogramm. Eng. Remote sensing, 51 : 463 – 466.

Francis, R. E. 1970. Ground Markers Aid in Procurement and Interpretation of Large Scale 70 mm Aerial Photography. J. Range Manage. 23 : 66 – 68.

Graetz, R. D., D. M. Carneggie, R. Hacker, C. Lendon and D. G. Wilcox, 1976. A Quantitative Evaluation of Landsat Imagery of Australian Rangelands. Australian Rangeland J. 1 : 53 – 59.

Hacker, R. B. 1980. Prospects for Satellite Applications in Australian Rangelands. Tropical Grasslands, 14 : 288 – 295.

Harris, R. W. 1951. Use of Aerial Photographs and Sub-Sampling in Range Inventories. J. Range Manage. 4 :270 – 278.

Huete, A. R. and R. D. Jackson, 1987. Suitability of Spectral Indices for Evaluating Vegetation Characteristics on Arid Rangelands. Remote Sensing Environ. 23 : 213 – 232.

Krupin, P. J. 1980. Aerial Photo Exploration for Open Range Water Supplies. Rangelands, 2 : 192.

Lyon, J. G., J. F. McCarthy and J. T. Heinen, 1986. Video Digitization of Aerial Photographs for Measurement of Wind Erosion Damage on Converted Rangeland. Photogramm. Eng. Remote Sensing, 52 : 373 – 377.

Maxwell, E. L. 1976. A Remote Rangeland Analysis System. J. Range Manage. 26 : 66 – 73.

McDaniel, K. C. and R. H. Haas, 1982. Assessing Mesquite-Grass Vegetation Condition from Landsat. Photogramm. Eng. Remote Sensing, 48 : 441 – 450.

Mouat, D. A., J. B. Bale, K. E. Foster and B. D. Treadwell, 1981. The Use of Remote Sensing for an Integrated Inventory of a Semi-Arid Area. J. Arid Environments, 4 : 169 – 179.

Pearson, R. L., C. J. Tucker and L. D. Miller, 1976. Spectral Mapping of Short Grass Prairie Biomass. Photogramm. Eng. 42 : 317 – 323.

Ripple, W. J. 1985. Asymptotic Reflectance Characteristics of Grass Vegetation. Photogramm. Eng. Remote Sensing, 51 : 1915 – 1921.

Robinov, C. J., P. S. Chavez, Jr., D. Gehring and R. Holmgren, 1981. Arid Land Monitoring using Landsat Albedo Difference Images. Remote Sensing. Environ. 2 : 153 – 156.

Tucker, C. J., L. D. Miller and R. L. Pearson, 1975. Short-grass Prairie Spectral Measurements. Photogramm. Eng. 41 : 1157 – 1162.

Wells, K. F. 1971. Measuring Vegetation Changes on Fixed Quadrats by Vertical Ground Stereo Photography. J. Range Manage. 24 : 233 – 236.

Williams, R. E., B. W. Allred, R. N. Denis and H. A. Paulsen, Jr., 1968. Conservation Development and Use of the World Rangelands. J. Range Manage. 21 : 355 – 360.

Wilson, R. O. and P. T. Tueller, 1987. Aerial and Ground Spectral Characteristics of Rangeland Plant Communities in Nevada. Remote Sensing Environ. 23 : 177 – 191.

Engineering Applications

Books

American Society of Photogrammetry, 1983. Manual of Remote Sensing, Vol. II, 2nd ed., R. N. Colwell (Ed.), Falls Church, Va.

Barrett, E. C. and L. F. Curtis, 1982. Introduction to Environmental Remote Sensing, 2nd ed., Halsted Press, Wiley, New York.

Estes, J. E. and L. W. Senger (Eds.), 1974. Remote Sensing : Techniques for Environmental Analysis. Hamilton, Santa Barbara, California.

Karale, R. L. (Ed.), 1992. Natural Resources Management – A New Perspective. NNRMS, ISRO, Department of Space, Government of India, Bangalore.

Kennie, T. J. M. and M. C. Matthews (Ed.), 1985. Remote Sensing in Civil Engineering. Halsted Press, Wiley, New York.

Lobeck, A. K. 1939. Geomorphology. McGraw-Hill Book Co. Inc., New York.

Lueder, D. R. 1959. Aerial Photographic Interpretation : Principles and Applications. McGraw-Hill, New York.

Taylor, John L. and David C. Williams, 1981. Urban Planning Practice in Developing Countries. Pergamon Press.

Thorbury, W. D. 1969. Principles of Geomorphology, 2nd ed., John Wiley & Sons, New York.

Turner, Alam (Ed.), 1980. The Cities of the Poor : Settlement Planning in Developing Countries. Crown Helm.

Way, D.S. 1978. Terrain Analysis. McGraw-Hill, New York.

Way, D. 1978. Terrain Analysis : A Guide to Site Selection using Aerial Photographic Interpretation, 2nd ed., Dowden, Hutchinson & Ross, Stroudsburg, Pa.

Woods, K. B. (Ed.), 1960. Highway Engineering Hand-Book. McGraw-Hill, New York.

Research Papers

Barr, D. J. and M. D. Hensey, 1974. Industrial Site Study with Remote Sensing. Photogramm. Eng. 40 : 79 – 86.

Belcher, D. J. 1945. Engineering Significance of Soil Patterns. Photogramm. Eng. 11 : 115 – 148.

Chaves, J. R. 1968. Color Photos for Highway Engineering. Photogramm. Eng. 34 : 375 – 379.

Dill, H. W., Jr., 1963. Airphoto Analysis in Outdoor Recreation-Site Inventory and Planning. Photogramm. Eng. 29 : 67 – 70.

Frost, R. E. 1953. Factors limiting the use of Aerial Photographs for Analysis of Soil and Terrain. Photogramm. Eng. 19 : 427 – 436.

Geraci, A. L. 1981. Remote Sensing Techniques Aid in Water Pollution Evaluation. Sea Technology, 10 : 20 – 21.

Hittle, J. E. 1946. The Application of Aerial Strip Photography to Highway and Airport Engineering. Highway Research Board Proc. 26 : 226 – 235.

Hittle, J. E. 1949. Airphoto Interpretation of Engineering Sites and Materials. Photogramm. Eng. 15 : 589 – 603.

Khorram, S. 1981. Use of Ocean Color Scanner Data in Water Quality Mapping. Photogramm. Eng. Remote Sensing, 47 : 667 – 676.

Khorram, S. 1981a. Water Quality Mapping from Landsat Digital Data. Int. J. Remote Sensing, 2 : 145 – 153.

Mintzer Olin, 1983. Engineering Applications. In : Manual of Remote Sensing, Vol. II, 2nd ed., R. N. Colwell (Ed.). American Society of Photogrammetry, Falls Church, Va. p. 1955 – 2109.

Murthy, Y. V. N. Krishna, V. Raghavswamy, S. K. Pathan and K. L. Majumdar, 1991. IRS-1A Applications for Urban Planning. Current Science (Special Issue), 61 (No.3 & 4) : 243 – 246.

Nanda, R. L. 1978. Use of Photo Interpretation Techniques for Survey of Highway Material Resources. J. Indian Soc. Photo Interpretation, 6 : 67 –74.

Raghavswamy, V., S. K. Pathan, P. Ram Mohan, R. J. Bhandari and Padma Priya, 1996. IRS-1C Applications for Urban Planning and Development. Current Science (Special Issue), 70 (7) : 582 – 588.

Rao, D. P., G. Behera, R. R. Navalgund, R. L. Karale and R. S. Rao, 1991. IRS-1A Applications for District Level Planning. Current Science (Special Issue), 61 (No. 3 & 4) : 260 – 265.

Rao, D. P., N. C. Gautam, R. Nagaraja and P. Ram Mohan, 1996. IRS-1C Applications in Land-Use Mapping and Planning. Current Science (Special Issue), 70 (7) : 575 – 581.

Rib. H. T. and R. D. Miles, 1969. Automatic Interpretation of Terrain Features. Photogramm. Eng. 25 : 153 – 164.

Scarpace, F. L., K. Holmquist and L. T. Fisher, 1979. Landsat Analysis of Lake Quality. Photogramm. Eng. Remote Sensing, 45 : 623 – 633.

Srivastava, P. K., B. Gopala Krishna and K. L. Majumdar, 1996. Cartography and Terrain Mapping using IRS-1C Data. Current Science (Special Issue), 70 (7) : 562 – 567.

Stafford, D. B. and J. Langfelder, 1971. Airphoto Survey of Coastal Erosion. Photogramm. Eng. 37 : 565 – 575.

Strandberg, C. H. 1964. An Aerial Water Quality Reconnaissance System. Photogramm. Eng. 30 : 46 – 54.

Strandberg, C. H. 1966. Water Quality Analysis. Photogramm. Eng. 32 : 234 – 248.

Turner, A. K. 1978. A Decade of Experience in Computer Route Selection. Photogramm. Eng. Remote Sensing, 44 : 1561 – 1576.

Wagner, R. R. 1963. Using Airphotos to measure Changes in Land-Use around Highway Interchanges. Photogramm. Eng. 29 : 645 – 649.

Geological Applications

Books

American Society of Photogrammetry, 1983. Manual of Remote Sensing, Vol. II, 2nd ed., R. N. Colwell (Ed.), Falls Church, Va.

Avery, T. E. 1968. Interpretation of Aerial Photographs, 2nd ed., Burgess Publishing Co., Minneapolis.

Barrett, E. C. and L. F. Curtis, 1982. Introduction to Environmental Remote Sensing, 2nd ed., Halsted Press, Wiley, New York.

Civella, L., P. Gasparini, G. Luongo and A. Rapolla (Eds.), 1974. Physical Volcanology. Elsevier Scientific Publishing Co., New York.

Druary, S. A. 1987. Image Interpretation in Geology. Allen and Unwin Ltd., London.

Embleton Clifford and C. A. M. King, 1975. Glacial Geomorphology. John Wiley & Sons, New York.

Estes, J. E. and L. W. Senger (Eds.), 1974. Remote Sensing : Techniques for Environmental Analysis. Hamilton, Santa Barbara, California.

Fernch, H. M. 1976. The Periglacial Environment. Longman Inc., New York.

Karale, R. L. (Ed.), 1992. Natural Resources Management – A New Perspective. NNRMS, ISRO, Department of Space, Government of India, Bangalore.

Lueder, D. R. 1959. Aerial Photographic Interpretation : Principles and Applications. McGraw-Hill, New York.

Nefedov, K. E. and T. A. Popova, 1972. Deciphering of Ground Water from Aerial Photographs. Amerind Publishers, New Delhi.

Paterson, W. S. B. 1981. Physics of Glaciers. Pergamon Press, New York.

Richter, C. F. 1958. Elementary Seismology. W. H. Freeman, San Francisco.

Scheidegger, A. E. 1970. Theoretical Geomorphology, 2nd ed., Springer-Verlag, Berlin.

Siegal, B. S. and A. R. Gillespie (Eds.), 1986. Remote Sensing in Geology. Wiley, New York.

Smith, W. L. (Ed.), 1977. Remote Sensing Applications for Mineral Exploration. Dowden, Hutchinson & Ross, Stroudsburg, Pa.

Thorbury, W. D. 1969. Principles of Geomorphology, 2nd ed., John Wiley & Sons, New York.

Verstappen, H. 1977. Remote Sensing in Geomorphology. Elsevier Scientific Publications, The Netherlands.

Research Papers

Allen, C. R. 1975. Geological Criteria for Evaluating Seismicity. Geological Soc. Am. Bull. 86 : 1041 – 1057.

Belcher, D. J. 1948. Determination of Soil Conditions from Aerial Photographs. Photogramm. Eng. 14: 482 – 488.

Brock, B. B. 1957. World Patterns and Lineaments. Trans. Geological Soc. South Africa, 60 : 127 – 160.

Brooks, R. L., W. J. Campbell, R. O. Ramseier, H. R. Stanley and H. J. Zwally, 1978. Ice Sheet Topography by Satellite Altimetry. Nature, 274 : 539 – 543.

Carey, S. W. 1962. Scale of Geotectonic Phenomena. Geological Soc. India J. 3 : 97 – 105.

Clark, M. M. 1978. Finding Active Faults using Aerial Photographs. Earthquake Information Bull. 10: 169 – 173.

Elson, J. A. 1980. Glacial Geology. In : Remote Sensing in Geology, B. S. Siegal and A. R. Gillespie (Eds.). John Wiley & Sons, New York. p. 505 – 551.

Jayaweera, K. O. L. F., R. Seifert and G. Wendler, 1976. Satellite Observations of the Eruption of Tolbachik Volcano. Trans. Am. Geophys. Union, 57 : 196 – 200.

Krimmel, R. M. and M. F. Meier, 1975. Glacier Applications of ERTS Images. J. Glaciology 15 : 391 – 401.

Lattman, L. H. 1958. Technique of Mapping Geologic Fracture Traces and Lineaments on Aerial Photographs. Photogramm. Eng. 24 : 568 – 576.

Meier, M. F. and Austin Post, 1969. What are Glacier Surges ? Canadian J. Earth Sci. 6 (Part-2) : 807 – 817.

Moody, J. D. and M. J. Hill, 1956. Wrench-Fault Tectonics. Geological Soc. Am. Bull. 67 : 1207 – 1246.

Morgan, V. I. And W. E. Budd, 1975. Radio Echosounding of the Lambert Glacier Basin. J. Glaciology, 15 : 103 – 111.

Moxham, R. M. 1967. Changes in Surface Temperature at Taal Volcano, Philippines 1965 – 1966. Bull. Volcanologique, 31 : 215 – 234.

Moxham, R. M. 1970. Thermal Features at Volcanoes in the Cascade Range as observed by Aerial Infrared Surveys. Bull. Volcanologique, 34 : 77 – 106.

Muller Fritz, 1962. Zonation in the Accumulation Area of the Glaciers of Axel Heiberg Island, N.W.T., Canada. J. Glaciology 4 : 302 – 313.

Ostrem Gunnar, 1975. ERTS Data in Glaciology – An Effort to monitor Glacier Mass Balance from Satellite Imagery. J. Glaciology 15 : 403 – 414.

Rabchevsky, G. A. 1979. Coal Fires. Earth Sci., Summer Issue : 12 – 15.

Rabchevsky, G. A. 1982. Coal Fires in Mines, Refuse Banks and Unmined deposits. Earth Sci., Summer Issue : 16 –19.

Ray, L. L. 1940. Glacial Chronology of the Southern Rocky Mountains. Geological Soc. Am Bull. 51 : 1851 – 1917.

Robin, G. deQ., 1975. Radio Echosounding : Glaciological Interpretations and Applications. J. Glaciology 15 : 49 – 63.

Rosenfeld, C. A. 1980. Observations on the Mount St. Helens Eruption. Am. Scientist, 68 : 494 – 509.

Tapponnier Paul and Peter Molnar, 1977. Active Faulting and Cenozoic Tectonics in China. J. Geophys. Res. 82 : 2905 – 2930.

Tapponnier, P. and P. Molnar, 1979. Active Faulting and Cenozoic Tectonics of the Tien Shan, Mongolia and Baykal Regions. J. Geophys. Res. 84 : 3425 – 3459.

Thorarinsson Sigurdur, 1969. Glacier Surges in Iceland with special reference to the Surges of Bruarjokull. Canadian J. Earth Sci. 6 (Part-2) : 875 – 882.

Williams, Richard S.,Jr., 1983. Geological Applications. In : Manual of remote Sensing, Vol. II, 2nd ed., R. N. Colwell (Ed.). American Society of Photogrammetry, Falls Church, Va. p. 1667 – 1953.

Williams, R. S.,Jr., J. G. Ferrigno, T. M. Kent and J. W. Schoonmaker Jr., 1982. Landsat Images and Mosaics of Antarctica for Mapping and Glaciological Studies. Ann. Glaciology, 3 : 321 – 326.

Wright, L. A., J. K. Otton and B. W. Troxel, 1974. Turtleback Surfaces of Death Valley viewed as a Phenomenon of Extensional Tectonics. Geology, 2 : 53 – 54.

Appendix : I

Indian Space Odyssey : The Milestones*

1962 : Indian National Committee for Space Research (INCOSPAR) formed and it worked on establishing Thumba Equatorial Rocket Launching Station (TERLS).

1963 : First sounding rocket launch from TERLS started (November 21, 1963).

1965 : Space Science & Technology Centre (SSTC) established in Thumba.

1967 : Satellite Telecommunication Earth Station set up at Ahmedabad.

1968 : TERLS dedicated to the United Nations (February 2, 1968).

1969 : Indian Space Research Organisation (ISRO) formed (August 15, 1969).

1972 : Space Commission and Department of Space set up. ISRO brought under DOS (June 1, 1972).

1972-76 : Air-borne remote sensing experiments.

1975 : ISRO becomes Government Organisation (April 1, 1975).

First Indian Satellite, Aryabahtta, launched (April 19, 1975).

1975-76 : Satellite Instructional Television Experiment (SITE) conducted.

1977 : Satellite Telecommunication Experiments Project (STEP) carried out.

1979 : Bhaskara-l, an experimental satellite for earth observations, launched (June 7, 1979).

First Experimental launch of SLV-3 with Rohini Technology Payload on board (August 10, 1979). Satellite could not be placed in orbit.

1980 : Second Experimental launch of SLV-3 Rohini satellite successfully placed in orbit. (July 18, 1980).

1981 : First developmental launch of SLV-3. RS-D1 placed in orbit (May 31, 1981).

APPLE, an experimental geo-stationary communication satellite successfully launched (June 19, 1981).

Bhasakara-II launched (November 20, 1981).

1982 : INSAT-1A launched (April 10, 1982). Deactivated on September 6, 1982.

1983 : Second developmental launch of SLV-3 RS-D2 placed in orbit (April 17, 1983).

INSAT-1B, launched (August 30, 1983).

1984 : Indo-Soviet manned space mission (April 1984).

1987 : First developmental launch of ASLV with SROSS-1 satellite on board (March 24, 1987). Satellite could not be placed in orbit.

1988 : Launch of first operational Indian Remote Sensing satellite, IRS-1A (March 17, 1988).

INSAT-1C launched (July 22, 1988). Abandoned in November 1989.

Second developmental launch of ASLV with SROSS-2 on board (July 13, 1988). Satellite could not be placed in orbit.

1990 : INSAT-1D launched (June 12, 1990).

1991 : Launch of second operational Remote Sensing satellite, IRS-1B (August 29, 1991).

1992 : Third developmental launch of ASLV with SROSS-C on board (May 20, 1992). Satellite placed in orbit.

INSAT-2A, the first satellite of the indigenously-built second-generation INSAT series, launched (July 10, 1992).

1993 : INSAT-2B, the second satellite in the INSAT-2 series, launched (July 23, 1993).

First developmental launch of PSLV with IRS-1E on board (September 20, 1993). Satellite could not be placed in orbit.

1994 : Fourth developmental launch of ASLV with SROSS-C2 on board (May 4, 1994). Satellite placed in orbit.

Second developmental launch of PSLV with IRS-P2 on board (October 15, 1994). Satellite successfully placed in polar sun synchronous orbit.

1995 : INSAT-2C, the third satellite in the INSAT-2 series, launched (Deccember 7, 1995).

Launch of third operational Indian Remote Sensing Satellite, IRS-1C (December 28, 1995).

1996 : Third developmental launch of PSLV with IRS-P3 on board (March 21, 1996). Satellite placed in polar sun-synchronous orbit.

1997 : INSAT- 2D, fourth satellite in the INSAT series, launched (June 4, 1997). Becomes inoperable on Octobeer 4, 1997. (An in-orbit satellite, ARABSAT-1C, since renamed INSAT-2DT, was acquired in November 1997 to partly augment the INSAT system).

First operational launch of PSLV with IRS-1D on board (September 29, 1997). Satellite placed in orbit.

1998 : INSAT system capacity augmented with the readiness of INSAT-2DT acquired from ARABSAT. (January 1998).

1999 : INSAT-2E, the last satellite in the multipurpose INSAT-2 series, launched by Ariane from Kourou, French Guyana (April 3, 1999).

Indian Remote Sensing Satellite, IRS-P4 (OCEANSAT), launched by polar Satellite Launch Vehicle (PSLV-C2) along with Korean KITSAT-3 and German DLR-TUBSAT from Sriharikota (May 26, 1999).

2000 : INSAT-3B, the first satellite in the third generation INSAT-3 series, launched by Ariane from Kourou, French Guyana (March 22, 2000).

2001 : Successful flight test of Geo-synchronous Satellite Launch Vehicle (GSLV), (April 18, 2001) with an experimental satellite GSAT-1 on board.

Successful launch of PSLV-C3 (October 22, 2001) placing three satellites India's TES, Belgian PROBA and German BIRD, into Polar sun-synchronous orbit.

2002 : Successful launch of INSAT-3C by Ariane from Kourou, French Guyana (January 24, 2002).

Seccessful launch of KALPANA-1 by ISRO's PSLV from SDSC, SHAR (September 12, 2002).

2003 : Successful launch of INSAT-3A by Ariane from Kourou, French Guyana (April 10, 2003).

Successful launch of second developmental test flight of GSLV (GSLV-D2) with GSAT-2 on board from SDSC SHAR, (May 8, 2003).

Successful launch of INSAT-3E by Ariane from Kourou, French Guyana (September 28, 2003).

Successful launch of IRS-P6 (RESOURCESAT-1) by ISRO's PSLV-C5 from SDSC SHAR, (October 17, 2003).

---

* *Source* : Annual Report, DOS, Govt. of India, 2003-2004.

Appendix : II

Acronyms

ADEOS	Advanced Earth Observation Satellite
AIS	Airborne Imaging Spectrometer
ALPS	Automated Launch Processing System
AMI	Active Microwave Instrumentation
ANN	Artificial Neural Network
AOCI	Airborne Ocean Color Imager
AOCS	Attitude and Orbit Control System
APT	Automatic Picture Transmission
ASAS	Advanced Soild-state Array Spectroradiometer
ASLV	Augmented Satellite Launch Vehicle
ASTER	Advanced Spaceborne Thermal Emission and reflection Radiometer
ATI	Apparent Thermal Inertia
AVHRR	Advanced Very High Resolution Radiometer
AVIRIS	Airborne Visible and Infrared Imaging Spectrometer
AWiFS	Advanced Wide Field Sensor
BSQ	Band Sequential
BIL	Band-Interleaved-by-Lines
BIP	Band-Interleaved-by-Pixel
BIS	Band-Interleaved-by-Sample
Bit	Binary Digit
BSS	Broadcast Satellite Service
CD	Compact Disk
CCD	Charge-Coupled Device (or Detector)
CCT	Computer Compatible Tape
CCU	Central Control Unit
CDF	Cumulative Distribution Function
CIR	Color Infrared
CLS	Constrained Least Squares
C/N	Carrier-to-Noise Ratio
CNES	(French) Centre National d' Etudes Spatiales
COPUOS	Committee on Peaceful Uses of Outer Space
CPU	Central Processing Unit

CRT	Cathode Ray Tube
CST	Color Space Transform
CZCS	Coastal Zone Color Scanner
DCP	Data Collection Platform
DCS	Data Collection System
DCST	Data Collection Storage and Transmission
DEM	Digital Elevation Model
DES	Delhi Earth Station
DMSP	Defense Meteorological Satellite Program
DN	Digital Number
DOD	Department of Ocean Development
DOMSAT	Domestic Satellite
DOS	Department of Space (Government of India)
DRT	Data Relay Transponder
DTG	Dynamically Tuned Gyroscope
DTM	Digital Terrain Model
DWS	Disaster Warning System
ECIL	Electronic Corporation of India Limited
EDC	EROS Data Center
EFL	Effective Focal Length
EGC	Engine Gimbal Control
EMC	Electromagnetic Compatibility
EMI	Electromagnetic Interference
EMR	Electromagnetic Radiation
EOC	Edge of Coverage
EOS	Earth Observation System
EOSAT	Earth Observation Satellite
EPA	Environmental Protection Agency
EREP	Earth Resources Experiment Package
EROS	Earth Resource Observation System
ERS	European Resource Satellite
ERTS	Earth Resource Technology Satellite, (now called Landsat)
ESA	European Space Agency
ETM	Enhanced Thematic Mapper
FANS	Future Air Navigation System
FCC	False Color Composite
FEAST	Finite Element Analysis for Structural Test
FLIR	Forward Looking Infrared
FMECA	Failure Mode Effects and Criticality Analysis
FM	Frequency Modulation
FOV	Field of View
FSS	Fixed Satellite Services

GAC	Global Area Coverage
GCP	Ground Control Point
GFOV	Ground-projected Field of View
GIFOV	Ground-projected Instantaneous Field of View
GIS	Geographic Information System
GL	Gray Level
GOES	Geostationary Operational Environmental Satellite
GPS	Global Positioning System
GRB	Gamma Ray Burst
GSD	Ground Sample Distance
GSI	Ground Sample Interval
GSLV	Geosynchronous Satellite Launch Vehicle
GSOC	German Space Operations Center
GTS	Global Telecommunications System
HAL	Hindustan Aeronautics Limited
HAT	High Altitude Test
HBF	High-Boost Filter
HCMM	Heat Capacity Mapping Mission
HILS	Hardware-in-Loop Simulation
HPF	High-Pass Filter
HRIR	High Resolution Infrared
HRIS	High Resolution Imaging Spectrometer
HRV	High Resolution Visible
HSI	Hyperspectral Imager
HYDICE	Hyperspectral Digital Imagery Collection Experiment
IAF	International Astronautical Federation
IARI	Indian Agricultural Research Institute
ICAR	Indian Council of Agricultural Research
IFOV	Instantaneous Field of View
IGS	International Ground Stations
IHS	Intensity, Hue and Saturation
IISU	ISRO Inertial Systems Unit
IIT	Indian Institute of Technology
IMD	Indian Meteorological Department
IMDPS	INSAT Meteorological Data Processing System
IMSD	Integrated Mission for Sustainable Development
INMARSAT	International Maritime Satellite
INSAT	Indian National Satellite
INTELSAT	International Telecommunication Satellite
IRS	Indian Remote-sensing Satellite
IRIS	Integrated Radar Imaging System
ISAC	ISRO Satellite Center
ISRO	Indian Space Research Organization

ISTRAC	ISRO Telemetry Tracking and Command (Network)
ITU	International Telecommunications Union
JNSDA	Japan National Space Development Agency
JERS	Japanese Earth Resources Satellite
JPL	Jet Propulsion Laboratory
JSC	Johnson Space Center
LAC	Local Area Coverage
LACIE	Large Area Crop Inventory Experiment
LED	Light Emitting Diode
LEO	Low-Earth Orbit
LEOS	Laboratory for Electro-Optic Sensors (a Unit of ISAC)
LFC	Large Format Camera
LIDAR	Light Detection And Ranging
LIF	Laser-Induced Fluorescence
LISS	Linear Imaging Self Scanner
LMSS	Land Mobile Satellite Services
LNA	Low Noise Amplifier
LPF	Low-Pass Filter
LPSC	Liquid Propulsion System Center
LRM	Local Range Modification
LSSC	Large Space Simulation Chamber
LUT	Look-Up Table / Local User Terminal
LWIR	Long Wave Infrared
MAS	Modis Airborne Simulator
MCC	Mission Control Center
MCF	Master Control Facility
MDUC	Meteorological Data Utilization Center
MEIS	Multispectral Electro-optical Imaging System
MEOSS	Monocular Electro-optic Stereo Scanner
MISR	Multi-angle Imaging Spectro Radiometer
MNF	Maximum Noise Fraction
MODIS	Moderate-resolution Imaging Spectroradiometer
MOMS	Modular Optoelectronic Multispectral Scanner
MOS	Metal Oxide Semiconductor
MSS	Multispectral Scanner/Mobile Satellite Service
MST	Mobile Service Tower (for PSLV)
MTF	Modulation Transfer Function
MWIR	Mid-Wave Infrared
MWR	Multi-Wavelength Radiometer
NAA	National Airport Authority
NADAMS	National Agricultural Drought Assessment and Monitoring System

NAL	National Aerospace Laboratory
NASA	National Aeronautics and Space Administration
NBSS&LUP	National Bureau of Soil Survey and Land Use Planning
NCIC	National Cartographic Information Center
NDC	NRSA Data Center
NDVI	Normalized Difference Vegetation Index
NHAP	National High Altitude Photography
NICNET	National Informatics Center Network
NIO	National Institute of Oceanography
NIR	Near Infrared
NMRF	National Mesosphere-stratosphere-troposphere Radar Facility
NNRMS	National Natural Resources Management System
NOAA	National Oceanic and Atmospheric Administration
NRDC	National Research and Development Corporation
NRIS	National Resources Information System
NRSA	National Remote Sensing Agency
NSSDC	National Space Science Data Center
NWDB	National Wasteland Development Board
OBTR	On Board Tape Recorder
OMS	Orbital Maneuvering System
ONGC	Oil and Natural Gas Commission
PAN	Panchromatic (Camera)
PCB	Printed Circuit Board
PCMC	Precision Coherent Monopulse C-band Radar
PCT	Principal Component Transformation
PEACE	Protection of Environment for Assuring Clean Earth
Pixel	Picture Element
PLB	Personal Locator Beacons
PRL	Physical Research Laboratory
PSF	Point Spread Function
PSLV	Polar Satellite Launch Vehicle
PWPFM	Pulse Width Pulse Frequency Modulation
RABMN	Remote Area Business Message Network
RADAR	Radio Detection and Ranging
RAM	Random Access Memory
RAR	Real Aperture Radar
RBV	Return Beam Vidicon
RCC	Rescue Coordination Center
RCS	Reaction Control System
RDSS	Radio Determination Satellite Services
RESINS	Redundant Strap-down Inertial Navigation System
RESPOND	Research Sponsored (by ISRO, program)

RF	Representative Fraction
RGB	Red Green Blue
RLE	Run-Length Encoding
RN	Radio Networking
RPA	Retarded Potential Analyzer
RRSSC	Regional Remote Sensing Service Center
RTD	Real Time Decision
SAC	Space Applications Center (of ISRO)
SADA	Solar Array Drive Assembly
SAMIR	Satellite Microwave Radiometer
SAR	Synthetic Aperture Radar
SAS&R	Satellite-Aided Search and Rescue
SAVI	Soil-adjusted Vegetation Index
SBRTN	Satellite-based Rural Telegraphy Network
SCC	Spacecraft Control Center
SDUC	Secondary Data Utilization Center
SHAR	Sri Harikota Range
SIR	Shuttle Imaging Radar
SITVC	Secondary Injection Thrust Vector Control system
SISEX	Shuttle Imaging Spectrometer Experiment
SLAR	Side Looking Airborne Radar
SM	Structural Model
SMS	Synchronous Meteorological Satellites
S/N, SNR	Signal-to-Noise Ratio
SOM	Space Oblique Mercator
SONAR	Sound Navigation And Ranging
SPC	Standardized Principal Components
SPCT	Standardized Principal Component Transformation
SPINS	Stabilized Platform Inertial Navigation System
SPL	Space Physics Laboratory
SPOT	Systeme Probatoire d' Observation de la Terre
SROSS	Stretched Rohini Satellite Series
SR	Simple Ratio
SSDA	Sequential Similarity Detection Algorithm
SSM/I	Special Sensor Microwave / Imager
SST	Sea Surface Temperature
STFSDC	Standard Time and Frequency Signal Dissemination Service
SVI	Spectral Vegetation Index
SWIR	Short Wave Infrared
TC	True Color
TCT	Tasseled-Cap Transformation
TDRS	Tracking and Data Relay Satellite
TERLS	Thumba Equatorial Rocket Launching Station

TIMS	Thermal Infrared Multispectral Scanner
TIR	Thermal Infrared
TM	Thematic Mapper
TMS	Thematic Mapper Simulator
TTC	Telemetry Tracking and Command (Network)
TV	Television
TVI	Transformed Vegetation Index
TVRO	Television Receive Only
TWT	Travelling Wave Tube
UHF	Ultra High Frequency
USO	Udaipur Solar Observatory
UV	Ultraviolet
UTM	Universal Transverse Mercator
VBT	Vainu Bappu Telescope
VHRR	Very High Resolution Radiometer
VI	Vegetation Index
VIS	Visible (spectrum)
VNIR	Visible and Near Infrared
VSSC	Vikram Sarabhai Space Center
WARC	World Administrative Radio Conference
WiFS	Wide Field Sensor
WMO	World Meteorological Organization
WRS	World Reference System

Appendix : III

Glossary

Absorptance	A measure of the ability of a body to absorb incident energy. Spectral absorptance is used to indicate with respect to a specific wavelength band.
Absorption	The process by which radiant energy is absorbed by a body (and subsequent radiation transformation by it).
Absorption Band	A range of wavelength interval in the electromagnetic spectrum within which radiant energy is absorbed by a body.
Absorptivity	An intrinsic property of the material showing its capacity to absorb incident radiant energy. Spectral absorptivity is used to indicate with respect to a specific wavelength/ wavelength band.
Accuracy	The success in / closeness of estimating the true value in a measurement process.
Across-Track Scanner	A remote sensing system that scans the swath normal to the direction of flight like a whisk broom (Alternately called a Whisk-broom scanner).
Active Remote Sensing	Remote sensing technique that uses its own (artificial) source of radiation to illuminate the target, such as in Radar.
Albedo	Ratio of electromagnetic energy reflected by an earth's surface feature to the amount of energy incident upon it.
Algorithm	(1) A fixed step-by-step procedure to accomplish a given result. (2) A computer oriented procedure for resolving a problem.
Alphanumeric	A character set composed of letters, integers, punctuation marks and special symbols.
Altitude	Height above a datum (the mean sea level).
Amplifier	A device capable of increasing the power output of an electrical or electromagnetic radiation signal.
Analog	A form of data display in which values are shown in graphic form, such as curves. Also a form of computing in which values are represented by directly measurable quantities, such as voltages or resistances. Analog computing methods contrast with digital methods in which values are treated numerically.

Glossary

Ancillary Data	In remote sensing, secondary data pertaining to the area, such as topographical, demographic or climatological data. Ancillary data may be digitized and used in the analysis process in conjunction with the primary remote sensing data.
Angle of Depression	In SLAR usage, the angle between the horizontal plane passing through the antenna and the line connecting the antenna and the target.
Angle of Incidence	(1) The angle between the direction of incoming electromagnetic radiation and the normal to the intercepting surface. (2) In SLAR system this is the angle between the vertical and a line connecting the antenna and the target.
Angle of Reflection	The angle that the electromagnetic radiation reflected from a surface makes with the normal to the surface.
Anomaly	An area of an image that differs from the surrounding normal area. For example, a concentration of vegetation within a desert scene constitute an anomaly.
Antenna	The device that radiates electromagnetic radiation from a transmitter and receives electromagnetic radiation from other antennas / other sources.
Antenna (SAR)	The effective antenna produced by storing and comparing the Doppler signals received while the aircraft travels along its flight path. This synthetic antenna (or array) is many times longer than the physical antenna actually used, thus sharpening the effective beam width and improving the azimuth resolution.
Aperture	The opening in a lens diaphragm through which light passes.
Angular Resolution	Minimum separation between two resolvable targets expressed in radian measure.
ATI	Apparent Thermal Inertia. An approximation of thermal inertia calculated as one minus albedo divided by the difference between daytime and night-time radiant temperatures.
Atmospheric Windows	Those wavelength ranges in which radiation can pass through the atmosphere with relatively little attenuation.
Attenuation	Any process in which the flux density of a parallel beam of energy decreases with increasing distance from the energy source.
Attitude	The angular orientation of a remote sensing system with respect to a geographical reference system.
Azimuth	The geographical orientation of a line given as an angle measured clockwise from north.
Background	Area on an image or the terrain that surrounds an area of interest or target.
Backscatter	(Also called backscattering) The scattering of radiant energy into

	the hemisphere of space bounded by a plane normal to the direction of the incident radiation and lying on the same side as the incident ray. The opposite of forward scatter.
Backscattering Coefficient	In Radar, a quantitative measure of the intensity of energy returned to a radar antenna from the terrain.
Band	A wavelength interval in the electromagnetic spectrum.
Band-Pass Filter	A wave filter that has single transmission band extending from a lower cutoff frequency greater than zero to a finite upper cutoff frequency.
Batch Processing	The method of data processing in which data and programs are entered into a computer that carries out the entire processing operation with no further instructions.
Beam	A focused pulse of energy.
Bit	(1) An abbreviation of binary digit, which in digital computing represents an exponent of the base 2.
Blackbody	An ideal substance that absorbs all the radiant energy incident on it and emits radiant energy at the maximum possible rate per unit area at each wavelength for any given temperature. No actual substance behaves as a true blackbody.
Blackbody Radiation	The electromagnetic radiation emitted by an ideal blackbody by virtue of its temperature.
Brightness Temperature	(1) The temperature of a blackbody radiating the same amount of energy per unit area at the wavelengths under consideration as the observed body. Also called the effective temperature. (2) The apparent temperature of a gray body (non-blackbody) determined by measurement with an optical pyrometer or radiometer.
Byte	A group of eight bits of digital data.
Calibration	The process of comparing certain specific measurements in an instrument with a standard.
Cartography	Map and chart construction.
Cascade (Electronics)	To arrange a series of elements or devices such that the output of one feeds directly into the input of another. The cascade series usually serves to amplify the effect.
CRT	Cathode Ray Tube. A vacuum tube capable of producing a black-and-white or color image by beaming electrons onto a sensitized screen. As a component of a data processing system, the CRT can be used to provide rapid, pictorial access to numerical data.
Change-Detection Images	Images prepared by digitally comparing two original images acquired at different times. The gray tones of each pixel on a change-detection image portray the amount of difference between the original images.

Glossary

Class
A surface characteristic type that is of interest to the investigator, such as forest by type and condition, or water by sediment load.

Classification
The process of assigning individual pixels of a multispectral image to categories, generally on the basis of spectral reflectance characteristics.

Clipping
The shearing off of the peak of a signal. This may affect either the positive or negative peaks, or both.

Clustering
The analysis of a set of measurement vectors to detect their inherent tendency to form clusters in Multidimensional measurement space.

Color Composite (Multiband Photography)
A color picture produced by assigning a color to a particular spectral band. In Landsat, blue is ordinarily assigned to MSS band 4 (0.5- 0.6 μm), green to band 5 (0.6 – 0.7 μm) and red to band 7 (0.8 – 1.1 μm), to form a picture closely approximating a color-infrared photograph.

Color Temperature
An estimate of the temperature of an incandescent body, determined by observing the wavelength at which it is emitting with peak intensity (its color), determined by applying Wien's Law.

Complementary Colors
(1) Two spectral colors are complementary if, when added together (as by projection), they produce neutral-hue light. (2) Colors of pigment which when mixed produce a gray.

Contrast Stretching
Improving the contrast of images by digital processing. The original range of digital values is expanded to utilize the full contrast range of the recording film or display device.

Contrast Ratio
On an image, the ratio of reflectances between the brightest and the darkest parts of the image.

Convergence of Evidence
The bringing together of several types of information in order that a conclusion may be drawn in the light of all available data. In remote sensing, often implies increase in scale to obtain more information about a smaller area.

Coordinates, Geographical
A ystem of spherical coordinates for describing the position of points on earth. The declinations and polar bearings in this system are the latitudes and longitudes respectively.

Corner Reflector (Dihedral)
A dihedral (two sided) corner reflector is formed by two intersecting flat surfaces perpendicular to each other. Radar energy striking one of these surfaces is reflected back to the antenna via the other surface. Frequently used on control points in radar survey.

Correlator, Optical (Radar)
A device that uses the original synthetic aperture radar signal film recording of Doppler phase histories to make the radar image by methods that are similar to those used in optical Fourier transforms.

Covariance
The measure of how two variables change in relation to each other (covariability). If the larger values of Y tend to be associated with the larger values of X, then covariance will be positive. If larger values

	of Y are associated with smaller values of X, the covariance will be negative. When there is no particular association between X and Y, the covariance value will approach zero.
Coverage, Stereoscopic	Aerial photographs taken with sufficient overlap to permit complete stereoscopic examination.
Cross-Polarized	Describes a radar pulse in which the polarization direction of the return beam is normal to the polarization direction of the transmitted beam. Cross-polarized images may be HV (horizontal transmission, vertical return) or VH (vertical transmission, horizontal return).
Crown Diameter, Visible	The apparent diameter of a tree crown imaged on a vertical aerial photograph.
Cultural Features	All map details representing man made elements of the landscape.
Cursor	Aiming device, such as a lens with cross-hairs, on a digitizer or an interactive computer display.
CZCS	Coastal Zone Color Scanner. A satellite-carried multispectral scanner designed to measure chlorophyll concentrations in the ocean.
Data Acquisition System	The collection devices and media that measures physical variables and records them prior to input to the data processing system. Also called Data Collection System : **DCS**.
Databank	A well-defined collection of data, usually of the same general type, which can be accessed by a computer.
Data Dimensionality	The number of variables (channels) present in the data set. The term *intrinsic dimensionality* refers to the smallest number of variables that could be used to represent the data set accurately.
Data Link	Any communication channel or circuit used to transmit data from a computer, a readout device, or a storage device.
Data Processing	Application of procedures – mechanical, electrical, computational or others, whereby data are changed from one form into another.
Data Reduction	Transformation of observed values into useful, ordered, or simplified information.
Decision Rule (Classification Rule)	The criterion used to establish discriminant functions for classification (e.g., nearest-neighbor rule, minimum-distance-to-means rule, maximum-likelihood rule).
Definition (Photography)	The degree of sharpness, that is, distinctness of small detail in the picture, image, negative or print.
Density (D)	A measure of the degree of blackening of an exposed film, plate or paper after development, or of the direct image (in case of a printout material). It is defined strictly as the logarithm of the optical opacity, where the opacity is the ratio of the incident to the transmitted (or reflected) light or transmissivity, T, as $D = \log(I/T)$.

Density Slicing	The process of converting the continuous gray tone of an image into a series of density intervals, or slices, each corresponding to a specific digital range.
Detection	A unit is said to be detected if the decision rule is able to assign it as belonging only to some given subset of categories from the set of all categories. Detection of a unit does not imply that the decision rule is able to identify the unit as specifically belonging to one particular category.
Detector (Radiation)	A device providing an electrical output that is a useful measure of incident radiation. It is broadly divisible into two groups : thermal (sensitive to temperature changes), and photodetectors (sensitive to changes in photon flux incident on the detector), or it may also include antennas and film. Typical thermal detectors are thermocouples, thermopiles and thermisters (also termed bolometers).
Dielectric Constant	Electrical property of matter that influences radar returns (also referred to as complex dielectric constant).
Diffraction	The propagation of electromagnetic radiation around the edges of opaque objects into the shadow region. A point of light seen or projected through a circular aperture will always be imaged as a bright center surrounded by light rings of gradually diminishing intensity in the shadow region. Such a pattern is called a diffraction disk, Airy disk or centric.
Diffuse Reflector	Any surface that reflects incident rays in many directions, either because of irregularities in the surface. Or because the material is optically inhomogeneous, as a paint, the opposite of a specular reflector. Ordinary writing papers are good examples of diffuse reflectors, whereas mirrors or highly polished plates are examples of specular reflectors in the visible portion of the electromagnetic spectrum. Almost all terrestrial surfaces (except calm water) act as diffuse reflectors of incident solar radiation. The smoothness or roughness of a surface depends on the wavelength of the incident electromagnetic radiation.
Digitization	The process of converting an image recorded originally on a photographic material into numerical format.
Discriminant Function	One of a set of mathematical functions which in remote sensing are commonly derived from training samples and a decision rule, and are used to divide the measurement space into decision regions.
Displacement	Any shift in the position of an image on a photograph which does not alter the perspective characteristics of the photograph (i.e., shift due to tilt of the photograph, scale change in the photograph, and relief of the objects photographed).
Display	An output device that produces a visible representation of a data set for quick visual access; usually the primary hardware component is a cathode ray tube.

Distribution Function	The relative frequency with which different values of a variable occur.
DN	Digital Number. The value of reflectance recorded for each pixel on the CCTs.
Doppler Effect	A change in the observed frequency of electromagnetic or other waves caused by relative motion between the source and the observer.
Dynamic Range	The ratio of maximum measurable signal to the minimum detectable signal.
Depolarized	Refers to a change in polarization of a transmitted radar pulse as a result of various interactions with the terrain surface.
Detectability	Measure of the smallest object that can be discerned on an image.
Digital Image Processing	Computer manipulation of the digital number values of an image.
Digitizer	Device for scanning an image and converting it into numerical format.
Directional Filter	Mathematical filter designed to enhance on an image those linear features oriented in a particular direction.
Dwell Time	Time required for a detector IFOV to sweep across a ground resolution cell.
Edge Enhancement	Image processing technique that emphasizes the appearance of edges and lines.
Ektachrome	A Kodak color positive film.
Electromagnetic Radiation	Energy propagated through space or through material media in the form of an advancing interaction between electric and magnetic fields.
Electromagnetic Spectrum	The ordered array of known electromagnetic radiations extending from the shortest wavelength (γ-rays) to the longest wavelength (electrical disturbances).
Elevation	(1) Vertical distance from the datum, usually mean sea level, to a point or object on the earth's surface. Not to be confused with altitude, which refers to points or objects above the earth's surface. (2) Architectural : An orthographic projection of any object onto a vertical plane.
Emission	With respect to electromagnetic radiation : The process by which a body emits electromagnetic radiation usually as a consequence of its temperature only.
Emissivity (ε)	The ratio of radiation given off by a surface to the radiation given off by a blackbody at the same temperature; a blackbody has an emissivity of 1, other objects between 0 and 1.
Emittance	The radiant flux per unit area emitted by a body, (exitance).
Equivalent Blackbody Temperature	The temperature measured radiometrically corresponding to that which a black body would have.

False Color	The use of one color to represent another. For example, the use of red colour to represent infrared light in color infrared film.
Far Infrared	A term for the longer wavelengths of the infrared region, from 25 µm to 1 mm, the generally accepted shorter wavelength limit of the microwave part of the electromagnetic spectrum. This band is severely limited in terrestrial use, as the atmosphere transmits very little radiation between 25 µm and the millimeter regions.
Far Range	Refers to the portion of the SLAR image farthest from the aircraft or spacecraft flight path.
Feature Extraction	The process in which an initial measurement pattern or some subsequence of measurement patterns is transformed to a new pattern feature.
FOV	Field of View. The solid angle through which an instrument is sensitive to radiation. Owing to various effects, diffractions etc., the edges are not sharp. In practice they are defined as the *half-power* points, i.e. the angle outwards from the optical axis, at which the energy sensed by the radiometer drops to half its on-axis value.
Filter (Optical)	(1) Any material which, by absorption or reflection, selectively modifies the radiation transmitted through an optical system. (2) To remove a certain component or components of electromagnetic radiation usually by means of a filter, although other devices may be used.
Filter (Digital)	Mathematical procedure for modifying values of numerical data.
Filtering	In analysis, the removal of certain spectral or spatial frequencies to highlight features in the remaining image.
Format	The arrangement of descriptive data in descriptors, identifiers or labels. The arrangement of data in bit, byte and word form in the CPU.
Frequency	Number of oscillations per unit time, or number of wavelengths that pass a point per unit time.
Gain	(1) A general term used to denote an increase in signal power in transmission from one point to another. Gain is usually expressed in decibels. (2) An increase or amplification.
Gaussian	A statistical term that refers to a normal distribution of values.
GCP	Ground Control Point. A geographical feature of known location that is recognizable on image and can be used to determine geometrical corrections.
Geocoding	Geographical referencing or coding of location of data items.
Geometric Correction	Image processing procedure that corrects spatial distortions in an image.
Geometrical Transformations	Adjustments made in the image data to change its geometrical character, usually to improve its geometrical consistency or cartographic utility.

Geostationary Satellite	Satellites travelling at the angular velocity at which the earth rotates in the same direction; as a result, they remain stationary above the same point on earth at all times.
GOES	Geostationary Operational Environmental Satellite. A NOAA satellite that acquires visible and thermal infrared images for meteorological purposes.
Gray Body	A radiating surface whose radiation has essentially the same spectral energy distribution as that of a blackbody at the same temperature, but whose emissive power is less. Its absorptivity is nonselective.
Gray Scale	A calibrated sequence of gray tones ranging from black to white.
Grid Line	One of the lines in a grid system; a line used to divide a map into squares. East-west lines in a grid system are x-lines, and north-south lines are y-lines.
Ground Control	Accurate data on the horizontal and/or vertical positions of identifiable ground points.
Ground Data	Supporting data collected on the ground, and information derived therefrom, as an aid to the interpretation of remotely recorded surveys, such as airborne imagery etc. Generally this should be performed concurrently with the airbornme surveys. Data as to weather, soils and vegetation types and conditions are typical.
Ground Information	Information derived from ground data and surveys to support interpretation of remotely sensed data.
Ground Range	The distance from the ground track (nadir) to a given object.
Ground Receiving Station	Facility that records data transmitted by a satellite, such as IRS, Landsat and so on.
Ground Resolution Cell	The area on the terrain that is covered by the instantaneous field of view (IFOV) of a detector. The size of the ground resolution cell is determined by the altitude of the remote sensing system and the instantaneous field of view of the detector.
Ground Swath	Width of the strip of terrain that is imaged by a scanner system.
Ground Track	The vertical projection of the actual flight path of an aerial or space vehicle onto the surface of the earth or other body.
Ground Truth	Term coined for data and information obtained on surface or subsurface features to aid in interpretation of remotely sensed data. Ground data and ground information are preferred terms.
Hardware	The physical components of a computer and its peripheral equipment. Contrasted with software.
Heat Capacity (c)	Ratio of heat absorbed or released by a material to the corresponding temperature rise or fall. Expressed in calories per gram per degree centigrade. Also called thermal capacity.

Glossary

HCMM	Heat Capacity Mapping Mission. NASA satellite orbited in 1978 to record day-time and night-time visible and thermal infrared images of large areas.
Highlights	Areas of bright tone on an image.
Histogram	The graphical display of a set of data which shows the frequency of occurrence (along the vertical axis) of individual measurements or values (along the horizontal axis); a frequency distribution.
Hue	The attribute of a color by virtue of which it differs from gray of the same brilliance, and which allows it to be classed as red, yellow, green, blue or intermediate shades of these colors.
Image Enhancement	Any one of a group of operations that improve the detectability of the targets or categories. These operations include, but are not limited to, contrast improvement, edge enhancement, spatial filtering, noise suppression, image smoothing and image sharpening.
Image Processing	Encompasses all the various operations that can be applied to photographic or image data. These include, but are not limited to, image compression, image restoration, image enhancement, preprocessing, quantization, spatial filtering and other image pattern recognition techniques.
Image Restoration	A process by which a degraded image is restored to its original condition. Image restoration is possible only to the extent that the degradation transform is mathematically invertible.
Imagery	The products of image forming instruments (analogous to photography).
Infrared	Pertaining to energy in the 0.7 – 100 μm wavelength region of the electromagnetic spectrum.
Insolation	Incident solar radiation.
IFOV	Instantaneous Field of View. A term specifically denoting the narrow field of view designed into detectors, particularly scanning radiometer systems, so that, while as much as 120° may be under scan, only electromagnetic radiation from a small area is recorded at any one instant.
Interactive Image Processing	The use of an operator or analyst at a console that provides the means of assessing, preprocessing, feature extracting, classifying, identifying and displaying the original imagery or the processed imagery for his subjective evaluations and further interactions.
Interpretation Key	Characteristic or combination of characteristics that enable an interpreter to identify an object on an image.
Irradiance	The measure, in power unit, of radiant flux incident upon a surface. It has the dimensions of energy per unit time (namely, watts).
Irradiation	The impinging of electromagnetic radiation on an object or surface.

Jitter (1) Intensity of the signal or trace of a cathode ray tube. (2) Small rapid variations in a waveform due to deliberate or accidental electrical or mechanical disturbances or to changes in the supply voltages, in the characteristics of components, etc.

Kernel Two-dimensional array of digital numbers used in digital filtering.

Kinetic Temperature The internal temperature of an object, which is determined by the molecular motion. Kinetic temperature is measured with a contact thermometer, and differs from radiant temperature, which is a function of emissivity and internal temperature.

Lambertian Surface An ideal, perfectly diffusing surface, which reflects energy equally in all directions.

Langley A unit of luminous intensity, defined as 4.184×10^4 joules / m^2.

Laplacian Filter A form of nondirectional digital filter.

Layover In radar images, the geometric displacement of the top of objects towards the near range relative to their base.

Lineament A linear topographical or tonal feature on the terrain and on images and maps, which may represent a zone of structural weakness.

Line-pair Pair of light and dark bars of equal widths. The number of such line pairs aligned side by side that can be distinguished per unit distance expresses the resolving power of an imaging system.

Look Angle (Radar) The direction of the look, or direction in which the antenna is pointing when transmitting to and receiving from a particular cell.

Look Direction Direction in which pulses of microwave energy are transmitted by a SLAR system. Look direction is normal to the azimuth direction. Also called range direction.

Low-Sun-Angle Photograph Aerial photograph acquired in the morning, evening, or winter when the sun is at a low elevation above the horizon.

Luminance Quantitative measure of the intensity of light from a source.

Map A representation in a plane surface, at an established scale, of the physical features (natural, artificial or both) of a part of the earth's surface, with the means of orientation indicated.

Map (Thematic) A map designed to demonstrate particular features or concepts. In conventional use this term excludes topographical maps.

Maximum Likelihood Rule A statistical decision criterion to assist in the classification of overlapping signatures; pixels are assigned to the class of highest probability.

Microwave Electromagnetic waves between 1mm and 1m in wavelength, or 300 GHz to 0.3 GHz in frequency. Microwave region is bounded on the

	shorter wavelength side by the far infrared (at 1mm) and on the long wavelength side by very high-frequency radio waves (at 1m).
Mie Scattering	Multiple reflection of light waves by atmospheric particles that have the appropriate dimensions of the wavelength of light.
Minimum-Distance Classifier	A classification technique that assigns raw data to the class whose mean falls the shortest Euclidean distance from it.
Minimum Ground Separation	Minimum distance on the ground between two targets at which they can be resolved on an image.
Minus-Blue Photographs	Black-and-white photographs acquired using a filter that removes blue wavelengths to produce higher spatial resolution.
Modulation	To vary the frequency, phase or amplitude of electromagnetic waves.
Modulation Transfer Function (MTF)	A method of describing spatial resolution.
Mosaic	An assemblage of overlapping aerial or space photographs or images whose edges have been matched to form a continuous pictorial representation of a portion of the earth's surface.
Mosaicking	The assembling of photographs or other images whose edges are cut and matched to form a continuous photographic representation of a portion of the earth's surface.
Multiband System	A system for simultaneously observing the same (small) target with several filtered bands, through which data can be recorded. Usually applied to cameras; may be used for scanning radiometers that use dispersant optics to split wave-length bands apart for viewing by several filtered detectors.
Multichannel System	Usually used for scanning systems capable of observing and recording several channels of data simultaneously, preferably through the same aperture.
Multispectral	Generally used for remote sensing in two or more spectral bands, such as visible and infrared.
Multispectral Classification	Identification of terrain categories by digital processing of data acquired by multispectral scanners.
Multispectral Scanner	Scanner system that simultaneously acquires images of the same scene at different wavelengths.
Multivariate Analysis	A data analysis approach that makes use of multi-dimensional interrelations and correlations within the data for effective discrimination.
Nadir	(1) That point on the celestial sphere vertically below the observer, or 180° from the zenith. (2) That point on the ground vertically beneath the perspective center of the camera lens.
Near Range	Refers to the portion of a SLAR image closest to the aircraft flight path.

Noise Random or regular interfering effects in the data which degrade its information-bearing quality.

Nondirectional Filter Mathematical filter that treats all orientations of linear features equally. Geometric irregularities on images that are not constant and can not be predicted from the characteristics of the imaging system.

Nonsystematic Distortion Geometric irregularities on images that are not constant and can not be predicted from the characteristics of the imaging system.

Orbit (1) The path of a body or particle under the influence of a gravitational or other force. For instance, the path of a satellite around a celestial body such as the earth, under the influence of gravity. (2) To go around the earth or other body in a trajectory.

Orthophotograph Photograph having the properties of an orthographic projection. It is derived from a conventional perspective photograph by simple or differential rectification so that image displacements caused by camera tilt and terrain relief are removed.

Overlap (1) The area common to two successive photographs or images along the same flight path, expressed as the percentage of the photograph or image area. (2) Extent to which adjacent images or photographs cover the same terrain, expressed as a percentage.

Overlay (1) A transparent sheet giving information to supplement that shown on maps. When the overlay is laid over the map on which it is based, its details will supplement the map. (2) A tracing of selected details on a photograph, mosaic or map to present the interpreted features and the pertinent detail.

Panchromatic Used for films that are sensitive to a broadband electromagnetic radiation (e.g., the entire visible part of the spectrum), and for broadband photograph.

Panchromatic Film Black-and-white film that is sensitive to all visible wavelengths.

Parallax The apparent change in the position of one object or point, with respect to another, when viewed from different angles. As applied to aerial photos, the term refers to the apparent displacement of two points along the same vertical line when viewed from a point (the exposure station) not on the same vertical line.

Parallel-Polarized Describes a radar pulse in which the polarization of the return beam is the same as that of the transmission. Parallel-polarized images may be HH (horizontal transmission, horizontal return) or VV (vertical transmission, vertical return).

Pass In digital filters, refers to the spatial frequency of data transmitted by the filter. High-pass filters transmit high-frequency data, low-pass filters transmit low-frequency data.

Glossary

Passive Remote Sensing — Remote sensing of energy naturally reflected or emitted from the terrain.

Pattern — Regular repetition of tonal variations on an image or photograph.

Pattern Recognition — Concerned with, but not limited to, problems of : (1) pattern identification, (2) pattern discrimination, (3) pattern classification, (4) pattern segmentation, (5) cluster identification, (6) feature selection, (7) feature extraction, (8) preprocessing, (9) filtering, (10) screening and (11) enhancement.

Periodic Line Dropout — Defect on satellite images in which no data are recorded for every m^{th} or n^{th} scan line, causing a black line on the image.

Periodic Line Striping — Defect on satellite images in which every m^{th} or n^{th} scan line is brighter or darker than the others, caused by the sensitivity of one detector being higher or lower than the others.

Perihelion — For an elliptic orbit about the sun, the point closest to the sun.

Perspective — Representation, on a plane or curved surface, of natural objects as they appear to the eye.

Photodetector — Device for measuring energy in the visible light band.

Photogrammetry — The art / science of obtaining reliable measurements by means of photography.

Pitch — Rotation of an aircraft about the horizontal axis normal to its longitudinal axis, which causes a nose-up nose-down attitude.

Pixel — Contraction of picture element. A data element having both spatial and spectral aspects. The spatial variable defines the apparent size of the resolution cell (i.e., the area on the ground represented by the data values), and the spectral variable defines the intensity of the spectral response for that cell in a particular channel.

Polarization — The direction of vibration of the electrical field vector of the electromagnetic radiation. In SLAR systems polarization is either horizontal or vertical.

Positive — (1) A photographic image having approximately the same rendition of light and shade as the original object / terrain. (2) A film, plate or paper containing such an image.

Previsual Symptom — A vegetation anomaly that is recognizable on infrared film before it is visible to the naked eye or on normal color photographs. It results when stressed vegetation loses its ability to reflect photographic infrared energy and is recognizable on infrared color film by a decrease in brightness of the red hues.

Primary Colors — A set of three colors that in various combinations will produce the full range of colors in the visible spectrum. There are two sets of primary colors : additive and subtractive.

Principal-Component Image — Digitally processed image produced by a transformation that recognizes maximum variance in multispectral images.

Printout	Display of computer data in alphanumeric format.
Pseudoscopic View	A reversal of normal stereoscopic effect, causing valleys to appear as ridges and ridges as valleys.
Pulse	Short burst of electromagnetic radiation transmitted by a radar antenna.
Pulse Length	Duration of a burst of energy transmitted by a radar antenna, measured in microseconds.
Pushbroom Scanner	An alternate term for the along-track scanner.
Radar	Acronym for radio detection and ranging. Radar is an active form of remote sensing that operates in the microwave and radio wavelength regions.
Radar Beam	A fan shaped beam of electromagnetic energy produced by the radar transmitter.
Radar, Brute Force	An imaging real aperture radar employing a long physical antenna to achieve a narrow beam width for improved resolution.
Radar Shadow	A dark area of no return on a radar image that extends in the far-range direction from an object on the terrain that intercepts the radar beam.
Radar, Synthetic Aperture	A radar in which a synthetically long apparent or effective aperture is constructed (mathematically) by integrating multiple returns from the same ground cell, taking advantage of the Doppler effect to produce a phase history film (hologram) or tape that may be optically or digitally processed to reproduce an image of much higher azimuth resolution.
Radiance	The accepted term of radiant flux in power units (e.g., W) and not for flux density per solid angle (e.g., $W\ cm^{-2}\ sr^{-1}$).
Radiant Flux	The time rate of flow of radiant energy; radiant power.
Radiant Power	Rate of change of radiant energy with time. May be further qualified as spectral radiant power, at a given wavelength.
Radiant Temperature	Concentration of the radiant flux from a material. Radiant temperature is the product of the kinetic temperature and the emissivity to the power one-fourth.
Radiometer	An instrument for quantitatively measuring the intensity of electromagnetic radiation in some band of wavelengths in any part of the electromagnetic spectrum (e.g., infrared radiometer, microwave radiometer)
Radiometric Correction	Correcting gain and offset variations in satellite data. Procedure calibrates and corrects the radiation data provided by the satellite sensor detectors.

Glossary

Random Line Dropout — In scanner images, the loss of data from individual scan lines in a nonsystematic fashion.

Range Direction — For radar images, this is the direction in which energy is transmitted from the antenna and is normal to the azimuth direction. Also called look direction.

Range, Dynamic — The ratio of maximum measurable signal to the minimum detectable signal. The upper limit usually is set by saturation and the lower limit by noise.

Range Resolution — In radar images, the spatial resolution in the range direction, which is determined by the pulse length of the transmitted microwave energy.

Raster — The scanned (illuminated) area of the Cathode Ray Tube.

Raster Pattern — Pattern of horizontal lines swept by an electron beam across the face of a Cathode Ray Tube that constitute the image display.

Ratio Image — An image prepared by processing digital multispectral data as follows : for each pixel, the value of one band is divided by that of another. The resulting digital values are displayed as an image.

Rayleigh Criterion, in Radar — The relationship between surface roughness, depression angle and wavelength that determines whether a surface will respond in a rough or smooth fashion to the radar pulse.

Rayleigh Scattering — The wavelength-dependent scattering of electromagnetic radiation by particles in the atmosphere much smaller than the wavelengths scattered.

Real-Aperture Radar — Side looking airborne radar in which azimuth resolution is determined by the physical length of the antenna and by the wavelength. The radar returns are recorded directly to produce images. Al;so called brute force radar.

Real Time — Time in which reporting on events or recording of events is simultaneous with the events.

Recognizability — Ability to identify an object on an image.

Rectification — The process of projecting a tilted or oblique photograph onto a horizontal reference plane, the angular relation between the photography and the plane being determined by ground reconnaissance. Transformation is the special process of rectifying the oblique images from a multiple-lens camera to equivalent vertical images by projection onto a plane that is perpendicular to the camera axis. In this case, the projection is onto a plane determined by the angular relations of the camera axis and not necessarily onto a horizontal plane. Differential rectification : The process of removing the effects of tilt, relief and other distortions from imagery by correcting small portions of the imagery independently.

Registration — Process of superposing two or more images or photographs so that equivalent geographic (ground control) points coincide.

Relief	Vertical irregularities of a surface.
Relief Displacement	A shift in position of the optical image of an object caused by height of the object above or depth below a datum plane.
Remote Sensing	In the broadest sense, the measurement or acquisition of information of a remote object or phenomenon, by a recording device (the remote sensor).
Representative Fraction	The relation between map or photo distance and ground distance expressed as a fraction (e.g., 1/25,000), or often as a ratio (1 : 25,000). Also called scale.
Resolution	Ability to separate closely spaced objects on an image or photograph. It may be expressed as the most closely spaced line-pairs per unit distance that can be distinguished. Also called spatial resolution.
Resolution Cell	The smallest area in a scene considered as a unit of data.
Resolving Power	A mathematical expression of lens definition, usually stated as the maximum number of line-pairs per millimeter that can be resolved in an image.
Return, in Radar	A pulse of microwave energy reflected by the terrain and received at the radar antenna. The strength of the return is referred to as the return intensity.
Roll	Rotation of an aircraft about the longitudinal axis to cause a wing-up or wing-down attitude.
Roll Compensation System	Component of an airborne scanner system that measures and records the roll of the aircraft. This information is used to correct the imagery for distortion due to roll.
Rough Criterion	In radar, the relationship between surface roughness, depression angle and wavelength that determines whether a surface will be considered by the incident radar pulse as a rough or intermediate fashion.
Roughness	For radar images, this term describes the average vertical relief of small-scale irregularities of the terrain surface. Also called surface roughness.
Sample	A subset of a population selected to obtain information concerning the characteristics of the population.
Satellite	An object in orbit around a celestial body.
Saturation	In IHS system, represents the purity of color. Saturation is also the condition where energy flux exceeds the sensitivity range of a detector.
Scale	The ratio of a distance on an image (or photograph or map) to its corresponding distance on the ground. The scale of a photograph varies from point to point because of displacements caused by tilt and relief, but is usually taken as f/H, where f is the focal length

Glossary

of the camera and H is the height of the camera above the mean ground elevation.

Scan Line The narrow strip on the ground that is swept by the instantaneous field of view of a detector in a scanner system.

Scanner (1) Any device that scans, and thus produces an image.
(2) A radar set incorporating a rotatable antenna, or radiator element, motor devices, mounting etc. for directing a searching radar beam through space and imparting target informatin to an indicator.

Scanning Radiometer A radiometer, which by the use of a rotating or oscillating plane mirror, can scan a path normal to the movement of the radiometer.

Scattering Multiple reflections of electromagnetic waves by particles and surfaces.

Scatterometer Non-imaging radar device that quantitatively records backscatter of terrain as a function of incidence angle.

Scene Area on the ground that is covered by an image or photograph.

Seasat NASA unmanned satellite that acquired L-band radar images of the sea surface in 1978.

Sensitivity The degree to which a detector responds to electromagnetic energy incident upon it.

Sensor Device that receives electromagnetic radiation and converts it into a signal that can be recorded and displayed as either numerical data or an image.

Shuttle Imaging Radar L-band radar system deployed on the Space Shuttle.

Sidelap Extent of lateral overlap between images acquired on adjacent flight lines.

Side Looking Radar An all weather, day / night, side scanning remote sensor which is particularly effective in imaging large areas of terrain. It is an active sensor, as it generates its own energy which is transmitted and received to produce an image of the ground. Also referred to as side looking airborne radar.

Signal Carrier of information. Electromagnetic signals are received by the remote sensor from the scene and converted to another form for transmission to the processing system.

Signal-To-Noise Ratio The ratio of the level of the information-bearing signal power to the level of the noise power.

Signature Any characteristic or series of characteristics by which a material or an object may be recognized in an image, photo, or data set. See also spectral signature.

Signature Analysis Techniques Techniques that use the variation in the spectral reflectance or emittance of objects as a method of identifying the objects.

Signature Extension	The use of training statistics obtained from one geographical area to classify data from similar areas some distance away; includes consideration of changes in atmosphere, and other geographical and temporal conditions that can cause differences in signal level for signal classes of interest. See spectral signature.
Slant Range	For radar images, this term represents the distance measured along a line between the antenna and the target.
Smoothing	Averaging of densities in adjacent areas to produce more gradual transitions.
Smooth Criterion	In radar, the relationship between surface roughness, depression angle and wavelength that determines whether a surface will scatter the incident radar pulse in a smooth or intermediate fashion.
Software	The computer programs that drive the hardware components of a data processing system; includes system monitoring programs, programming language processors, data handling utilities and data analysis programs.
Spatial Filter	An image transformation, usually a one-to-one operator used to lessen noise or enhance certain characteristics of the image. For any particular (x,y) coordinate on the transformed image, the spatial filter assigns a gray shade on the basis of the gray shades of a particular spatial pattern near the coordinates (x,y).
Spatial Information	Information conveyed by the spatial variations of spectral response (or other physical variables) present in the scene.
Spectral Band	An interval in the electromagnetic spectrum defined by two wavelengths, frequencies or wave numbers.
Spectral Interval	The width, generally expressed in wavelength or frequency of a particular portion of the electromagnetic spectrum. A given sensor (e.g., radiometer or camera film) is designed to measure, or be sensitive to energy received at the satellite from that part of the spectrum. Also termed spectral band.
Spectral Reflectance	The reflectance of electromagnetic energy at specified wavelength intervals.
Spectral Regions	Conveniently designated ranges of wavelengths subdividing the electromagnetic spectrum; for example, the visible region, near infrared region, middle infrared region, microwave region.
Spectral Response	The response of a material as a function of wavelength to incident energy, particularly in terms of the measurable energy reflected from and emitted by the material.
Spectral Sensitivity	Response or sensitivity, of a film or detector to radiation in different spectral regions.

Glossary

Spectral Signature	Quantitative measurement of the properties of an object at one or several wavelength intervals.
Spectral Vegetation Index	An index of relative amount and vigor of vegetation. The index may be calculated from two or more spectral band information (e.g., NDVI, greenness index and so on).
Spectrometer	A device to measure the spectral distribution of electromagnetic radiation.
Spectrophotometer	A photometer which measures the intensity of electromagnetic radiation as a function of the frequency or wavelength of the electromagnetic radiation.
Specular Reflection	The reflectance of electromagnetic energy without scattering or diffusion, as from a surface that is smooth in relation to the wavelengths of incident energy. Also called mirror reflection.
Steradian	The unit of solid angle that cuts unit area from the surface of a sphere of unit radius centered at the vertex of the solid angle. There are 4π steradians in a sphere.
Stereo Base	A line representing the distance and direction between complementary image points (photo nadir points) on a streo-pair of photos correctly oriented and adjusted for comfortable stereoscopic vision under a given stereoscope, or with the unaided eyes.
Stereoscope	A binocular optical instrument for assisting the observer to view two properly oriented photographs or diagrams to obtain the mental impression of a three dimensional model.
Stereoscopic Image	That mental impression of a three dimensional object which results from stereoscopic vision (stereo viewing).
Stereo-Pair	Two overlapping images or photographs that may be viewed stereoscopically.
Subscene	A portion of an image that is used for detailed analysis.
Subtractive Primary Colors	Yellow, magenta and cyan. When used as filters for white light, these colors remove blue, green and red light respectively.
Sun Synchronous	An earth satellite orbit in which the orbital plane is near polar and the altitude such that the satellite passes over all places on earth having the same latitude twice daily at the same local sun time.
Supervised Classification	Digital information extraction technique in which the operator provides training-site information that the computer uses to assign pixels to classes or categories.
Swath Width (**Total Field of View**)	The overall plane angle or linear ground distance covered by a multispectral scanner in the across-track direction.
Synchronous Satellite, with Respect to Earth	An equatorial west-to-east satellite orbiting the earth at an altitude of 36,000 Km, at which altitude it makes one revolution in 24 hrs synchronous with the earth's rotation.

Synoptic View	The ability to see or otherwise measure widely dispersed areas at the same time and under the same conditions; e.g., the overall view of a large portion of the earth's surface which can be obtained from satellite altitudes.
Synoptic Stereo Image	A stereo model made by digital processing of a single image. Topographic data are used in calculating the geometric distortions. Also called digital elevation model or digital terrain model.
System	Structured organization of people, theory methods and equipment to carry out an assigned set of tasks (e.g., digital image processing task).
Systematic Distortion	Geometric irregularities on images that are caused by known and predictable characteristics.
Target	(1) An object on the terrain of specific interest in a remote sensing investigation. (2) The portion of the earth's surface that produces, by reflection or emission, the radiation measured by the remote sensing system.
Telemetry	The science of measuring quantities or data, transmitting the data to a distant station and there interpreting, indicating or recording the data.
Telemetry Link	The system for transmitting data over long distances using radio techniques.
Texture	In a photo image, the frequency of change and arrangement of tones. Some descriptive adjectives for textures are fine, medium or coarse; and stippled or mottled.
Thermal Band	A general term for middle infrared wavelengths which are transmitted through the atmospheric window at $8 - 14$ µm. Occasionally used for the windows around $3 - 6$ µm.
Thermal Capacity (C)	The ability of a material to store heat, expressed in cal g^{-1} $°C^{-1}$.
Thermal Conductivity (K)	The measure of the rate at which heat passes through a material, expressed in cal cm^{-1} s^{-1} $°C^{-1}$.
Thermal Crossover	On a plot of radiant temperature versus time, this refers to the point at which the temperature curves for two different materials intersect.
Thermal Diffusivity (κ)	Governs the rate at which temperature changes within a substance, expressed in cm^2 s^{-1}.
Thermal Inertia (P)	A measure of the response of a material to temperature changes, expressed in cal cm^{-2} $°C^{-1}$ $s^{-1/2}$.
Thermal Infrared	The preferred term for the middle wavelength range of the infrared region, extending roughly from 3 µm at the end of the near infrared, to about $15 - 20$ µm, where the far infrared begins. In practice the limits represent the envelop of energy emitted by the earth behaving as a gray body with a surface temperature around 300 K.

Glossary

Thematic Mapper	An across-track scanner deployed on Landsat that records seven bands of data from the visible through thermal infrared regions.
Threshold	The boundary in spectral space beyond which a data point, or pixel, has such a low probability of inclusion in a given class that the pixel is excluded from that class.
Tilt	The angle between the optical axis of the camera and the plumb line for a given photo.
Tone	Each distinguishable shade of gray from white to black on an image.
Tracking and Data Relay Satellite	Geostationary satellite used to communicate between ground recording stations and satellites such as Landsat.
Training	Informing the computer system which sites to analyze for spectral properties or signatures of specific land cover classes; also called signature extraction.
Training Samples	The data samples of known identity used to determine decision boundaries in the measurement or feature space prior to classification of the overall set of data vectors from a scene.
Training Sites	Recognizable areas on an image with distinct (spectral) properties useful for identifying other similar areas in supervised classification.
Transmissivity	Transmittance for a unit thickness of sample. May be further qualified as spectral transmissivity.
Transmittance	The ratio of radiant energy transmitted through a body to that incident upon it.
Transparency	(1) The light transmitting capability of a material. (2) Image on a transparent photographic material, normally a positive image.
Travel Time	In radar, the time interval between the generation of a pulse of microwave energy and its return from the terrain.
Unsupervised Classification	Digital information extraction technique in which the computer assigns pixels to categories with no instructions from the operator.
Vegetation Anomaly	Deviation from the normal distribution or properties of vegetation. Vegetation anomalies may be caused by faults, trace elements in soil, or other factors.
Value	Degree of lightness, one of the attributes, along with hue and saturation, that may be thought of as the dimensions of color.
Variance	Variance of a random variable is the expected value of square of the deviation between that variable and its expected value. It is a measure of the dispersion of the individual unit values about their mean.

Vertical Exaggeration	In a stereo model, the extent to which the vertical scale appears larger than the horizontal scale.
Vidicon	(1) A storage-type electronically scanned photoconductive television camera tube, which often has a response to radiations beyond the limits of the visible region. Particularly useful in space applications, as no film is required. (2) An image-plane scanning device.
Vignetting	A gradual reduction in density of parts of a photographic image caused by the stopping of some of the rays entering the lens.
Volume Scattering	In radar, interaction between electromagnetic radiation and the interior of a material.
White Noise	Noise whose spectral density is independent of frequency.
Window	A band of electromagnetic spectrum which offers maximum transmission and minimal attenuation through a particular medium with the use of a specific sensor.
X-Band	Radar wavelength region from 2.4 to 3.8 cm.
Yaw	Rotation of an aircraft about its vertical axis, causing the longitudinal axis to deviate from the flight line.
Zenith	The point in the celestial sphere that is exactly overhead; opposed to nadir.

Appendix : IV

Index

A

Absorber 11
Absorptance 15, 17
Absorption band 146
Absorptivity 17
Accuracy of positional
  measurement 179
Across-track scanner 62, 63
Active faults 223
Active sensor 159
Adjacency
  polygon-polygon 197
  polygon-polygon (arc) 197
  polygon-polygon (node) 198
  line-line 199
  line-node 200
Administrative boundaries 192
Advantageous geometry 181
Aerial photograph 73, 224
Aerial thermographs 225
Agricultural applications 212
Airborne multispectral scanner 76
Algebraic computation
  (3-D location) 178
Along-track scanner 63, 64
Amplitude modulation 90
Analysis :
  data/signal 1, 225
  image 100
Ancillary data/information 209
Anisotropic properties 49
Antenna 51
Antenna gain 51, 52

Apparent soil thermal inertia 214
Arc attribute table 194
Arc coordinates 194
Arch dam 220
Archival 95
Aridity index 214, 215
Artificial neural network classification 143
Aspect 39, 133, 218
Associated attribute data 193, 194
Association 133
Atmospheric effects :
  absorption bands 18, 19, 25
  scattering 18, 26, 27
  transmission 18, 19, 25
Atmospheric windows 28
Atomic clock 175
Attitude 87
Attribute data 194
Auroral activity 7
Automatic digitization 192
Azimuth resolution 165

B

Back scattering cross-section 39
Back slope 221
Band-interleaved-by-pixel 95
Band-interleaved-by-line 95
Band-Pass Filter 126
Band-Sequential 95, 101
Band width 28-34
Beam-width (radar) 166
Binary encoding 147
Bit 99, 100

Black-and-white film 56
Blackbody radiation curves 10, 11
Box filter 126
Brightness temperature 50, 51
Brute force radar 166
Buffers 205
Bulk density 162

C

C/A-code 179
Calving glacier 226
Camera (framing system) :
    large format camera 59
    Hasselblad camera 59
    Return beam vidicon 60
Cascaded linear filters 127
Cascade volcanoes 226
Cathode ray tube 39, 100, 160, 208
Cavity 10
Central processing unit 99
Channel gradient 222
Channel sinuosity 222
Charge-coupled device 63
Classification/classifiers
  (digital) :
    artificial neural network 143
    level slice/box 136
    maximum likelihood 138
    minumum-distance-to-means 136
    parallelepiped 136, 138
    true parallelepiped 138
Classification algorithm for
  hyperspectral data 147
Climatic classification 215
Clock bias error 179
Close canopy 219
Coal mine-fire 227
Coastal zone color scanner 32
Code synchronization 176
Collateral data 213, 216
Color film 56
Color infrared 56
Color resolution 100
Color space transform 118
Combination filter 127

Compiler 100
Complex dielectric constant 39, 43
Computer compatible tape 100, 101
Compton scattering 26
Computation of 3-D location :
    algebraic method 178
    method of triangulation 177
Connectivity 200
Containment 202
Continuous receiver (GPS) 181
Contrast 68, 70, 71, 112
Contrast stretch 113
Convolution filter 126
Corner reflector (radar) 42
Cosmic ray particles 5
Correlation 110
Coulomb 5
Covariance (of multispectral
  data) 110
Coverage (of images) 86, 224
Coverage type (GIS) :
    point coverage 193, 194
    line coverage 193, 194
    polygon coverage 193, 194
Crop area estimation 212
Crop production forecasting 212
Crop stress monitoring 212
Cultural environment 190

D

Dam :
    arch dam 220
    earthen dam 220
    concrete dam 220
    gravity dam 220
Dam failure 227
Dam site selection 220
Data acquisition 90
Data analysis 225
Data analysis function 204
Database management system 185
Data calibration/correction 131
    geometric correction 94
    radiometric correction 93
Data collection platform 91

Index

Data collection system 91
Data / image display 101
Data dissemination 95
Data formatting 95
Data input function 203
Data management function 204
Data manipulation function 204
Data loading 100
Data output function 207
Data representation 187
Data retrieval 185, 188
Decorrelation stretch
 (color images) 117
Deforestation 215
Density slicing 100
Depression angle (radar) 162
Depth of penetration (radar) 162
Descriptive data 185
Desertification 214
Detector :
 film (VIS/NIR, IR) 56
 linear array (LISS) 64, 65
 photoconductive 57
 photoemissive 57
 photon 57
 photovoltaic 57
 quantum 57
 thermal 56
Dielectric properties 43
Differential GPS 181, 183
Digital image processing 99
Digital elevation model 218
Digital number 106
Digital terrain model 167
Digitizer 186
Discriminant function 132
Display resolution 100
Directional filters 127
Distance from headwaters 223
Doppler frequency shifts 166
Drainage pattern 214
Duplexer 39

E

Earthquake 220, 223, 224
Ecological range condition 219
Edge enhancement 127
Effective resolution element 68
Effluent irrigation 221
Electric vector 7
Electrodynamics 8
Electromagnetic induction 7
Electromagnetic radiation :
 frequency 8
 generation of 8
 interaction with matter 17, 37
 spectrum 9
 wavelength 9, 11, 13
Electromagnetic signal 8
Emission of electromagnetic signal 10
Emissivity :
 black body 12
 gray body 12
 selective radiator 12
 of terrestrial materials 24
Emittance
 radiant 15
Engineering applications 220
Engineering factors 220, 221
Environmental impact 221
Environmental problems 220
Ephemeris of GPS satellites 175, 176
Errors in GPS 180
Establishment stage (crop) 212

F

False color composite 101
Fast movement/surge glaciers 226
Faults (geologic) 224
Feature class 132
Feature extraction 132
Feature extraction for
 hyperspectral data 146
Feromones 7
Field :
 electric 5
 electromagnetic 7
 gravitational 4

magnetic 5
pressure 5
Field of view 63
Filters 125
Flood flow 221
Flood peaks and fronts 224
Flood plains 227
Flow depth 221
Fluvial features 221
Foliar cover :
    close canopy 219
open canopy 219
Forage production 220
Forage quality 220
Forest 217
Forest fires 218
Forest fire damage assessment 218
Forest fire detection 218
Forest fire hazard zone mapping 218
Fourier transform 130
Fractures 225
Fracture patterns :
    radial 225
    ring types 225
Frequency modulation 9, 90
Fuzzy set classification 145

G

Gaussian (normal) distribution 109
Gaussian maximum likelihood
  classifier 138, 141
Geographies 185
Geographic information system 185
GIS files 187
GIS input to remote sensing system 209
Geologic hazard 221
Geometric correction 94
Geomorphic hazard 224
Geostationary satellite 78
Geosynchronous orbits 78
Glacier-outburst floods 226
Glaciologic hazard 226
GPS receiver 174
Global positioning system 174
Gradient filter 129

Grassland 220
Gray body 12
Gray level (GL) 113, 114
Gravitational constant 4
Gravity survey 4
Greenness index 123, 124, 125
Ground control point (GCP) 105, 106
Ground control segment 176
Ground cover (crop area) 212
Ground coverage (by satellite swath) 86
Ground failures 223, 224
Ground monitoring stations 176
Ground-projected instantaneous
  field of view 62-65
Ground resolution (radar) 167
Ground resolution cell 63, 65, 164
Ground truth collection 131
Ground water 214
Ground water supplies 221

H

Hard classification 133
Hasselblad camera 59
Hardware 99
Haze 27
Head scraps 224
Heat islands 216
Heavy grazing 219
Heterogeneous aquifers 214
High altitude photography 75
High altitude sounding rockets 76
High-Boost filter 127
High-gain antenna 90
High-Pass filter 126
Histogram 107
Histogram equalization 115
Historical data (on crop productivity) 213
Historical data (meteorological) 213
Human induced hazard 226
Hummocks 224
Hybrid supervised-unsupervised
  classification 143
Hydrology 187, 194, 220
Hyperspectral classification 147
Hyperspectral image analysis 145

I

Identifier 108, 187
Image registration 106
Image restoration/display 101
Image-based information system 209
Image classification (thematic) 132
Image enhancement 113, 125
Image magnification 103
Image processing 99
Image quality parameters :
    contrast 112
    modulation 112
    S/N 112
Image rectification 104
Image reduction 101
Image statistics : univariate 106
    kurtosis 107
    skewness 107
    variance 107
Image statistics : multivariate 109
    correlation 110
    covariance 110
Imaging radar 159
Imaging sensors 57
Imaging system 57
Impact assessment study 225
Improving efficiency (of GPS) 181
Industrialization 216
Information extraction 132
Infrared band 9, 28, 29
Infrared emission 14, 24
Infrasonic 6
Infrasonic arrays 225
Inhomogeneities 52
Instantaneous field of view 62-65
Interaction of microwave with
  bodies of known shapes :
    facets 39
    spheres 41
    cylinders 42
    dihedral corner reflectors 42
    trihedral corner reflectors 42
Interaction of radiation with
  matter (VIS/NIR regions) 17
Interaction of radiation with
  matter (microwave region) 37

IHS transformation 118
International space station 77
Interpretation keys : 133, 214
    tone, size, shape, texture, pattern, association
    location, aspect, shadow, resolution
Inverse square law 4
Ionospheric error 180
Irradiance 15

J

Jiggling charges/fields 8

K

K-means clustering algorithm 142
Kernel 127
Kirchhoff's law 12
Kurtosis 107

L

Land cover 187
Landsat 77, 79, 80
Landsat MSS and TM coefficients 124, 125
Landslides 224
Land-subsidence features 224
Land surface water 213
Land use/land cover
  classification system 216, 217
Land use/land cover changes 216
Land use zoning 192
Large format camera 59
Lava flow 225
Leaf area index 124
Level-slice classifier 136
Lineaments 214
Linear contrast stretch 114
Linear filter 126
Linear imaging self scanner 64, 65
Linkage between remote
  sensing and GIS 208
Livestock 220
Lobes 224
Look angle (radar) 159
Loss factor/loss tangent (δ) 45

Low sun angle photograph 224
Low-Pass filter 126
Low sun-elevation imagery 214, 224

M

Manned satellites 76, 77
Map dissolve 205
Map overlay 205
Mapping unit 72
Master control station (GPS) 175, 176
Maths coprocessor 99
Maximum flood inundation 224
Maximum likelihood classifier 138
Maximum vegetative phase 212
Mean 107
Median 107
Megapolish 216
Meteorological data 213, 216, 219, 220
Meteorological satellites 77
Method of triangulation (GPS) 177
Metropolish 216
Microwave 9, 28, 29
Microwave bands 10, 29, 161
Microwave emission 50
Microwave radiometry 49
Midwave infrared 29
Mie scattering 27
Minimum-distance-to-means classifier 136
Min-max stretch (color images) 117
Mode 107
Modulation 112
Modulation ratio 119
Moisture content :
    leaf 45, 48
    soil 45, 48, 219
Molecular recognition 7
Molecular signals 7
Multichannel receiver (GPS) 181
Multipath interference 180
Multiple volume scatterers 45, 46, 47
Multispectral images 101, 117, 118, 136
Multispectral remote sensing 216, 221
Multispectral scanner 57, 62, 63
Multitemporal imagery 132, 215, 217

Multitemporal remote sensing 215, 217
Multilevel sensing 73, 74, 76

N

Nadir view 65
Natural landscape 191, 192
Navigation 182
  Navigational aid 174
Negative slope 163
Near infrared 28, 56, 101
Noise 112, 113
Normal (Gaussian) distribution 108, 109
Non-imaging sensors 57
Non-imaging system 57
Non-linear contrast stretch 115
Non-selective scattering 27
Non-systematic remdom noise 113
Normalization stretch 116
Normalization stretch
  (color images) 117
Normalized antennae temperature 52
Normalized difference vegetation index 119

O

Oblique aerial photographs 225
Observation platforms :
    handheld 73
    airborne 73
    spaceborne 76
Ocean color monitor 31
Open canopy 219
Operational meteorology 6
Operating system 100
Optical region :
    visible 28, 56
    NIR/IR 28, 29
Opto-electronic detectors 56
Orbit :
    satellite 78
    shuttle 86
    sun-synchronous 86
    geosynchronous 78
Orbital configuration 174
Orbital period 175
Oscillation of electric charges 8

Index

Outcrop fires 227
Output 207

P

Panchromatic camera 29, 30
Panchromatic film 56
Parallelepiped classifier 138
Passes 86
Passive microwave emission 50
Passive microwave sensing 50
Passive sensors 2
Pattern recognition 132
Penetration depth (radar) 45
Permittivity of medium 45
Perpendicular vegetation index 120
Phase modulation 90
Photographic film 56, 90
Photon 14
Photon detectors 57
Picture element (pixel) 101, 102, 103
Planck's constant 13
Planck's law of radiation 13
Plant community spectral signatures 219
Plant species :
    increaser 219
    decreaser 219
    invader 219
Platform 73
Point attribute table 194
Point coordinates 194
Pointing accuracy 87
Polar satellite launch vehicle 78
Polarimeter 67
Polarized wave 8, 9
Polygon arc list 194
Polygon attribute table 195
Positive slope 163
Power spectrum 131
Preprocessing 92
Primary colors 101
Principal component transform 120
Probability density function 140, 141
Pseudo range (GPS) 178
Pulse length 164

Q

Quanta 6
Quantum detectors 57
Query (the database) 185

R

Radar :
    side looking 37
    real aperture 37
    synthetic aperture 37
Radar imagery 47, 226
Radar pulse 160
Radar return 39, 160
Radar-terrain interaction 47
Radiant energy 15
Radiant flux 15
Radiant photon emittance 14
Radiation terminology 15
Radiator 12
Radio communication 90
Radio-echo sounding 226
Radiometer 67
Radio-positioning 174
Radiometric correction 93
Radiometry :
    IR 10, 24
    Microwave 49
Raman scattering 26
Ranch 220
Random access memory 99
Range 219, 220
Range biomass 220
Range direction 159
Rangeland applications 217
Rangeland resource monitoring 219
Range resolution (radar) 164
    near range 164
    far range 164
Raster 187, 189
Ratio vegetation index 119
Ratioing (of image data) 118
Rayleigh's criterion for resolution 68
Rayleigh's criterion for
    surface roughness (Radar) 47, 162
Rayleigh scattering 26

Real aperture radar 37
Rectification 104
Reference stretch 116
Reflectance spectroscopy 2
Reflection 17, 24
Reflector 12, 42
Refractive index 14
Registration 106
Remote sensing :
    applications 2, 212-227
    natural 2
    artificial 2
    passive 2
    active 2
Remote sensing input to GIS 209
Representing reality 192
Resolution 67
    spatial 68, 164
    spectral 72
    radiometric 72
    temporal 72
Restoration (image) 101
Return beam vidicon 60
River channel contraction 221
River crossing site
  selection for bridges 221
River erosion 221
River sedimentation 221
Roam mechanism 102
Roughness (surface) 47, 162
Row direction (radar) 161

S

Salinization 161
Satellite constellation (GPS) 174
Satellite launch vehicle :
    PSLV 78
    GSLV 78
Satellite orbit :
    sunsynchronous 86
    geosynchronous 78
    shuttle 86
Satellite ranging 176
Satellite thermography 225

Saturation stretch 115
Scale 66
Scan lines 64
Scanner :
    across-track/whiskbroom 62
    along-track/pushbroom 63
Scanning system 62
Scatter diagram/scatterogram 136
Scattering :
    Rayleigh 26
    Mie 26, 27
    nonselective 26, 27
    Compton 26
    Raman 26
Scattering coefficient/cross-section 39-42
Scatterometer 67, 159, 167
Scrambling 181
Seasat 167, 168
Seasonal range (land) conditions 220
Sediment fining 222
Seismic hazard 223
Seismic sensors 225
Seismicity 223
Selective radiator 12
Sensing 1, 99
Sensors 56
Sensor materials 56
Sensor platforms :
    ground observation 73
    airborne observation 73
    spaceborne observation 76
Separability (feature) 134, 135
Sequential aerial photography 217, 225, 227
Sequential receiver (GPS) 181
Sequential topographic maps 225
Shadow 133
Shallow aquifers 214
Shallow coal seam fires 227
Shoreline uplift 223
Short wave infrared 28, 30
Shuttle imaging radar 169, 170
Side looking radar 37, 160
Side looking scanner 65
Signal
    acoustic 5

Index

attenuation 8
disturbance in force fields 4, 5
(gravitational, electric, magnetic, pressure)
electromagnetic 7, 8
particulate 6
reception 39, 63, 90
transmission 39, 63, 90
Signal-to-noise ratio 113
Signature (spectral) 17-24, 90
Single channel receiver (GPS) 181
Skin depth 45, 162
Slant range resolution (radar) 164
Slip displacements 224
Slope 163, 219
Slow movement glacier 226
Smoothness criterion (radar) 162
Soft classification 133
Software 100
Soils 123
Soil adjusted vegetation index 120
Soil brightness 123
Soil erosion 215
Soil line 122, 123
Soil moisture 131, 219
Soil texture 162, 214
Solar irradiance 18
Sounding rocket 6, 76
Space platform 73, 76, 77
Space segment (GPS) 175
Space station 77
Spatial data 187, 193, 204
Spatial pattern recognition 132
Spatial transforms 125-131
Speckle (radar noise) 160
Spectral angle mapping 147
Spectral bands 29
Spectral derivative ratio 146
Spectral finger prints 146
Spectral pattern recognition 132
Spectral radiant exitance 15
Spectral reflectance 15
Spectral signature 17-24, 90
Specular reflection 47, 49
Spillway discharge 227

Spontaneous ignition 227
SPOT satellite 81, 82, 86
Stability of slopes 221
Standard deviation 107
Standardized principal component transformation 123
Statistical filters 129
Stefan-Boltzmann law 13
Stereo image 221, 224
Stereoscopic viewing 221
Storage disk 100
Storage regulation 223
Streets 192
Sub-pixel classification 144
Subsurface water 213
Subsidence 223
Sun-synchronous satellite 86
Supervised classification 134
Surface refuse fires 227
Surface roughness 47, 49, 162
Surface roughness anomalies 47, 49
Surface scattering (radar) 45
Survey markers 227
Swath width 37, 62, 63
Synthetic aperture radar 166
System parameters (radar) 47

T

Target parameters (radar) 47
Target-signal interaction mechanisms 27
Tasseled-cap transformation 123
Tectonic uplift 223
Temperature 219
Temporal pattern recognition 132
Temporal ratio image 220
Tephra (ash falls) 224
Terrain moisture content 162
Thematic layers of data 187
Thematic mapper 216
Thermal :
    capacity 32
    inertia 214
Thermal anomalies 214, 225
Thermal detectors 56
Thermal infrared 219

Thermal infrared multispectral scanner 30, 31, 33
Thermal radiation 10
Thermography, aerial 225
Thresholding, gray level 116
Tillage practices 161
Tilt meter array 225
Time transfer (GPS) 174
Tone 133
Topographic depressions 214
Topography 187, 224
Topologic properties 194, 197, 200
Tracking and data relay satellite 91
Training for classification 146
Training sites 132
Transects 103
Transformed vegetation index 120
Transmittance 15, 17
True color 101

U

Ultrasonic 6
Ultraviolet 9
Unsupervised classification 135
Upper atmospheric studies 6
Urban planning applications 184, 207
Urbanization 216
User segment 176
Utilities 185, 190

V

Vector (GIS) 190

Vegetation anomaly 214
Vegetation vigor 122, 220
Verification 131
Visualization of database 185, 187
Visualization of data elements :
    points 187, 188
    lines 187, 188
    polygons 187, 188
    raster 187, 189
    vector 189, 190
Visualization of hyperspectral image 145
Volcanic eruption 225
Volcanic hazard 224
Volume scattering (radar) 45

W

Warping 223
Waste disposal 221
Water bodies 213
Water resources 213
Wavelength 161
Weed infestation 161
Wien's law 13
Wildlife 220

Y

Yellowness 123, 124

Z

Zoning 192